AF780727

Applied Geophysics in Hydrocarbon Exploration, Energy Storage and CCUS

Applied Geophysics in Hydrocarbon Exploration, Energy Storage and CCUS

Editor

Paulo T. L. Menezes

MDPI

Basel • Beijing • Wuhan • Barcelona • Belgrade • Novi Sad • Cluj • Manchester

Editor
Paulo T. L. Menezes
Universidade do Estado
do Rio de Janeiro
Rio de Janeiro
Brazil

Editorial Office
MDPI
St. Alban-Anlage 66
4052 Basel, Switzerland

This is a reprint of articles from the Special Issue published online in the open access journal *Minerals* (ISSN 2075-163X) (available at: https://www.mdpi.com/journal/minerals/special_issues/2B03UTQ0YL).

For citation purposes, cite each article independently as indicated on the article page online and as indicated below:

Lastname, A.A.; Lastname, B.B. Article Title. *Journal Name* **Year**, *Volume Number*, Page Range.

ISBN 978-3-7258-0545-7 (Hbk)
ISBN 978-3-7258-0546-4 (PDF)
doi.org/10.3390/books978-3-7258-0546-4

© 2024 by the authors. Articles in this book are Open Access and distributed under the Creative Commons Attribution (CC BY) license. The book as a whole is distributed by MDPI under the terms and conditions of the Creative Commons Attribution-NonCommercial-NoDerivs (CC BY-NC-ND) license.

Contents

About the Editor

Paulo T. L. Menezes

Paulo T. L. Menezes received a B.S. (1986) in geology from the Rio de Janeiro State University, Brazil, an M.S. (1990) in geophysics from the Federal University of Pará, Brazil, and a Ph.D. (1996) in geophysics from the National Observatory of Rio de Janeiro. From 1996 to 1997, he worked as a consultant at Geomag Aerolevantamentos Ltda. Since 1997, he has been a professor at Rio de Janeiro State University. He also works as a senior geophysicist at Petrobras. His research interests include interpreting potential field, seismic, and electromagnetic data.

Preface

Geophysical methods are powerful tools in the hydrocarbon industry, allowing subsurface imaging for reservoir characterization, carbon capture, and energy storage applications.

This reprint "Application of Geophysical Data Interpretation in Geological and Mineral Potential Mapping" published by MDPI, is a compilation of scientific papers on new interpretation results and technical developments in geophysical methods such as seismic and multiphysics approaches applied to hydrocarbon exploration, CCUS, and energy storage (including geothermal). More specifically, the papers in this reprint addressed three main problems: exploration case studies from a regional to a local scale; reservoir characterization and monitoring; and carbon capture and energy storage.

In conclusion, "Application of Geophysical Data Interpretation in Geological and Mineral Potential Mapping" results from a cooperative endeavor to compile and share knowledge from the geophysical field. All the scientific papers in this reprint are original contributions that provide a comprehensive understanding of applications of geophysical methods. The Guest Editor sincerely thanks the contributing authors who initially believed in this reprint project.

Paulo T. L. Menezes
Editor

Editorial

Editorial for the Special Issue "Applied Geophysics in Hydrocarbon Exploration, Energy Storage and CCUS"

Paulo T. L. Menezes [1,2]

1 Departamento de Geologia Aplicada, Faculdade de Geologia/UERJ, Rua São Francisco Xavier, 524, sala 2009, Bloco A, Rio de Janeiro 20550-900, Brazil; paulo.menezes@uerj.br
2 PETROBRAS-EXP/PEN/AB, Av. Henrique Valadares 28, 4 andar, Rio de Janeiro 20231-030, Brazil

Since its inception, applied geophysics methods have been crucial in the oil and gas exploration industry. These methods are now employed in all life stages of hydrocarbon reservoirs.

In recent years, advancements in applied geophysics technology have greatly improved our ability to image and understand underground reservoirs. This has reduced exploration costs, improved optimization, and supported efforts to create a more sustainable hydrocarbon industry with less environmental impact. To reduce emissions from the energy industry, companies worldwide are focusing on carbon capture, utilization, and storage (CCUS) projects. Additionally, energy storage in deep reservoirs, such as hydrogen or compressed air, is a key component of current environmental and energy governmental policies.

This Special Issue aims to provide a space for discussing and sharing progress in interpretation workflows and innovations in technology related to hydrocarbon exploration, CCUS applications, and energy storage, all with an eye towards the future.

Our focus in this Special Issue included geophysical method applications from regional exploration to reservoir characterization and monitoring, and carbon and energy storage solutions. We especially welcomed the submission of case studies, reviews, new developments, and the integration of methodologies. We have divided the themes into three sections:

1. Exploration case studies, from regional to local scales.
2. Reservoir characterization and monitoring.
3. Applied geophysics in carbon capture, utilization and storage (CCUS), and energy storage.

As a result, we received an engaged response from the geoscience community, with ten papers covering various important facets of the oil and gas and energy transition industries.

Dell'Aversana [1] presents an integrated deep learning framework that can be widely applied for image analysis and automatic classification in many Earth disciplines, including mineralogy, petrography, paleontology, well-log analysis, and geophysical imaging. When analyzing and classifying images, there are several types of deep learning models available, such as fully connected deep neural networks (FCNNs), convolutional neural networks (CNNs or ConvNet), and residual networks (ResNets). According to Dell'Aversana, all these methods effectively recognize and classify mineral images, with ResNets being the most accurate and precise. He also compared deep learning techniques to other machine learning algorithms, such as random forest, naive Bayes, adaptive boosting, support vector machine, and decision tree. He found that deep neural networks generally perform better in classification. Additionally, he discusses how this deep learning approach can be applied to other types of images and geo-data, making it a versatile and multipurpose methodology for analyzing and automatically classifying multidisciplinary information. This article also serves as a tutorial, providing a detailed explanation of all the key steps in the workflow.

Exploration companies worldwide have widely used marine-controlled source electromagnetics (CSEM) for exploration purposes. However, due to its perceived value and cost

Citation: Menezes, P.T.L. Editorial for the Special Issue "Applied Geophysics in Hydrocarbon Exploration, Energy Storage and CCUS". *Minerals* **2023**, *13*, 1335. https://doi.org/10.3390/min13101335

Received: 11 October 2023
Accepted: 14 October 2023
Published: 17 October 2023

Copyright: © 2023 by the author. Licensee MDPI, Basel, Switzerland. This article is an open access article distributed under the terms and conditions of the Creative Commons Attribution (CC BY) license (https://creativecommons.org/licenses/by/4.0/).

compared to seismic and the lack of realistic case studies, the industry has yet to show much interest in using it for time-lapse reservoir-monitoring (4D) applications. To increase the value of information and reduce survey costs, Menezes and coauthors [2] propose performing joint operations where seismic and CSEM data are acquired during the same survey and at equivalent spatial densities. Additionally, they propose a new multiphysics ocean-bottom nodes (OBN) concept that shows how CSEM can be a cost-efficient and effective integrator to 4D seismic projects. By performing a feasibility study, Ref.[2] demonstrated that horizontal magnetic field components could be used instead of horizontal electric field components to map the 3D resistivity distribution and 4D fluid change responses in a given reservoir. That would make engineering a new OBN class easier and cheaper, as various miniaturized magnetic fields and seismic sensors are available off-the-shelf or ready to operate.

Estimating rock properties accurately from seismic data is essential in the petroleum industry. Two vital properties are the compressional velocity (Vp) and the quality factor (Q), which measure waves' energy losses as they propagate in the subsurface. These properties are typically obtained from multichannel seismic acquisitions. Santos and coauthors [3] developed a method to estimate Q and Vp using single-channel seismic data by using the windowed discrete Fourier transform for a single seismic trace, then calculating the peak and dominant frequency that changes with time. Their method uses a linear equation to adjust the estimated effective quality factor derived from migrated seismic data. The purpose of this correction is to account for the influence of lower-frequency content from more distant events that may be affecting shorter events. The methodology was applied to the Joetsu Knoll massive gas hydrates (GH) site, an SW-NE anticline structure on the eastern margin of the Sea of Japan. Their outcomes indicate a progressive gas hydrate depletion northward along the dome.

In today's world, finding ways to reduce our carbon footprint and protect the environment is crucial. One way to accomplish this is by recycling or burying carbon dioxide in depleted petroleum reservoirs and shifting our exploration strategies to focus on hydrogen reservoirs. These resources may occur in the same or different reservoirs, so searching for both is more efficient. Meju and Saleh [4] proposed a CSEM-MT workflow for investigating reservoirs within a play-based exploration and production framework. That involves finding the right basin and block, selecting the right prospect, drilling the proper well, and looking for opportunities for sustainability and CO2 recycling or burial in the appropriate reservoirs. Recent methodological developments integrating 3D CSEM-MT imaging into the appropriate structural constraints are described, along with case studies demonstrating how this can help us understand deep geological processes and the distribution of potential hydrocarbon, geothermal, and hydrogen reservoirs. These advancements could play a critical role in helping us achieve net-zero emissions by 2050.

Menezes and coauthors [5] provide a comprehensive historical overview of the CSEM method in its 20-year usage in the Brazilian continental margin. The authors have shown progress in understanding CSEM resistivity data across various geological scenarios since 2003. Their review presents a roadmap of technical advancements in acquisition design and interpretation techniques. Accordingly, they have shown the expansion of CSEM usage beyond the lead ranking classic use to general applications, including salt imaging, gas hydrates, geohazard mapping, and reservoir characterization and monitoring. Ultimately, Ref. [5] discuss potential upcoming CSEM applications in new energy resources and carbon capture and storage.

Arelaro and coauthors [6] conducted a seafloor 4D gravity feasibility study for monitoring deep-water hydrocarbon reservoirs. They simulated the gravity effect by creating different density and pore pressure distributions using a fluid flow simulator in a realistic model of a turbiditic oil field in Campos Basin, offshore Brazil. These reservoirs are analogs of several other passive-margin turbiditic systems worldwide. Ref. [6] considered four reservoir scenarios with and without seafloor subsidence. Their research indicates that the gravity responses exceeded the acceptable value of 3 μGal 12 years after the first base survey. The maximum gravity anomaly area corresponds to the oil–water substitution in the production zones. A maximum seafloor subsidence of 0.6 cm was calculated, resulting in no detectable gravity effects. Their results support future 4D seafloor gravity acquisitions for monitoring oil production in the deep-water passive-margin turbiditic reservoirs.

Lyrio and Li [7] have developed an innovative approach to map the basement structures of sedimentary basins, which involves integrating surface gravity data, seismic imaging, and well-logging information. Their method depends on a nonlinear inversion algorithm that constructs the shape and depth of the basement from surface gravity data. Using the primal-logarithmic barrier method, Ref. [7] also incorporated depth constraints from wells. In addition, they use seismic data, where available, to image the basement depth, which serves as a reference model for the inversion algorithm. Combining these elements, Ref. [7] can simultaneously define basement structures that fit seismic and gravity data. They successfully applied their new methodology in the Recôncavo Basin, Brazil, a syn-rift onshore mature basin with a strong correlation between oil field distribution and tectonic framework. Their approach has improved the basement definition and highlighted new exploration targets in the studied area.

Accurately predicting the quality and occurrence of source rocks in a sedimentary basin is crucial to evaluating petroleum resources. Analyzing rock samples in a laboratory is the most precise method to obtain their geochemical properties, but rock information is often limited. Moreover, source rocks could be sampled at positions that may not represent the oil kitchens' average organic content and quality. To overcome these challenges, Reis and coauthors [8] propose a seismic interpretation workflow supported by machine learning methods such as random forest, DBSCAN, and NGBoost to automate the source rock characterization methodology from the seismic data. Their technique helps maximize available data, expand information, and reduce data analysis time. Automating the input data quality control, extrapolating laboratory measurements to continuous well logs of geochemical properties, and estimating these properties in 3D volumes of total organic carbon (TOC) can be generated using machine learning techniques. The authors argue that their approach provides more accurate predictions, reduces uncertainties in the characterization of source rocks, and assists in exploratory decision-making. The proposed method was applied to the pre-salt source rocks from Santos Basin (Brazil) and allowed for the quantification of the TOC distribution in the source rocks, improving the interpretation of the main source rock interval based solely on seismic amplitude data.

Jiang and coauthors [9] develop a supervised deep learning method to predict the low-frequency components of the inverted acoustic impedance, combining various seismic and geological attributes that contain low-frequency information, such as relative geological age, interval velocity, and integrated instantaneous amplitude. Based on the results obtained from synthetic and real data, Ref. [9] argue that the proposed method is capable of enhancing the prediction accuracy of low-frequency components, with a significant improvement in the actual data case with a 57.7% increase as compared to the impedance predicted through well-log interpolation.

The Controlled-Source Electromagnetic (CSEM) method is valuable for obtaining information about reservoir fluids and their spatial distribution. It can be utilized in various applications such as storage for carbon dioxide (CO_2), enhanced oil recovery (EOR), geothermal, and lithium exploration. One of the benefits of the CSEM method is its versatility, as it can be tailored to meet specific reservoir objectives by selecting the appropriate components of a multi-component system. Barajas-Olalde and coauthors [10] show the applicability of CSEM in a CO_2 storage site in North Dakota, USA. Their study describes the procedures involved and highlights how surface measurements can achieve log-scale sensitivity when appropriately upscaled. Furthermore, Ref. [10] also evaluate the sensitivity of CSEM in other case studies, such as EOR, geothermal, and lithium exploration applications.

I am grateful to all the authors who submitted their papers to this Special Issue. Additionally, I appreciate the contribution of the reviewers, who helped improve the quality of the manuscripts and contributed to the success of this Special Issue. Finally, I would like to thank the Minerals editors for their assistance in the publication process.

Conflicts of Interest: The author declares no conflict of interest.

References

1. Dell'Aversana, P. An Integrated Deep Learning Framework for Classification of Mineral Thin Sections and Other Geo-Data, a Tutorial. *Minerals* **2023**, *13*, 584. [CrossRef]
2. Menezes, P.; Zerilli, A.; Correa, J.; Menor, E.; Ferreira, S.; Labruzzo, T. New Concept Ocean-Bottom Multiphysics (OBMP) Nodes for Reservoir Monitoring. *Minerals* **2023**, *13*, 602. [CrossRef]
3. Santos, L.A.; Matsumoto, R.; Freitas, F.D.S.; Santos, M.A.C. Geophysical Characterization and Attenuation Correction Applied for Hydrate Bearing Sediments in Japan Sea. *Minerals* **2023**, *13*, 655. [CrossRef]
4. Meju, M.A.; Saleh, A.S. Using Large-Size Three-Dimensional Marine Electromagnetic Data for the Efficient Combined Investigation of Natural Hydrogen and Hydrocarbon Gas Reservoirs: A Geologically Consistent and Process-Oriented Approach with Implications for Carbon Footprint Reduction. *Minerals* **2023**, *13*, 745. [CrossRef]
5. Menezes, P.T.L.; Ferreira, S.M.; Correa, J.L.; Menor, E.N. Twenty Years of CSEM Exploration in the Brazilian Continental Margin. *Minerals* **2023**, *13*, 870. [CrossRef]
6. Arelaro, A.D.; Barbosa, V.C.F.; Oliveira, V.C., Jr.; Menezes, P.T.L. Feasibility of 4D Gravity Monitoring in Deep-Water Turbidites Reservoirs. *Minerals* **2023**, *13*, 907. [CrossRef]
7. Lyrio, J.C.S.O.; Li, Y. Basement Mapping Using Nonlinear Gravity Inversion with Borehole and Seismic Constraints. *Minerals* **2023**, *13*, 1173. [CrossRef]
8. Reis, M.A.A.d.A.d.; Damasceno, A.C.; Roriz, C.E.D.; Korenchendler, A.L.; da Silva, A.M.; da Silva Praxedes, E.; Silva, V.G. Source Rock Evaluation from Rock to Seismic Data: An Integrated Machine-Learning-Based Work Flow and Application in the Brazilian Presalt (Santos Basin). *Minerals* **2023**, *13*, 1179. [CrossRef]
9. Jiang, L.; Castagna, J.P.; Zhang, Z.; Russell, B. Prediction of Reflection Seismic Low-Frequency Components of Acoustic Impedance Using Deep Learning. *Minerals* **2023**, *13*, 1187. [CrossRef]
10. Barajas-Olalde, C.; Adams, D.C.; Curcio, A.; Davydycheva, S.; Klapperich, R.J.; Martinez, Y.; Paembonan, A.Y.; Peck, W.D.; Strack, K.; Soupios, P. Application of Electromagnetic Methods for Reservoir Monitoring with Emphasis on Carbon Capture, Utilization, and Storage. *Minerals* **2023**, *13*, 1308. [CrossRef]

Disclaimer/Publisher's Note: The statements, opinions and data contained in all publications are solely those of the individual author(s) and contributor(s) and not of MDPI and/or the editor(s). MDPI and/or the editor(s) disclaim responsibility for any injury to people or property resulting from any ideas, methods, instructions or products referred to in the content.

minerals

Article

Feasibility of 4D Gravity Monitoring in Deep-Water Turbidites Reservoirs

Andre D. Arelaro [1,2,*], Valeria C. F. Barbosa [2], Vanderlei C. Oliveira Jr [2] and Paulo T. L. Menezes [1,3]

1 Petroleo Brasileiro S.A. (Petrobras), Rio de Janeiro 20231-030, Brazil
2 Department of Geophysics, Observatório Nacional, Rio de Janeiro 20921-400, Brazil
3 Departamento de Geologia Aplicada, Faculdade de Geologia, Universidade do Estado do Rio de Janeiro (UERJ), Rio de Janeiro 20550-900, Brazil
* Correspondence: andre.arelaro@gmail.com

Abstract: We present a seafloor 4D gravity feasibility analysis for monitoring deep-water hydrocarbon reservoirs. To perform the study, we have simulated the gravity effect due to different density and pore pressure distributions derived from a realistic model of a turbiditic oil field in Campos Basin, offshore Brazil. These reservoirs are analogs of several other passive-margin turbiditic systems located around the world. We considered four reservoir scenarios including and not including seafloor subsidence. Our results indicate that the gravity responses are higher than the feasible value of 3 μGal 12 years following the base survey. The area of maximum gravity anomaly corresponds to where we suppose hydrocarbon extraction occurs. A maximum seafloor subsidence of 0.6 cm was estimated, resulting in no detectable gravity effects. Our results endorse the 4D seafloor gravity acquisition as a beneficial tool for monitoring deep-water passive-margin turbiditic reservoirs.

Keywords: reservoir monitoring; 4D gravity; feasibility test

Citation: Arelaro, A.D.; Barbosa, V.C.F.; Oliveira Jr, V.C.; Menezes, P.T.L. Feasibility of 4D Gravity Monitoring in Deep-Water Turbidites Reservoirs. *Minerals* **2023**, *13*, 907. https://doi.org/10.3390/min13070907

Academic Editor: Georgy Cherkashov

Received: 25 April 2023
Revised: 24 May 2023
Accepted: 29 May 2023
Published: 5 July 2023

Copyright: © 2023 by the authors. Licensee MDPI, Basel, Switzerland. This article is an open access article distributed under the terms and conditions of the Creative Commons Attribution (CC BY) license (https://creativecommons.org/licenses/by/4.0/).

1. Introduction

Currently, the oil industry is seeking ultra-deep marine exploratory opportunities that can reach water depths of over 2000 m with sedimentary overloads of a few kilometers. In these scenarios, using geophysical methods as an imaging tool is challenging because the distance between sources and the sensor can be substantial and impacts data quality. In these cases, an effective way to reduce the signal amplitude loss is to place the geophysical sensors on the seafloor closer to the desired targets. Concerning gravity measurements, high-resolution instruments set on the seafloor can allow for hydrocarbon production monitoring through 4D (time-lapse) acquisitions [1–3].

Gravity acquisitions can be conducted in almost every type of environment. Regarding aquatic environments, gravimetric surveys have been performed since the 1940s. First in lakes and later in marine regions with shallow depths [4]; then, going to deeper and more complicated sites over the years. Land 4D gravity acquisitions have been performed for some decades with various purposes, from geothermal field studies to volcano monitoring, aquifer water storage, mine subsidence, tectonic and post-glacial isostatic movements, and hydrocarbon exploration and production [3,5,6]. Since the end of the 1990s, seafloor 4D gravity acquisitions have been performed in the shallow waters of the North Sea to monitor the fluid dynamics in hydrocarbon fields [3,7,8] and seafloor subsidence due to hydrocarbon production.

Once the gravimetric method is sensitive to mass variations, seafloor 4D gravity measurements can improve our understanding of fluid movements inside a reservoir. Moreover, it directly impacts hydrocarbon recovery and reduction of geological uncertainties such as mass variation estimation, reservoir water influx characterization, gas/oil or oil/water contact estimation, and CO_2 injection and storage. It can also contribute to a better and optimized drilling plan by diminishing the number of wells, substantially reducing the costs of a hydrocarbon field development project [3,9–15]. Furthermore, if pressure gauges are settled with gravimeters on the seafloor, one can obtain information about seafloor movement due to oil production [3,8,11,16–18]. The measure of seafloor deformation is crucial to reduce the risks of human exposure and production facilities. At last, seafloor 4D gravity is a relatively fast, cost-effective, and environmentally friendly geophysical method that can be used complementary to 4D seismic surveys [3,5,8,9,15,18,19].

Several studies performed all over the world in the last two decades show the efficiency of the seafloor 4D gravity technique. These include feasibility studies, survey improvements, time-lapse gravity processing, application of inversion methods, use of borehole and gravity gradiometry information, and real data interpretation [7,10–12,16,17,19–27].

In deep and ultra-deep waters, as in the case of the largest Brazilian hydrocarbon fields, the seafloor 4D gravity survey is a technological challenge. In this scenario, it is necessary to evaluate the technical feasibility of using the seafloor 4D gravity surveys for hydrocarbon-field monitoring. Thus, this study focuses on the feasibility of the seafloor 4D gravity acquisition for monitoring hydrocarbon reservoirs and seafloor movement. By improving the work developed by [28], we have performed 4D forward gravity modeling of a typical passive-margin turbiditic reservoir at Campos Basin, offshore Brazil. We tested three scenarios with seafloor movement and another without this phenomenon. Using the feasible (threshold) limit of 3 μGal, which can be considered as a conservative value concerning recent seafloor 4D gravity acquisitions [15,18], results exhibit gravity responses higher than the feasible limit, which validate the seafloor 4D gravity survey for similar situations.

2. Materials and Methods

2.1. Gravity Anomalies and Seafloor Changes

Consider a marine sedimentary basin where the seafloor is the interface separating the water column from the basin sediments (Figure 1). We assume that the water column and the sediments are homogeneous, with constant densities ρ_w and ρ_s, respectively. Inside the sediment layer stands a hydrocarbon-producing reservoir, with known dimensions and time-variable bulk density $\rho_b(t)$ due to its function.

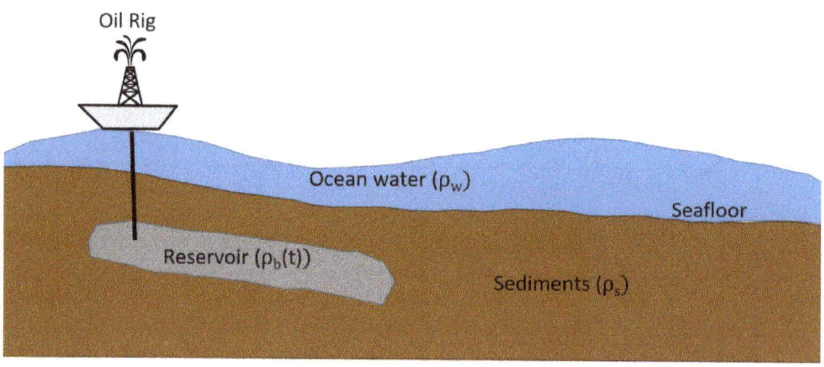

Figure 1. Sketch of a marine environment containing an ocean water layer (blue) with a constant density of ρ_w. Below the seafloor, there is a sedimentary layer (brown) with density ρ_s and a hydrocarbon-producing reservoir (gray) enclosed within the sediments, showing time-variable bulk density $\rho_b(t)$.

Let us simulate a seafloor 4D gravity survey over the model described in Figure 1. In this case, three phenomena can generate time-variable gravity anomalies: (1) water layer variation due to tides and waves; (2) reservoir mass variation due to hydrocarbon production; and (3) seafloor relief movement due to hydrocarbon production. The role of gravity modeling is to facilitate the study of the gravity response by isolating the gravity effects caused by these phenomena. This work focuses on the 4D gravity effect yielded by the reservoir's mass variation and the seafloor movement due to hydrocarbon production. Thus, the 4D gravity effect of the ocean water movement has not been taken into account.

The gravity effect of a 3D density distribution can be calculated by dividing the study region in a sequence of vertical prisms adjacent in the three spatial directions with the $x-$axis to the north, $y-$axis rising to the east, and $z-$axis falling downward. From now on, this mesh of 3D vertical, juxtaposed prisms in the horizontal and vertical directions is called the interpretation model. The gravitational acceleration in the vertical direction $g_z(P_i, t_k)$ evaluated in a point $P_i = (x_i, y_i, z_i(t_k))$ representing the seafloor in a time t_k can be calculated according to Blakely [29]:

$$g_z(P_i, t_k) = \gamma \sum_{j=1}^{N} \Delta\rho_j(t_k) f_{ij}(P_i, x_j, y_j, z_j(t_k)), \tag{1}$$

with

$$f_{ij}(P_i, x_j, y_j, z_j(t_k)) = \int_{z_{1j}(t_k)}^{z_{2j}(t_k)} \int_{y_{1j}}^{y_{2j}} \int_{x_{1j}}^{x_{2j}} \frac{z_j(t_k) - z_i(t_k)}{d_{ij}^3} dx_j dy_j dz_j, \tag{2}$$

where γ is the gravitational constant, $\Delta\rho_j(t_k)$ is the time-variable density contrast between the jth prism and its surroundings. N is the number of prisms in the interpretation model. The variable d_{ij} is the distance between the ith measurement point and the jth integration point within the jth prism $(x_j, y_j, z_j(t_k))$:

$$d_{ij} = [(x_i - x_j)^2 + (y_i - y_j)^2 + (z_i(t_k) - z_j(t_k))^2]^{1/2}. \tag{3}$$

Note that in Equations (1)–(3), the vertical coordinates $z_i(t_k)$ and $z_j(t_k)$ are also dependant on time because they can vary in the case of seafloor movement. In Equation (2), the integration is conducted in the variables x_j, y_j, and z_j denoting the x-, y-, and z-coordinates of an arbitrary point belonging to the interior of the jth prism. The integrals limits (Equation (2)) correspond to the jth prism borders in the following way: x_{1j} and x_{2j} are their south and north borders; y_{1j} and y_{2j} are their west and east borders; and z_{1j} and z_{2j} are their depths to the top and bottom. The analytical solution for Equation (1), for a mesh of rectangular prisms, was taken from Nagy et al. [30].

2.2. Reservoir Fluid Substitution and Seafloor Movement Effects

In a hydrocarbon-producing field (gas, oil, or both), the reservoir bulk density is time-variable because of the hydrocarbon removal or its substitution by water or other fluids inside the reservoir. As they have different densities, a time-variable gravity effect is generated. To calculate this effect in a specific position P_i and moment t_k, defined here as $g_z^r(P_i, t_k)$, we represent the reservoir as a mesh of vertical prisms (Figure 2) and use Equation (1). In this case, the density contrast is between the reservoir density at that time $(\rho_j(t_k))$ and the background sediment density (ρ_s).

Another important phenomenon is the seafloor movement caused by hydrocarbon exploitation. If the reservoir pore pressure decreases, it generates reservoir compaction, resulting in the seafloor sinking [16,31–33]. When it occurs, the region of subsidence, initially composed of seafloor sediments in time t_0 (Figure 3a), is replaced by water in time t_1 (Figure 3b) just after the subsidence. We assume that there is only vertical displacement on the seafloor. It means that the ith measurement point on the seafloor in t_0, $P_i = (x_i, y_i, z_i(t_0))$, changes to $P_i' = (x_i, y_i, z_i'(t_1))$ in t_1 (Figure 3). The opposite occurs when the pore pressure increases.

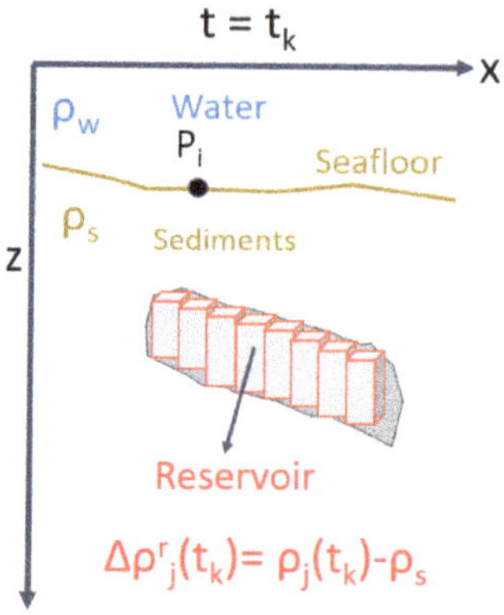

Figure 2. Sketch of a hydrocarbon-producing reservoir (gray prisms) in time $t = t_k$. Its density varies with time due to the substitution of hydrocarbon for water or other fluids inside the reservoir. The observation point P_i is on the seafloor. ρ_w is the water density and ρ_s is the sediment density. $\rho_j(t_k)$ is the density of the j−th prism representing the reservoir in $t = t_k$. $\Delta\rho_j^r(t_k)$ is the density contrast between the j−th prism in the reservoir and its surroundings (sediments) in $t = t_k$.

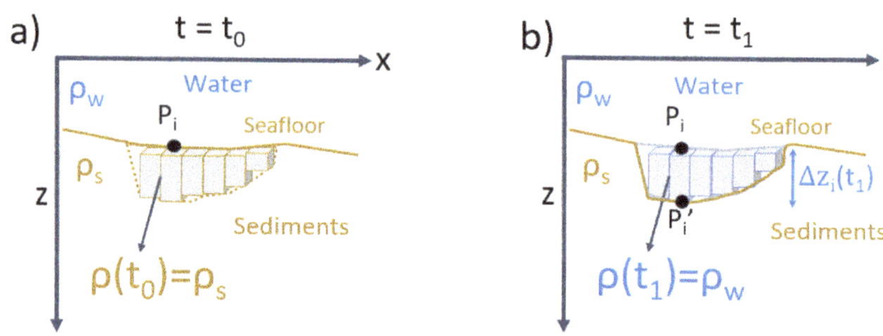

Figure 3. Sketch of a seafloor subsidence process (gray prisms) showing: (**a**) moment $t = t_0$ before subsidence and (**b**) moment $t = t_1$ after subsidence. In $t = t_0$, the observation point on the seafloor is P_i and the gray prisms are sediments. In $t = t_1$, this point P_i moved to P_i', at vertical distance Δz_i from P_i. The region represented by the gray prisms compounded by sediments in t_0 is substituted by water in t_1.

The change of the seafloor's vertical coordinates also causes an additional gravity effect that must be taken into account in the modeling calculations once the observation points are on the seafloor. The gravity variation Δg_z^s due to the change of the vertical position in the measurement points (Figure 3) is defined by:

$$\Delta g_z^s = 0.3086\Delta z_i = 0.3086(z_i'(t_1) - z_i(t_0)), \quad (4)$$

where $z_i(t_0)$ is the original vertical coordinate of the observation point P_i (Figure 3a), and $z_i'(t_1)$ is the vertical coordinate of the observation point after the bathymetric change at the point P_i' (Figure 3b). Equation (4) is the free-air gradient used to correct the gravity effect of the vertical distance between the measurement point and the difference in station elevation [29].

The total gravity effect due to seafloor movement, defined here as $g_z^s(P_i', t_1)$, is the sum of the gravity effect due to the substitution of sediments by water and the free-air gradient (Equation (4)). To calculate the first part, we discretize the new bathymetry region (i.e., the sinking volume after the seafloor movement) into a mesh of vertical prisms (blue prisms in Figure 3b). Then, the gravity effect due to this substitution can be calculated using Equation (1), where the density contrast is between the seafloor sediments (ρ_s) and the ocean water (ρ_w). If there is a subsidence movement, the density contrast is negative because the ocean water (less dense) substitutes the sediments (more dense). Moreover, if there is an uplift movement, the density contrast is positive.

2.3. Seafloor Movement Estimation

To simulate the seafloor subsidence due to the reservoir compaction in a semi-infinite elastic medium, we selected the strain nucleus concept originally applied in thermodynamics theory [34–37]. The seafloor subsidence Δz_i shown in Figure 3b can be calculated considering the subsurface as a set of strain nuclei that are elements of infinitesimal volume. Once we assume there is no compaction/expansion outside the reservoir, the total subsidence is due to the movement inside the reservoir. Moreover, we consider that there is no change in the reservoir volume, so the compaction is only due to the variation in the reservoir pore pressure. Then, total movement on the seafloor at P_i (Figure 3) is the summation of the movement occurring in the N_r strain nuclei in the reservoir, i.e.,

$$\Delta z(x_i, y_i, z_i(t_0)) = \sum_{j=1}^{N_r} \left[\int\int\int_{V_{1j}} w_{1ij} dV'_j + \int\int\int_{V_{2j}} w_{2ij} dV'_j \right] \tag{5}$$

where w_{1j} represents the infinitesimal vertical displacement in the jth strain nucleus in an infinite medium and w_{2j} is the correction of the vertical displacement considering a semi-space. The second integration in Equation (5) is called the "image nucleus solution" (Figure 4). The infinitesimal vertical displacement in the jth prism and its correction due to jth image nucleus are, respectively, given by:

$$w_{1ij} = \frac{A(1+\nu)}{E} \frac{\partial}{\partial z} \frac{1}{r_{1ij}} \Delta p_j, \tag{6}$$

and

$$w_{2ij} = \frac{A(1+\nu)}{E} \left[2z_i(t_0) \frac{\partial^2}{\partial z^2} \frac{1}{r_{2ij}} - (3-4\nu) \frac{\partial}{\partial z} \frac{1}{r_{2ij}} \right] \Delta p_j, \tag{7}$$

with

$$A = -\frac{c_m E}{4\pi(1+\nu)}, \tag{8}$$

where c_m is the uni-axial compaction coefficient given by:

$$c_m = \frac{(1+\nu)(1-2\nu)}{E(1-\nu)}. \tag{9}$$

E is the Young modulus, ν is the Poisson ratio, and Δp_j is the difference of pore pressure in the jth nucleus between the moments t_0 and t_1. r_{1ij} and r_{2ij} are the distances from the i-th point (x_i, y_i, z_i) to the j-th strain nucleus (x'_j, y'_j, z'_j) and the j-th image nucleus (x'_j, y'_j, z'_j), respectively (see Figure 4).

One can note the similarity between the integral functions that describe the gravity anomaly (Equation (1) and the vertical displacement in a semi-infinite elastic medium (Equation (5)). We solved these equations by using the approach proposed by [37] that have taken advantage of this similarity and used the closed expressions of the gravitational potential and its derivatives produced by the 3D right rectangular prism derived by [30,38] for calculating the displacement field on the seafloor.

Upon calculating the vertical displacement of the seafloor (Equation (5)), we obtain the new bathymetry, that is, the vertical coordinates $z'_i(t_1)$. Then, we use these coordinates in Equation (1) to obtain the gravity anomaly due to the vertical displacement of the seafloor.

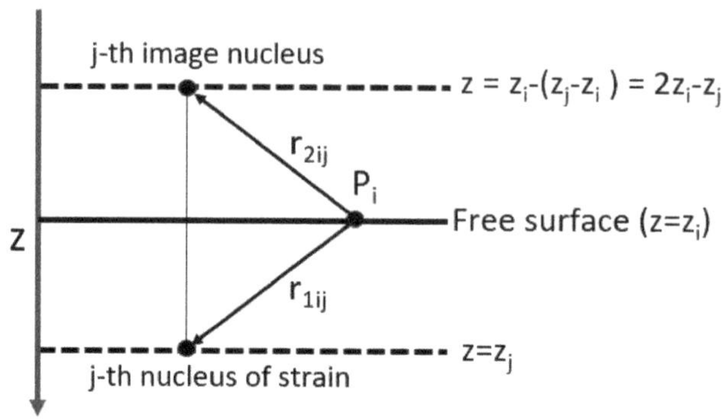

Figure 4. Schematic portrayal of the strain nucleus [31]. The free surface is in $z = z_i$ and the nucleus is in z_j. The observation point P_i is on the free surface and r_{1ij} and r_{2ij} correspond to the distances between the observation point to the strain nucleus and the image nucleus, respectively. According to [32], see chapter 12.

2.4. Four-Dimensional Gravity Anomaly

If there is no movement of the seafloor, the 4D gravity anomaly is the difference of the gravity effect at the station point P_i due to the reservoir production at different times (t_0 and t_1), i.e.,

$$\Delta g_z^{4D}(P_i, t_0, t_1) = g_z^r(P_i, t_1) - g_z^r(P_i, t_0), \tag{10}$$

where the function g_z^r represents the gravity effect of the reservoir region.

Moreover, if we have seafloor subsidence or uplifting (Figure 3b), it is necessary to consider the gravity effect due to these changes on the seafloor. Thus, the 4D gravity anomaly at the measurement point P_i' can be written as:

$$\Delta g_z^{4D}(P_i, P_i', t_0, t_1) = g_z^s(P_i', t_1) + g_z^r(P_i', t_1) - g_z^r(P_i, t_0), \tag{11}$$

where the functions g_z^s and g_z^r, respectively, represent the gravity effects due to seafloor movement and reservoir fluid substitution.

We stress that before the production starts, there is no seafloor movement; hence, the gravity effect in time t_o is only calculated at the original seafloor (point P_i). Equations (10) and (11) are the main point of this work. For all tested models, we adopted 3 μGal (equivalent to 3.10^{-8} ms^{-2} in SI) as a feasible value to detect the gravity anomalies. This is a conservative value comparing the precision achieved in the present surveys of this type [15,18].

2.5. Campos Basin Geological Setting

Several turbiditic oilfields are situated in the northeastern portion of the offshore Campos Basin, Brazil. The passive margin Campos Basin is one of the predominant Brazilian offshore oil provinces. Its tectonic-sedimentary evolution is associated with the breakup of the Gondwana supercontinent and the opening of the South Atlantic Ocean [39]. It comprises three main tectonic stages: rift, transitional, and drift (Figure 5).

The rift sedimentary sequence includes the Barremian lacustrine deposits of the Lagoa Feia Formation overlaying the Hauterivian (120–130 Ma) Cabiunas basalts. These volcanic rocks characterize the economic basement of the basin. The Lagoa Feia sediments are understood as the principal non-marine source rocks in the Campos Basin [40].

The transitional sequence encloses the Aptian sedimentation, from bottom to top: conglomerates, carbonates, and predominantly the evaporitic rocks deposited during a period of tectonic quiescence. This transitional stage defines the marine drift phase's antecedent, where the sediments are associated with the first seawater invasion via the Walvis Ridge [41].

The drift sequence starts with the Albian/Cenomanian shallow-water calcarenites and calcilutites of the Macaé Formation. They were followed by the Carapebus Formation, a marine Upper Cretaceous to Paleogene deep-water clastic section formed by shale, marls, and sandstone turbidite lenses. These sediments were deposited during general tectonic inactivity and thermal subsidence. The turbidite reservoir systems are sedimented in deep-water settings associated with slope and continental rise deposits [42] and form the most valuable post-salt petroleum reservoirs in the Campos Basin [40]. These turbidites are potential targets for several multi-physics appraisal and monitoring studies [43–45].

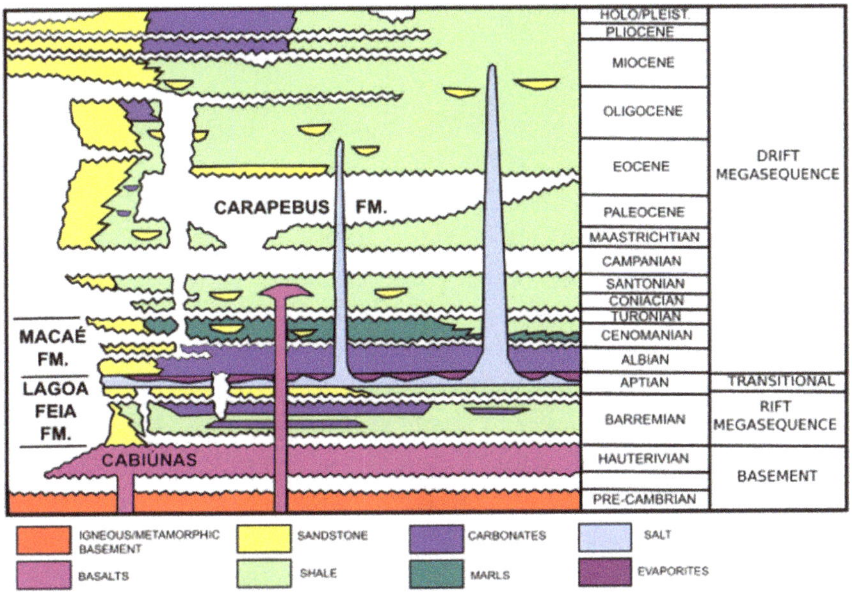

Figure 5. Simplified stratigraphic chart of Campos Basin, modified from [46].

The reservoir model comprises a typical turbiditic reservoir of the Campos Basin. The reservoir facies are thick, up to 300 m, formed by clean, massive, turbidite sandstones interbedded with shales and marls. The trap is of the structural/stratigraphic type. These reservoirs usually have high porosity values in the 26%–32% range [47,48].

2.6. Reservoir Model

The entire interpretation model comprises the extent of 14,050 m in the north axis, 13,250 m in the east axis, and 625 m in the vertical direction. The top and bottom depths are 2712 m and 3337 m deep, respectively. The data relating to the reservoir model (e.g., pore pressure, density, and Poisson's ratio) make up 1,950,312 measurements for each property, with a grid spacing of 50 m along the north and east directions and 25 m in the vertical direction. The data came from reservoir fluid flow simulations for the following years: 2002, 2013, 2014, 2015, and 2018. It is worth noting that the field produces only oil, not gas. The observation points consist of a regular grid of 57 $\times$ 54 points in the north and east directions, respectively, with a grid spacing of 250 m in both directions, totaling 3078 observations. The depth of the observation grid is 1338 m deep, the average bathymetry in the oil field region. Over the years, the variation in density has gone from 2.08 to 2.64 g/cm^3, and pore pressure data range from 33.2 to 34.2 MPa. The model background has a density of 2.64 g/cm^3 and a pore pressure of 0 Mpa, representing the region outside the reservoir. Poisson's ratio data values vary from 0.3237 to 0.3723. Figures 6, 7 and 8 show, respectively, the 3D distributions of density, pore pressure, and Poisson's ratio in 2002.

In order to obtain a better visualization, only the values different from the background are shown.

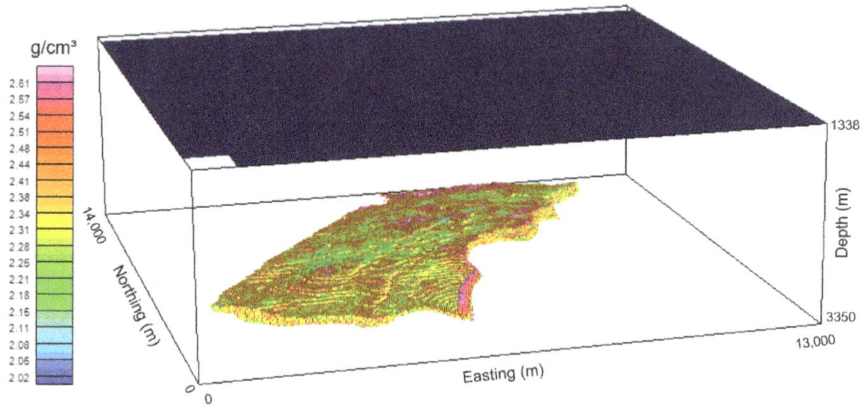

Figure 6. Three-dimensional density data distribution of the turbiditic reservoir model in 2002 (lower volume). The upper surface, with constant bathymetry is the place where the observation points are calculated (Equation (1)).

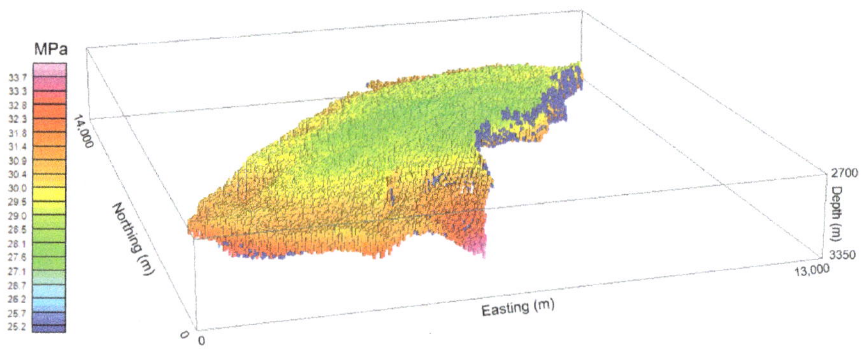

Figure 7. Three-dimensional pore pressure distribution of the turbiditic reservoir model in 2002.

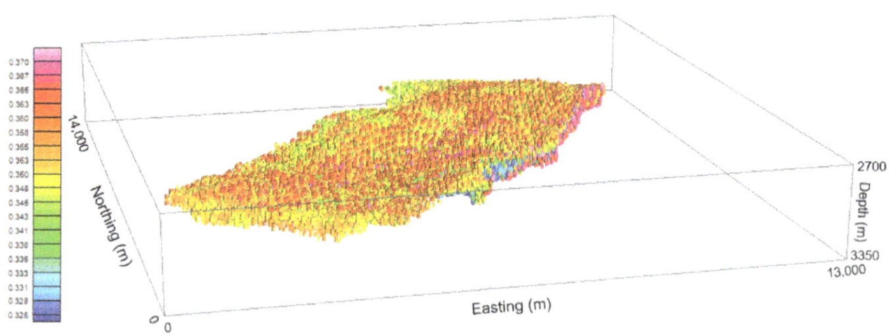

Figure 8. Three-dimensional Poisson ratio distribution of the turbiditic reservoir model in 2002.

3. Results

The results were divided into two groups regarding seafloor movement. The first group shows just one scenario with no seafloor movement, while the second group includes three scenarios where seafloor movement is considered.

3.1. Scenario without Seafloor Movement

In a scenario without seafloor movement, only density changes inside the reservoir are assumed to generate gravity anomalies. Thus, the data used in this scenario are the density differences over the years (Figure 9). Although we do not have information provided by boreholes, we can see an increase in density in the central-northeast region of the area (red arrows in Figure 9), which probably indicates where the production was occurring. This increase is due to the oil replaced with a denser fluid in the reservoir, probably formation water. Only values that are different from zero are shown in Figure 9, which means the colored cells represent where density has changed over the years.

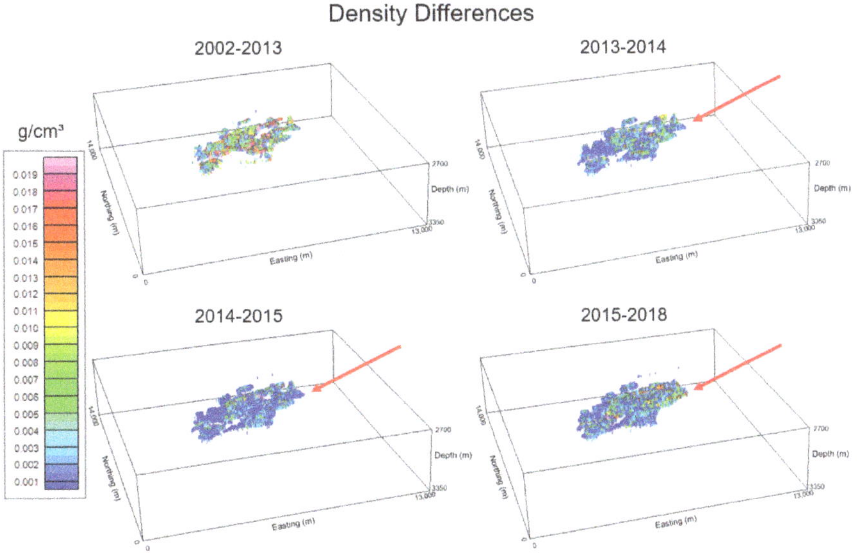

Figure 9. Three-dimensional perspective views of the density differences over the years within the reservoir model (lower volume in Figure 6). Red arrows indicate the region where the density is increasing. Only values that are different from zero are shown.

Using Equation (1), we calculate the gravity anomaly in each year produced by a set of N_r = 1,861,625 prisms simulating the reservoir model (lower volume in Figure 6) at the grid of observations located on the seafloor (upper surface in Figure 6). Since the reservoir is less dense than its surroundings, gravity anomalies are negative in all years, reaching a maximum amplitude of more than 1000 μGal. The resulting gravity anomalies in each year are not shown because the anomalies are very similar.

Since 2002 is the first year for which we have data, the differences over the years (2013 to 2018) due to density changes within the reservoir (Figure 9) were related to it, producing 4D gravity anomalies (Equation (10)). Figure 10 shows the 4D gravity anomalies over the production years, where we can see that the feasible limit of 3 μGal (dashed red line) is surpassed in 2014 and reaches the amplitude of about 7 μGal in 2018. Thus, according to this scenario, the 4D gravity effect due to oil production (Equation (10)) could be detected after 12 years from the base year.

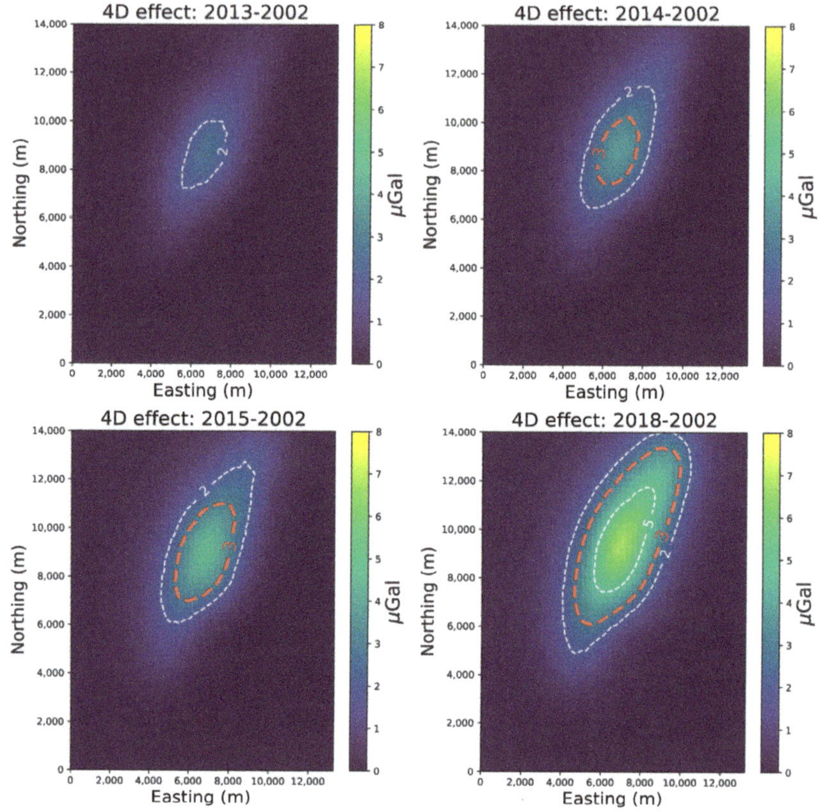

Figure 10. Four-dimensional gravity effect calculated using the reservoir model (lower volume in Figure 6) without seafloor movement. The feasible limit of 3 μGal is represented by the red lines. Observation points are located on the seafloor with a regular spacing of 250 m.

3.2. Scenarios with Seafloor Movement

We also modeled scenarios with seafloor movement calculated using Equation (5), which demands pore pressure and Poisson's ratio data from the reservoir flow simulator. In the case of Poisson's ratio, we choose three possible values according to the original data: 0.33, 0.35, and 0.37. Each value generates different scenarios. We defined Young's modulus as the constant value of 5 GPa related to reservoir sandstone rocks.

The differences over the years in pore pressure distribution (Figure 11) have a more complex pattern than the density differences (Figure 9). For ease of visualization, we split the pore pressure differences (Figure 11) into two parts: negative (left panels) and positive (right panels) parts. Equivalent to Figure 9, Figure 11 shows only cells with values that are different from zero. Note that between 2002 and 2013, the pore pressure differences are only negative and more intense in the same region where the density differences increase (red arrow in Figure 9). From 2013 to 2018, the pore pressure decreased in this region (red arrows in the left panels in Figure 11), but with less intensity and in smaller volume (i.e., a smaller number of model cells). However, pore pressure increased in the southwest region of the model (blue arrows in the right panels in Figure 11) between 2013 and 2018. We believe that this rise in pore pressure is related to fluid injection in the reservoir as a strategy to avoid severe depletion during the production life of the oil field.

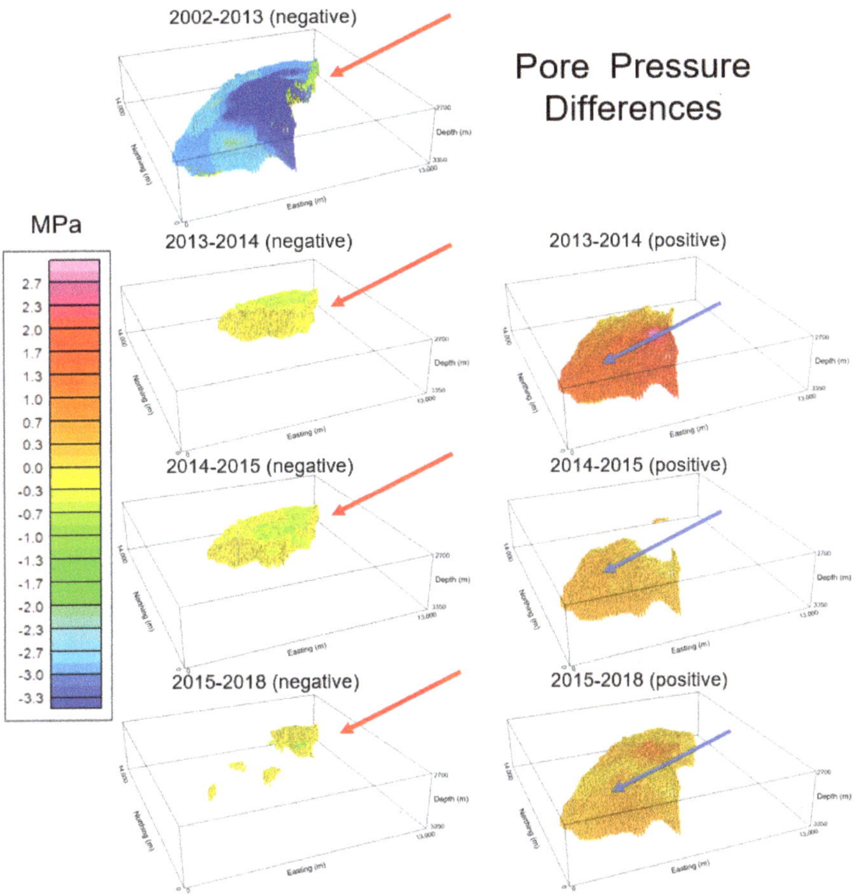

Figure 11. Three-dimensional perspective views of the pore pressure differences over the years from the reservoir model (lower volume in Figure 6). The left and right panels show, respectively, the negative and positive pore pressure differences. Note the pressure has decreased in the central–northeast area since 2013 (red arrows) and has increased in the southwest area since 2014 (blue arrows). Only values that are different from zero are shown.

The modeled seafloor movement in 2013 using three different Poisson ratios can be seen in Figure 12. When Poisson's ratio rises by 0.02, the amplitude of the seafloor displacement (subsidence or uplift) decreases by about 5 %. Figure 12 also shows small changes in the seafloor relief in the order of a few millimeters. Figure 13 shows that even the lowest value of Poisson's ratio ($\nu = 0.33$) does not produce seafloor movement greater than 0.6 cm. Between 2013 and 2015, the subsidence (positive values in Figure 13) grows in the central-northeast area while it diminishes in the southwest area. However, between 2015 and 2018, the subsidence in the central-northeast and the southwest regions decreases. As expected, these results are linked to the pore pressure dynamics over the years (Figure 11).

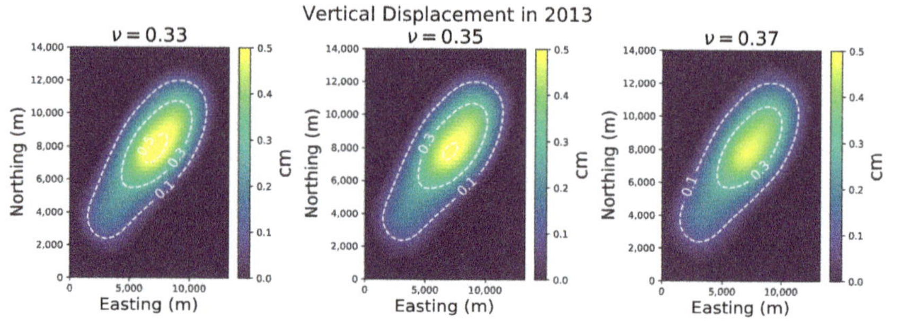

Figure 12. Seafloor movement simulated using Equation (5) for the 2013 year and using three different Poisson ratios (ν). The base year is 2002. Positive values represent the subsidence of the seafloor.

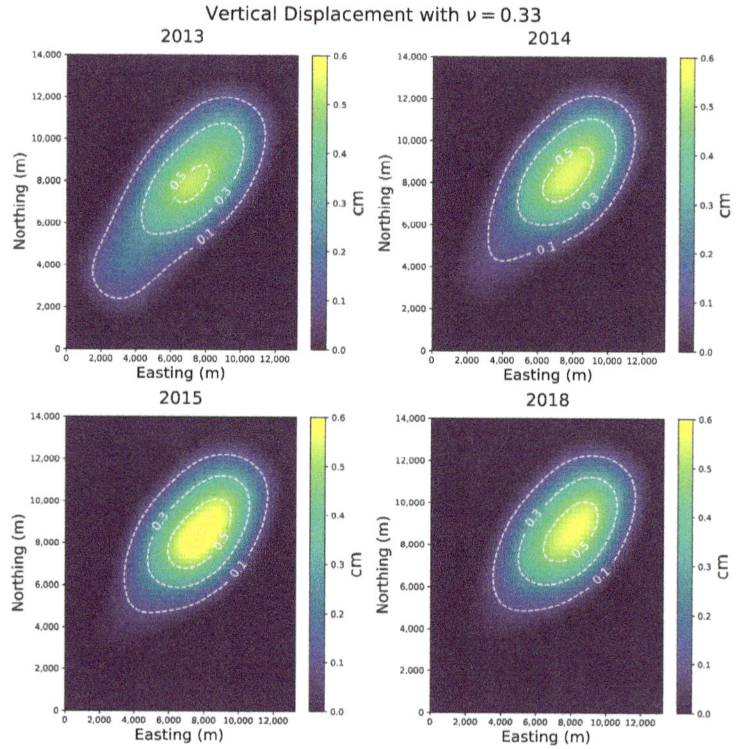

Figure 13. Seafloor vertical displacement simulated by using Equation (5) for years 2013, 2014, 2015, and 2018, for Poisson's ratio of 0.33. The base year is 2002. Seafloor subsidence is represented by positive values.

Upon estimating the seafloor movement we calculated for each year, the gravity effect due to this phenomenon, which is the sum of the gravity effects from the replacement of sediments by water on the seafloor (Equation (1)) and the free-air correction (Equation (4)). Because 2002 is the base year, we do not have subsidence effect for this year. Figure 14 exemplifies the changes in gravity effect in 2013 due to seafloor movement with $\nu = 0.33$. The shapes of the gravity anomalies concerning the rock/water substitution effect (Figure 14a) and of the vertical correction effect (Figure 14b) have a high correlation with the geometry of the seafloor movement (Figure 13). The rock/water substitution effect

(Figure 14a) is one order of magnitude lower than the vertical correction effect (Figure 14b). In isolation, these two gravity effects (Figure 14a,b) are below the feasible limit of 3 μGal, but the second one is in the same order of magnitude as this limit. The same procedure was repeated for years 2014, 2015, and 2018 with similar results, so they are not shown.

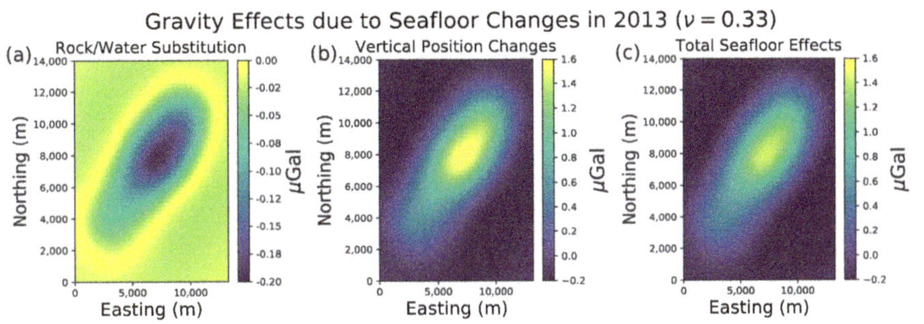

Figure 14. Gravity effect in 2013 due to: (**a**) rock/water substitution on the seafloor; (**b**) change of observations' vertical position and; (**c**) seafloor changes, which is the sum of (**a**,**b**). Observation points are located on the seafloor with a regular spacing of 250 m with $\nu = 0.33$.

Using the modified bathymetry, we updated the reservoir gravity effect over the years (with $\nu = 0.33$) to compare with the seafloor change gravity effects. These effects for 2013 are shown in Figure 15, where the total gravity is the sum of the reservoir and seafloor effects. The reservoir effect (Figure 15a) dominates the total gravity effect (Figure 15c) because it is three orders of magnitude greater than the gravity effect due to the seafloor changes (Figures 14c and 15b). The same is valid for other scenarios in different years and with different Poisson ratios.

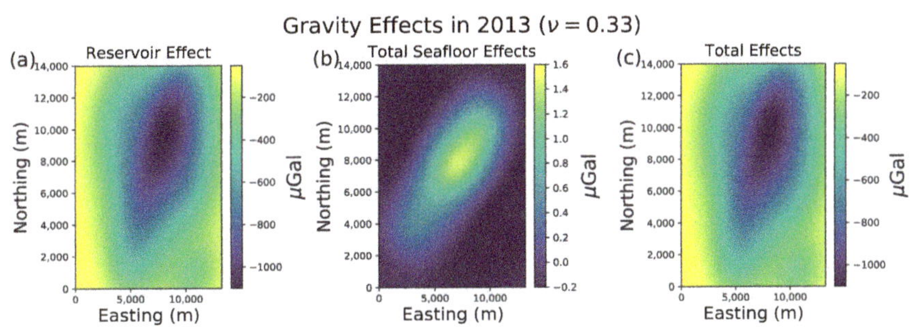

Figure 15. Gravity effect due to: (**a**) reservoir production; (**b**) seafloor changes; and (**c**) the total gravity effect, which is the sum of (**a**,**b**). Observation points are located on the seafloor with a regular spacing of 250 m.

At last, we calculated the 4D gravity anomaly over the years (using Equation (11) and $\nu = 0.33$) for the scenarios with seafloor changes. Figure 16 shows that the resulting 4D gravity anomaly is very similar in shape to that of the no-subsidence scenario (Figure 10). However, there is an increase in the maximum anomaly amplitude of about 14 %, which also occurs with the other two Poisson ratios (0.35 and 0.37), although with less intensity. In addition, the 4D gravity anomaly exceeds the feasible limit in 2013 when considering the vertical movement one year earlier than the case with no vertical movement (Figure 10). This result shows that even changes in the seafloor of a few millimeters cannot be neglected. In the Supplementary Material, we conducted tests on synthetic noise-corrupted data to investigate the sensitivity of our method to deal with distinct noise levels (Figures S1–S6).

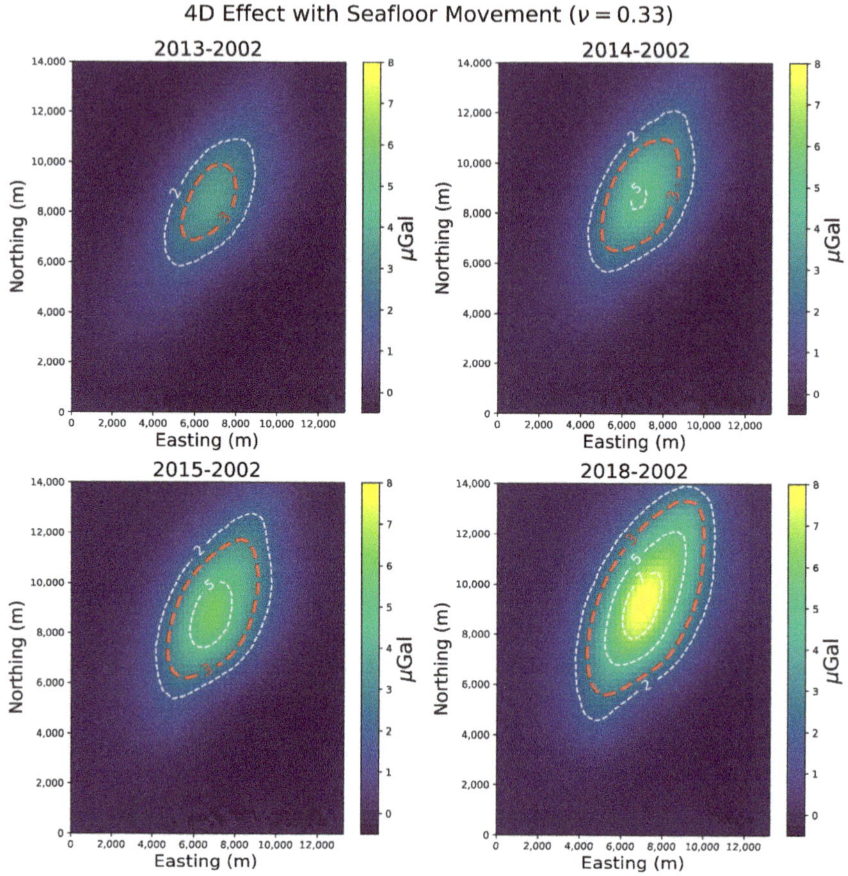

Figure 16. Four-dimensional gravity effect calculated using the reservoir model (lower volume in Figure 6) with seafloor movement (Figure 13). The feasible limit of 3 μGal is represented by the red lines. Observation points are located on the seafloor with a regular spacing of 250 m.

4. Discussion

Density, pore pressure, and Poisson's ratio from a reservoir flow simulator based on a Brazilian hydrocarbon field were the cornerstone information from which we modeled gravity anomalies and seafloor movement between 2002 and 2018. Negative pore pressure and positive density differences in the central-northeast area of the reservoir model suggest the occurrence of oil production there. Moreover, the increase in pore pressure in the southwest area since 2013 could be explained by the beginning of reservoir re-injection as a strategy for oil production. Once there is no density difference in the southwest between 2013 and 2018, the fluid used in this supposed injection could be the formation water or other fluid with the same density.

In this realistic reservoir model, between 2002 and 2018, the 4D gravity anomalies surpass the feasible limit of 3 μGal in the four tested scenarios: one without seafloor movement and three scenarios with subsidence and uplift, varying the Poisson ratios. These modeling results validate the use of 4D gravity measures for monitoring oil production and seafloor movement in considerable depths. In addition, The area of maximum gravity anomaly corresponds to where we suppose hydrocarbon extraction occurs. They also showed that seafloor subsidence leads to a 14 % increase in the 4D gravity anomaly amplitude. According to the methodology, gravity anomalies related to seafloor movement do not reach the detectability limit (3 μGal) but are in its order of magnitude (between −0.2 and 1.6 μGal, Figure 16). The order of magnitude of the time-lapse gravity anomalies calculated in all scenarios of some μGal agrees with the 4D anomalies obtained in real measurements in the North Sea [8,16,19]. We draw the readers' attention to the fact that several parameter combinations related to the reservoir in production could result in detectable anomalies due to seafloor changes. Some examples could be the variations in the Young modulus or the arrangement of density and pore pressure distributions with reservoir volume and depth.

5. Conclusions

Following various and diversified tests using a realistic geophysical model that includes 3D density, pore pressure, and Poisson's ratio distributions, we conclude that the seafloor 4D gravity survey should be beneficial for monitoring the reservoirs of the Brazilian turbiditic fields in deep waters, with or without seafloor movement due to production. There are many active fields already in the Brazilian offshore post-salt that could benefit from this technique. The extension of this work to analyze the feasibility of using 4D gravity acquisitions in the monitoring of reservoirs in Brazilian pre-salt fields has no methodological obstacles.

Supplementary Materials: The following supporting information can be downloaded at: https://www.mdpi.com/article/10.3390/min13070907/s1, Figure S1: Noise-corrupted 4D gravity anomaly with zero mean and standard deviation of 0.5 μGal in the scenario without subsidence. Compare with Figure 10 in the article showing the noise-free 4D gravity anomaly. The feasible limit of 3 μGal is represented by the red lines; Figure S2: Noise-corrupted 4D gravity anomaly with zero mean and standard deviation of 0.3 μGal in the scenario without subsidence. Compare with Figure 10 in the article showing the noise-free 4D gravity anomaly. The feasible limit of 3 μGal is represented by the red lines; Figure S3: Noise-corrupted 4D gravity anomaly with zero mean and standard deviation of 0.1 μGal in the scenario without subsidence. Compare with Figure 10 in the article showing the noise-free 4D gravity anomaly. The feasible limit of 3 μGal is represented by the red lines; Figure S4: Noise-corrupted 4D gravity anomaly with zero mean and standard deviation of 0.5 μGal in the scenario with subsidence. Compare with Figure 16 in the article showing the noise-free 4D gravity anomaly. The feasible limit of 3 μGal is represented by the red lines; Figure S5: Noise-corrupted 4D gravity anomaly with zero mean and standard deviation of 0.3 μGal in the scenario with subsidence. Compare with Figure 16 in the article showing the noise-free 4D gravity anomaly. The feasible limit of 3 μGal is represented by the red lines; Figure S6: Noise-corrupted 4D gravity anomaly with zero mean and standard deviation of 0.1 μGal in the scenario with subsidence. Compare with Figure 16 in the article showing the noise-free 4D gravity anomaly. The feasible limit of 3 μGal is represented by the red lines.

Author Contributions: Conceptualization, V.C.F.B. and V.C.O.J.; Methodology, A.D.A., V.C.F.B., V.C.O.J. and P.T.L.M.; Formal analysis, A.D.A.; Investigation, P.T.L.M.; Writing original draft, A.D.A., V.C.F.B., V.C.O.J. and P.T.L.M. All authors have read and agreed to the published version of the manuscript.

Funding: Valeria C.F. Barbosa was supported by fellowships from: CNPQ (grant 309624/2021-5) and FAPERJ (grant E-26/202.582/2019). Vanderlei C. Oliveira Jr was supported by fellowships from CNPQ (grant 315768/2020-7) and FAPERJ (grant E-26/202.729/2018).

Data Availability Statement: Reservoir flow simulator used in this research is classified information.

Acknowledgments: Andre D. Arelaro and Paulo T. L. Menezes thank Petrobras for the permission to publish this work. Valeria C. F. Barbosa and Vanderlei C. Oliveira Jr thank Observatório Nacional, CNPQ, and FAPERJ. All authors thank Julio C.S.O Lyrio, Mariana M. L. Silva, and Luiz A. Santos for the revision of this paper.

Conflicts of Interest: The authors declare no conflict of interest.

References

1. Sasagawa, G.S.; Crawford, W.; Eiken, O.; Nooner, S.; Stenvold, T.; Zumberge, M.A. A new sea-floor gravimeter. *Geophysics* **2003**, *68*, 544–553. [CrossRef]
2. Eiken, O.; Zumberge, M.; Hildebrand, J. Method for Monitoring Seafloor Subsidence and for Gravity Monitoring an Underground Hydrocarbon Reservoir. US Patent 6,813,564, 2 November 2004.
3. Stenvold, T. Offshore Gravimetric and Subsidence Monitoring. Ph.D. Thesis, Norwegian University of Science and Technology, Trondheim, Norway, 2008.
4. Pepper, T. The Gulf underwater gravimeter. *Geophysics* **1941**, *6*, 34–44. [CrossRef]
5. Biegert, E.; Ferguson, J.; Li, X. 4D gravity monitoring—Introduction. *Geophysics* **2008**, *73*, WA1–WA2. [CrossRef]
6. Crossley, D.; Hinderer, J.; Riccardi, U. The measurement of surface gravity. *Rep. Prog. Phys.* **2013**, *76*, 046101. [CrossRef] [PubMed]
7. Zumberge, M.; Alnes, H.; Eiken, O.; Sasagawa, G.; Stenvold, T. Precision of seafloor gravity and pressure measurements for reservoir monitoring. *Geophysics* **2008**, *73*, WA133–WA141. [CrossRef]
8. Alnes, H.; Stenvold, T.; Eiken, O. Experiences on seafloor gravimetric and subsidence monitoring above producing reservoirs. In Proceedings of the 72nd EAGE Conference and Exhibition Incorporating SPE EUROPEC 2010, Barcelona, Spain, 14–17 June 2010; European Association of Geoscientists & Engineers: Utrecht, The Netherlands, 2010; pp. cp–161.
9. Alnes, H.; Eiken, O.; Stenvold, T. Monitoring gas production and CO 2 injection at the Sleipner field using time-lapse gravimetry. *Geophysics* **2008**, *73*, WA155–WA161. [CrossRef]
10. Stenvold, T.; Eiken, O.; Landrø, M. Gravimetric monitoring of gas-reservoir water influx—A combined flow-and gravity-modeling approach. *Geophysics* **2008**, *73*, WA123–WA131. [CrossRef]
11. Eiken, O.; Stenvold, T.; Zumberge, M.; Alnes, H.; Sasagawa, G. Gravimetric monitoring of gas production from the Troll field. *Geophysics* **2008**, *73*, WA149–WA154. [CrossRef]
12. Krahenbuhl, R.A.; Martinez, C.; Li, Y.; Flanagan, G. Time-lapse monitoring of CO2 sequestration: A site investigation through integration of reservoir properties, seismic imaging, and borehole and surface gravity data. *Geophysics* **2015**, *80*, WA15–WA24. [CrossRef]
13. Røste, T.; Ke, G. Overburden 4D time shifts—Indicating undrained areas and fault transmissibility in the reservoir. *Lead. Edge* **2017**, *36*, 423–430. [CrossRef]
14. Van Camp, M.; Viron, O.; Watlet, A.; Meurers, B.; Francis, O.; Caudron, C. Geophysics From Terrestrial Time-Variable Gravity Measurements. *Rev. Geophys.* **2017**, *55*, 938–992. [CrossRef]
15. Ruiz, H.; Agersborg, R.; Hille, L.; Lien, M.; Lindgård, J.; Vatshelle, M. Monitoring Offshore CO2 Storage Using Time-lapse Gravity and Seafloor Deformation. In Proceedings of the EAGE/SEG Research Workshop 2017, Trondheim, Norway, 28–31 August 2017; European Association of Geoscientists & Engineers: Bunnik, The Netherlands, 2017; pp. cp–522.
16. Vevatne, J.; Alnes, H.; Eiken, O.; Stenvold, T.; Vassenden, F. Use of field-wide seafloor time-lapse gravity in history matching the Mikkel gas condensate field. In Proceedings of the 74th EAGE Conference and Exhibition Incorporating EUROPEC 2012, Copenhagen, Denmark, 4–7 June 2012; European Association of Geoscientists & Engineers: Bunnik, The Netherlands, 2012; pp. cp–293.
17. Ruiz, H.; Agersborg, R.; Hille, L.; Lindgård, J.E.; Lien, M.; Vatshelle, M. Monitoring offshore reservoirs using 4D gravity and subsidence with improved tide corrections. In *SEG Technical Program Expanded Abstracts 2016*; Society of Exploration Geophysicists: Tulsa, OK, USA, 2016; pp. 2946–2950.
18. Agersborg, R.; Hille, L.; Lien, M.; Lindgård, J.; Ruiz, H.; Vatshelle, M. Density Changes and Reservoir Compaction from In-situ Calibrated 4D Gravity and Subsidence Measured at the Seafloor. In Proceedings of the SPE Annual Technical Conference and Exhibition, San Antonio, TX, USA, 9–11 October 2017; Society of Petroleum Engineers: Richardson, TX, USA, 2017.
19. Vatshelle, M.; Glegola, M.; Lien, M.; Noble, T.; Ruiz, H. Monitoring the Ormen Lange field with 4D gravity and seafloor subsidence. In Proceedings of the 79th EAGE Conference and Exhibition 2017, Paris, France, 12–15 June 2017.
20. Sarkowi, M.; Kadir, W.G.; Santoso, D. Strategy of 4D microgravity survey for the monitoring of fluid dynamics in the subsurface. In Proceedings of the Proceedings World Geothermal Congress, Antalya, Turkey, 24–29 April 2005; pp. 1–5.
21. Ferguson, J.; Chen, T.; Brady, J.; Aiken, C.; Seibert, J. The 4D microgravity method for waterflood surveillance: Part II—Gravity measurements for the Prudhoe Bay reservoir, Alaska. *Geophysics* **2007**, *72*, I33–I43. [CrossRef]
22. Krahenbuhl, R.A.; Li, Y. Time-lapse gravity: A numerical demonstration using robust inversion and joint interpretation of 4D surface and borehole data. *Geophysics* **2012**, *77*, G33–G43. [CrossRef]
23. Krahenbuhl, R.A.; Reitz, A.; Li, Y.; Rim, H. Improved recovery of fluid movement through time-lapse borehole vector gravity. In *SEG Technical Program Expanded Abstracts 2014*; Society of Exploration Geophysicists: Tulsa, OK, USA, 2014; pp. 1348–1353.
24. Young, W.M.; Lumley, D. Feasibility analysis for time-lapse seafloor gravity monitoring of producing gas fields in the Northern Carnarvon Basin, offshore Australia. *Geophysics* **2015**, *80*, WA149–WA160. [CrossRef]
25. Reitz, A.; Krahenbuhl, R.; Li, Y. Feasibility of time-lapse gravity and gravity gradiometry monitoring for steam-assisted gravity drainage reservoirs. *Geophysics* **2015**, *80*, WA99–WA111. [CrossRef]
26. Elliott, E.J.; Braun, A. Gravity monitoring of 4D fluid migration in SAGD reservoirs—Forward modeling. *CSEG Rec.* **2016**, *41*, 16–21.
27. Balza, A.; Li, Y. Integration of time-lapse gravity and production data to monitor gas reservoirs. In Proceedings of the 79th EAGE Conference and Exhibition 2017-Workshops, Paris, France, 12–15 June 2017.

28. Arelaro, A.D.; Barbosa, V.C.; Oliveira Jr., V.C. Seafloor 4D gravity sensitivity analysis of a deep-water Brazilian oil field. In First International Meeting for Applied Geoscience & Energy Expanded Abstracts; Society of Exploration Geophysicists: Tulsa, OK, USA, 2021; pp. 884–888.
29. Blakely, R.J. *Potential Theory in Gravity and Magnetic Applications*; Cambridge University Press: Cambridge, UK, 1996.
30. Nagy, D.; Papp, G.; Benedek, J. The gravitational potential and its derivatives for the prism. *J. Geod.* **2000**, *74*, 552–560. Corrected in *J. Geod.* **2002**, *76*, 475. [CrossRef]
31. Geertsma, J. Land subsidence above compacting oil and gas reservoirs. *J. Pet. Technol.* **1973**, *25*, 734–744. [CrossRef]
32. Fjær, E.; Holt, R.M.; Raaen, A.; Horsrud, P. *Petroleum Related Rock Mechanics*; Elsevier: Amsterdam, The Netherlands, 2008.
33. van Thienen-Visser, K.; Pruiksma, J.; Breunese, J. Compaction and subsidence of the Groningen gas field in the Netherlands. *Proc. Int. Assoc. Hydrol. Sci.* **2015**, *372*, 367–373. [CrossRef]
34. Mindlin, R.D.; Cheng, D.H. Thermoelastic stress in the semi-infinite solid. *J. Appl. Phys.* **1950**, *21*, 931–933. [CrossRef]
35. Sen, B. Note on the stresses produced by nuclei of thermo-elastic strain in a semi-infinite elastic solid. *Q. Appl. Math.* **1951**, *8*, 365–369. [CrossRef]
36. Sharma, B.D. Stresses In an Infinite Slab due to a Nucleus of Thermoelastic Strain in it. *ZAMM-J. Appl. Math. Mech. Z. Angew. Math. Mech.* **1956**, *36*, 75–78. [CrossRef]
37. Barbosa, V.C.F.; Oliveira Jr., V.C.; Arelaro, A.D.; Borges, F.A.S. 3D Displacement and Stress Fields of Compacting Reservoir: Alternative Solutions. *Braz. J. Geophys.* **2022**, *40*, 1–12. [CrossRef]
38. Fukushima, T. Speed and accuracy improvements in standard algorithm for prismatic gravitational field. *Geophys. J. Int.* **2020**, *222*, 1898–1908. [CrossRef]
39. Asmus, H.E.; Ponte, F.C. The Brazilian marginal basins. In *The South Atlantic*; Springer: New York, NY, USA, 1973; pp. 87–133.
40. Bruhn, C.; T. Gomes, J.; Lucchese, C., Jr.; Johann, P. Campos Basin: Reservoir Characterization and Management-Historical Overview and Future Challenges. In Proceedings of the Offshore Technology Conference, Houston, TX, USA, 5–8 May 2003.
41. Leyden, R.; Asmus, H.; Zembruscki, S.; Bryan, G. South Atlantic diapiric structures. *AAPG Bull.* **1976**, *60*, 196–212.
42. Freire, W. Campos Basin Deepwater Giant Fields. Paper presented at the Offshore Technology Conference, Houston, TX, USA, 1–4 May 1989. . [CrossRef]
43. Johann, P.; Sansonowski, R.; Oliveira, R.; Bampi, D. 4D seismic in a heavy-oil, turbidite reservoir offshore Brazil. *Lead. Edge* **2009**, *28*, 718–729. [CrossRef]
44. Buonora, M.P.P.; Correa, J.L.; Martins, L.S.; Menezes, P.T.L.; Pinho, E.J.C.; Silva Crepaldi, J.L.; Ribas, M.P.P.; Ferreira, S.M.; Freitas, R.C. mCSEM data interpretation for hydrocarbon exploration: A fast interpretation workflow for drilling decision. *Interpretation* **2014**, *2*, SH1–SH11. [CrossRef]
45. Correa, J.L.; Menezes, P.T. Marlim R3D: A realistic model for controlled-source electromagnetic simulations—Phase 2: The controlled-source electromagnetic data set. *Geophysics* **2019**, *84*, E293–E299. [CrossRef]
46. Guardado, L.R.; Spadini, A.R.; Brandão, J.S.L.; Mello, M.R. Petroleum System of the Campos Basin, Brazil. In *Petroleum Systems of South Atlantic Margins: AAPG Memoir 73*; Mello, M.R., Katz, B.J., Eds.; AAPG: Tulsa, OK, USA 2000; Chapter 22; pp. 317–324.
47. Rodriguez-Suarez, C.; de Souza, J.A.B.; Sarzenski, D.J.; Ida, M.; Elias, V.L.G. Reservoir characterization of Roncador, Campos Basin. In Proceedings of the 8th International Congress of the Brazilian Geophysical Society, Rio de Janeiro, Brazil, 14–18 September 2003; European Association of Geoscientists & Engineers: Utrecht, The Netherlands, 2003; pp. cp–168.
48. Nascimento, T.M.; Menezes, P.T.; Braga, I.L. High-resolution acoustic impedance inversion to characterize turbidites at Marlim Field, Campos Basin, Brazil. *Interpretation* **2014**, *2*, T143–T153. [CrossRef]

Disclaimer/Publisher's Note: The statements, opinions and data contained in all publications are solely those of the individual author(s) and contributor(s) and not of MDPI and/or the editor(s). MDPI and/or the editor(s) disclaim responsibility for any injury to people or property resulting from any ideas, methods, instructions or products referred to in the content.

minerals

MDPI

Article

An Integrated Deep Learning Framework for Classification of Mineral Thin Sections and Other Geo-Data, a Tutorial

Paolo Dell'Aversana

Eni S.p.A., San Donato Milanese, 20097 Milan, Italy; paolo.dell'aversana@eni.com

Abstract: Recent studies have demonstrated the potential of machine learning methods for fast and accurate mineral classification based on microscope thin sections. Such methods can be extremely useful to support geoscientists during the phases of operational geology, especially when mineralogical and petrological data are fully integrated with other geological and geophysical information. In order to be effective, these methods require robust machine learning models trained on pre-labeled data. Furthermore, it is mandatory to optimize the hyper-parameters of the machine learning techniques in order to guarantee optimal classification accuracy and reliability. Nowadays, deep learning algorithms are widely applied for image analysis and automatic classification in a large range of Earth disciplines, including mineralogy, petrography, paleontology, well-log analysis, geophysical imaging, and so forth. The main reason for the recognized effectiveness of deep learning algorithms for image analysis is that they are able to quickly learn complex representations of images and patterns within them. Differently from traditional image-processing techniques based on handcrafted features, deep learning models automatically learn and extract features from the data, capturing, in almost real-time, complex relationships and patterns that are difficult to manually define. Many different types of deep learning models can be used for image analysis and classification, including fully connected deep neural networks (FCNNs), convolutional neural networks (CNNs or ConvNet), and residual networks (ResNets). In this paper, we compare some of these techniques and verify their effectiveness on the same dataset of mineralogical thin sections. We show that the different deep learning methods are all effective techniques in recognizing and classifying mineral images directly in the field, with ResNets outperforming the other techniques in terms of accuracy and precision. In addition, we compare the performance of deep learning techniques with different machine learning algorithms, including random forest, naive Bayes, adaptive boosting, support vector machine, and decision tree. Using quantitative performance indexes as well as confusion matrixes, we demonstrate that deep neural networks show generally better classification performances than the other approaches. Furthermore, we briefly discuss how to expand the same workflow to other types of images and geo-data, showing how this deep learning approach can be generalized to a multiscale/multipurpose methodology addressed to the analysis and automatic classification of multidisciplinary information. This article has tutorial purposes, too. For that reason, we will explain, with a didactical level of detail, all the key steps of the workflow.

Keywords: deep learning; minerals; thin sections; multidisciplinary images' fast classification

Citation: Dell'Aversana, P. An Integrated Deep Learning Framework for Classification of Mineral Thin Sections and Other Geo-Data, a Tutorial. *Minerals* **2023**, *13*, 584. https://doi.org/10.3390/min13050584

Academic Editor: Stanisław Mazur

Received: 20 March 2023
Revised: 17 April 2023
Accepted: 20 April 2023
Published: 22 April 2023

Copyright: © 2023 by the author. Licensee MDPI, Basel, Switzerland. This article is an open access article distributed under the terms and conditions of the Creative Commons Attribution (CC BY) license (https://creativecommons.org/licenses/by/4.0/).

1. Introduction

The recognition and analysis of images are critical tasks in various Earth disciplines, including seismic facies classification, well-log analysis, microfossil and mineralogical species' recognition, among others [1–6]. Recent studies have explored the use of machine learning methods for fast mineral classification based on microscope thin sections in field geology operations. For instance, one approach involves using digital image analysis to extract features from microscope thin-section images, such as grain size, shape, and color. Machine learning algorithms can then be trained on these features to automatically identify and classify minerals in real-time. One example of this approach is a study by She et al. [7], who used machine learning to classify different minerals in thin-section images of ore samples. The authors found that their method was able to classify minerals with an overall accuracy of 89.3%. Another approach involves using deep learning techniques, such as convolutional neural networks (CNNs), to directly learn the features of microscope thin-section images and classify minerals based on these features. For instance, Liu et al. [8] used a CNN to classify different minerals in thin-section images of rock samples. The authors found that their method was able to classify minerals with an overall accuracy of 93.7%. Overall, these studies demonstrate the potential of machine learning methods for fast mineral classification based on microscope thin sections in field geology operations. However, it is important to remark that these methods may require to be trained on pre-labeled data, in order to speed up the classification workflow. Furthermore, careful hyper-parameters' optimization is mandatory in order to achieve optimal accuracy and reliability through fast machine learning workflows.

In recent years, deep neural networks (DNNs) have made significant improvements in image classification, with the hierarchical models of the human visual system serving as a useful conceptual basis for building effective artificial networks distributed with a layered topology [9]. DNNs' hierarchical organization enables them to effectively share and reuse information, making them suitable for solving complex non-linear functions. The term "deep learning" is often used to describe multilayer neural networks with many hidden neuronal layers between the input and output layers. The simplest DNN architecture is the fully connected neural network (FCNN) or multilayer perceptron [10,11], which connects all nodes in one layer to all neurons in the next layer. However, this architecture is susceptible to overfitting and other problems that can limit the accuracy of the classification results. As anticipated above, Convolutional neural networks (CNNs or ConvNets) [12,13] represent a more sophisticated and effective DNN approach, especially in computer vision. They have shown excellent performance in solving complex image classification and pattern recognition problems. The first significant difference between ConvNets and FCNNs is the concept of "local processing". Neurons belonging to two successive layers in ConvNets are connected only locally, reducing the number of connections and the computation complexity. The connection weights are shared in groups, significantly reducing the number of weights. Additionally, ConvNets alternate between convolutional and pooling layers, which reduce the dimensions of data.

Unfortunately, the vanishing gradient problem can limit the effectiveness of DNNs with many hidden layers. This problem occurs when the gradient becomes vanishingly small, preventing the weight from changing its value during the backpropagation process, which can degrade the network's learning capabilities. Nonetheless, solutions have been proposed and successfully applied to address this problem, such as convolutional deep residual networks [14] (convolutional ResNet or, briefly, ResNet). As we will explain in detail in the methodological section, this type of deep neural network architecture includes residual connections between layers. These connections allow the network to bypass certain layers and learn residual functions, making it easier to train very deep networks. ResNets have been shown to achieve state-of-the-art performance on a variety of computer vision tasks, such as image classification and object detection. They can find interesting applications in several Earth disciplines, such as petrography, mineralogy, paleontology, sedimentology, and in all those fields where image recognition/classification plays a crucial role in the data interpretation workflow.

In this paper, we discuss examples of the application of different deep learning techniques to mineralogical classification problems through image analysis of microscope thin sections, by expanding our previous work on a similar subject [5]. The following is the structure of the paper:

- First, we introduce, briefly, the main methodological aspects of the techniques applied in our study, including fully connected, convolutional, and residual neural networks. We highlight benefits and limitations of these different deep learning methods.

- Next, in the example section, we start with an application based on the fully connected deep neural network. We show how this network is able to classify images of mineral thin sections, although with some uncertainties and classification mistakes.
- Then, we discuss another example using convolutional deep residual networks with a varying number of hidden layers. We show how the classification results can be improved with a ResNet architecture, with an accuracy that depends on the number of hidden layers.
- Next, we will compare the performance of the different deep neural networks, highlighting benefits and limitations of the various types of architecture.
- Finally, we compare the performances of deep neural networks with those of different types of algorithms, such as random forest, naive Bayes, adaptive boosting, support vector machine, and decision tree.

As stated earlier, this paper has tutorial purposes, too. For that reason, we explain in detail the main pragmatic aspects of the workflow addressed to image analysis and classification. These aspects involve image-embedding techniques, image pre-processing, feature engineering, hyper-parameters' optimization of network architecture, training and cross-validation techniques, classification methods, and representation of the results for each one of the various types of deep neural networks applied here.

2. Methodological Overview

In this section, we summarize the main characteristics of the different types of deep neural networks that we tested on the same experimental dataset. Our goal is to remark on the differences, limitations, and benefits of the various deep learning techniques. Additional technical/mathematical details, as well as a brief history about the developments of the neural network architecture over the past decades, can be found in a previous work that we dedicated to the description of the various types of deep learning architectures [15].

2.1. Fully Connected Neural Network (FCNN)

The fully connected neural network, also known as the dense neural network or multilayer perceptron, is a type of artificial neural network that is widely used in machine learning and deep learning applications. In a FCNN, each neuron in one layer is connected to every neuron in the next layer. The input layer receives the input data, and the output layer produces the output prediction. There can be one or more hidden layers between the input and output layers, which are responsible for extracting relevant features from the input data. Each neuron in a fully connected layer receives a weighted sum of inputs from the previous layer, adds a bias term, and applies a non-linear activation function to produce its output. The weights and biases of the network are learned during training using optimization techniques such as backpropagation.

FCNN are called "deep" when they have multiple hidden layers. Deep neural networks are capable of learning complex and hierarchical representations of input data, making them suitable for a wide range of applications, such as image recognition, natural language processing, and speech recognition. However, FCNN can suffer from overfitting, especially when dealing with high-dimensional input data. Regularization techniques such as L2 regularization can help to alleviate overfitting. Additionally, the large number of parameters in deep neural networks can make training slow and computationally expensive. Techniques such as batch normalization and weight initialization can help speed up training and improve performance.

Let us summarize the key features, benefits, and limitations of fully connected neural networks (FCNNs):

- FCNNs are a type of artificial neural network, where each neuron in one layer is connected to every neuron in the next layer.
- One of the benefits of FCNNs is that they can learn complex non-linear relationships between input and output data.

- They are also relatively simple to understand and implement, making them a popular choice for many machine learning tasks.
- However, FCNNs can be prone to overfitting, especially if the dataset is small or noisy.
- Additionally, training larger FCNNs can be computationally expensive and time-consuming.
- Regularization techniques, such as dropout, can be used to mitigate overfitting.

2.2. Convolutional Neural Network (CNN or ConvNet)

Convolutional neural networks are a type of deep neural network that are primarily used for image- and video-processing tasks. CNNs are inspired by the structure of the visual cortex in animals and are designed to automatically and adaptively learn spatial hierarchies of features from input data. The main building blocks of a CNN are convolutional layers, pooling layers, and fully connected layers. Convolutional layers perform a series of convolutions on the input data using a set of learnable filters or kernels. Each filter slides across the input data, computing dot products with the input data and producing a feature map that highlights a particular pattern or feature in the input.

Pooling layers down-sample the feature maps produced by the convolutional layers by computing a summary statistic (e.g., maximum or average) within a small region of the feature map. Pooling helps reduce the spatial size of the feature maps and control overfitting. The output of the convolutional and pooling layers is then fed into one or more fully connected layers, which perform classification or regression on the high-level features extracted from the input data. The filters or kernels in the convolutional layers are learned during training using backpropagation and stochastic gradient descent. The training process involves minimizing a loss function that measures the discrepancy between the network's predictions and the ground truth labels.

CNNs have several advantages over traditional image-processing techniques (including FCNNs). They can automatically learn a hierarchy of features from raw pixel data, eliminating the need for manual feature engineering. Additionally, CNNs are robust to small variations in the input data, such as translations and rotations, making them suitable for a wide range of real-world applications. CNNs have been successfully applied to a wide range of computer vision tasks, including image classification, object detection, face recognition, and image segmentation.

In summary, the following are the key concepts, pros, and cons of convolutional neural networks (ConvNets, or CNNs):

- CNNs are a type of deep neural network that are well-suited for image- and video-processing tasks.
- The key benefit of CNNs is their ability to automatically detect features or patterns in images, without the need for manual feature engineering.
- CNNs use convolutional layers to process input images, where each layer extracts specific features from the input image.
- The limitation of CNNs is that they can be prone to overfitting, especially if the dataset is small or noisy.

2.3. Deep Convolutional Residual Neural Network (ResNet)

Deep convolutional residual networks are a type of neural network architecture designed to address the problem of degradation in deep neural networks by allowing the networks to learn residual functions instead of directly learning the desired mapping.

The idea behind ResNets is based on the observation that as the depth of a neural network increases, the performance of the network can degrade, meaning that the accuracy on the training set starts to decrease. This degradation is because it becomes increasingly difficult for the network to learn the underlying mapping as the depth increases. ResNets address this problem by introducing a new type of residual block, which is a building block of the network (Figure 1).

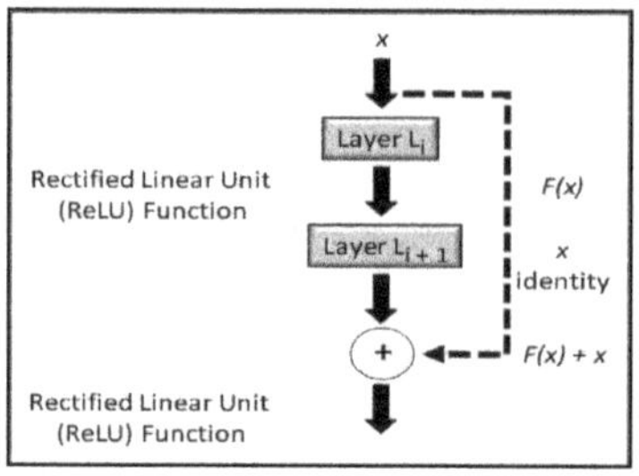

Figure 1. Shortcut connection [14].

To better understand the concept of residual blocks in neural networks, it can be useful to compare how it works through comparison with the basic structure of a "standard" neural network. In a typical neural network, each layer learns to transform its input data into a higher-level representation that can be used by the subsequent layers. Each layer is composed of a set of learnable parameters that are used to compute a set of outputs from the input data. The output of one layer is then passed as input to the next layer, and this process continues until the final output is produced.

A residual block, on the other hand, introduces a shortcut connection that bypasses one or more layers in the network (Figure 1).

The purpose of this shortcut connection is to allow the network to learn residual functions. A residual function is the difference between the input to the block and the output of the block. With reference to Figure 1, a building block is defined through identity mapping by shortcuts, as follows:

$$y = F(x, \{W_i\}) + x \tag{1}$$

Here, x and y are, respectively, the input and output vectors of the layers considered. The function $F(x, \{W_i\})$ is the residual mapping to be learned. For example, in case of a two-layer building block, we have:

$$F = W_2\sigma\cdot(W_1 x) \tag{2}$$

where W_1 and W_2 represent the weights of the two neuron layers, σ represents the "ReLU" (or "relu", as in Figure 1) activation function (see below for an explanation of "ReLU"), and the biases are omitted for the sake of simplicity. If F has only a single layer, Equation (1) corresponds to a linear layer, for which we have no observed advantage:

$$y = W_1 x + x \tag{3}$$

Finally, if we are considering convolutional layers of convolutional neural networks, the function $F(x, \{W_i\})$ can represent multiple convolutional layers. More specifically, in a residual block, the input data are first passed through a set of convolutional layers to transform them into a higher-level representation. This output is then added back to the original input data, resulting in the residual function. The residual function is then passed through another set of convolutional layers to produce the final output of the block.

The addition of the residual function to the original input data effectively allows the network to learn the difference between the desired output and the current output at that layer. This is important because in deep networks, it can be difficult for the network to directly learn the desired mapping. By allowing the network to learn residual functions, the network can more easily learn the desired mapping and improve its performance.

Overall, the use of residual blocks in neural networks is a powerful technique for improving the performance of very deep networks. By allowing the network to learn residual functions and use shortcut connections to reuse learned features from previous layers, residual networks are able to achieve state-of-the-art performance on a variety of tasks. The architecture of a ResNet typically consists of several layers of residual blocks, followed by a global average pooling layer and a fully connected layer for classification. The shortcut connections allow the network to be very deep, with over 1000 layers in some cases. The residual blocks also have the advantage of being able to reuse learned features from previous layers, which can lead to more efficient training and better performance. The main benefits of using ResNets are improved performance on very deep networks, faster convergence during training, and the ability to train networks with many layers. ResNets have been shown to outperform other state-of-the-art architectures on a variety of tasks, including image classification, object detection, and semantic segmentation.

In summary, deep residual networks are a powerful neural network architecture that address the problem of degradation in very deep networks by allowing the networks to learn residual functions. By introducing shortcut connections and residual blocks, ResNets are able to efficiently learn features from previous layers and achieve better performance on a variety of tasks.

Schematically, the following are the main advantages and disadvantages of ResNets:

- ResNets are a type of CNN that use residual blocks to enable the training of very deep models.
- The key concept of ResNets is the shortcut connections that allow information to bypass certain layers and be directly propagated to deeper layers.
- One benefit of ResNets is that they can achieve higher accuracy than traditional CNNs when working with very deep networks.
- ResNets can also mitigate the problem of vanishing gradients that can occur in very deep networks.
- However, ResNets have a higher computational cost than traditional CNNs, and can require more memory and longer training times.
- Additionally, the extra shortcut connections can make the network prone to overfitting if the dataset is small.

3. Examples

This section outlines the use of the various deep learning methods briefly explained in the previous section, starting with simpler fully connected neural networks and progressing to more advanced deep residual networks. We discuss how we applied these different architectures to classify images of mineral thin sections obtained from real samples. We applied these techniques to a test dataset with a didactical purpose, too. For that reason, in the following section, we explain the details of the main steps of the classification workflow, clarifying why and how we selected each specific hyper-parameter of our network architecture. Our goal is to show how it is possible to test and to evaluate the effectiveness of the different approaches in classifying the various mineral species through their microscope images.

3.1. Classification of Mineralogical Thin Sections Using FCNN

In this test, we utilized a dataset consisting of about 200 thin sections of rocks and minerals (link to the dataset: http://www.alexstrekeisen.it/index.php, accessed on 20 February 2023; courtesy of Alessandro Da Mommio). We collected these images for creating a labeled dataset for training the deep learning models used, successively, for classifying new images. The thin sections are presented as low-resolution colored (RGB) JPEG images with a resolution of 96 dpi and a size of 275 × 183 pixels. The objective of this study was to classify the thin sections into four distinct classes: augite, biotite, olivine, and plagioclase. Although the classification may seem straightforward, it was relatively challenging due to the similarities in the geometric features of different minerals and the potential effects of corrosion and alteration. Furthermore, we performed an additional classification test of four types of sedimentary rocks, using a set of jpeg images of thin sections. In this second type of test, we applied a suite of machine learning techniques (decision tree, random forest, adaptive boosting, support vector machine, and naive Bayes), with the goal of comparing different classification approaches with deep learning methods.

In the following part, we describe in detail all the steps of the workflow (as in the scheme of Figure 2).

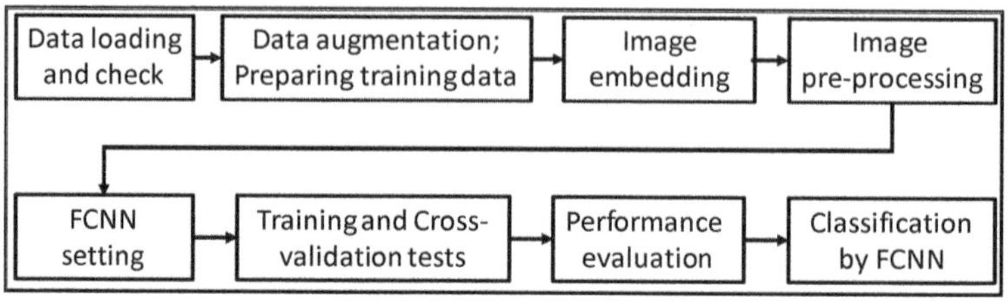

Figure 2. Block diagram of the classification workflow using FCNN. The workflow begins with data loading and data augmentation aimed at increasing the training dataset's size. It then moves on to image embedding to convert images into feature vectors. The following step is image pre-processing to normalize instances of the features and prepare the data for training. The subsequent steps are setting the hyper-parameters of the FCNN, training the network, and performing cross-validation tests. These are followed by performance evaluation using a suite of performance indexes and the confusion matrix method. Finally, the deep neural network classifies the unlabeled data.

3.1.1. Data Augmentation and Preparation of the Training Dataset

After data loading and check, we created a set of examples to train the FCNN. In this phase, the assistance of an expert in mineralogy was crucial for labeling a significant number of representative images of the various mineralogical classes. Due to the limited number of available images, the training dataset used in this study was relatively small. Defining a minimum number of training examples is a challenging task because it varies depending on the complexity of the classification problem, image quality and heterogeneity, and algorithms used. To estimate the appropriate number of training data required, specific tests can be performed, as explained below in the section dedicated to cross-validation tests. To address the issue of the small training dataset, we applied dataset augmentation techniques to add "artificial" images. These techniques allow for the application of transformation operators to the original data, including flipping (vertically and horizontally), rotating, zooming and scaling, cropping, translating the image (moving along the x or y axis), and adding Gaussian noise (distortion of high-frequency features). For this pre-processing, we used Python libraries, available in the Tensorflow package. The underlying concept behind this data augmentation approach is that the accuracy of the neural network model can be significantly improved by combining different operators across the original dataset.

3.1.2. Image Embedding

Image embedding is a technique used in computer vision and machine learning to convert an image into a feature vector, which can then be used for tasks such as image classification, object detection, and image retrieval. It involves transforming an image into a set of numbers that can be easily processed by machine learning algorithms.

We tested various algorithms and techniques for embedding our images (jpeg files of mineralogical thin sections), and finally, we adopted the SqeezeNet technique. This is an algorithm with a neural network architecture that uses a combination of convolutional layers and modules to extract features from images. The reason for our selection was because SqeezeNet was more computationally efficient than the other methods, while still preserving a high accuracy. Other techniques that we tested are the following:

VGG-16 and VGG-19: These are convolutional neural networks that were developed by the Visual Geometry Group at Oxford University. They consist of 16 or 19 layers, respectively, and are known for their effectiveness in image classification tasks.

Painters: This is an algorithm developed by Google that creates an embedding of an image by synthesizing a new image from it. It works by training a neural network to generate a painting that is similar to the original image, and then using the internal representation of the network as the image embedding.

DeepLoc: This is an algorithm developed by the University of Oxford that creates an embedding of an image by combining information from multiple layers of a convolutional neural network. It is designed specifically for the task of protein subcellular localization, which involves determining the location of proteins within cells based on microscopic images.

3.1.3. Pre-Processing

Before training the FCNN and using it for image classification, it was necessary to pre-process all the images of the datasets. The following are the main pre-processing steps applied to our data.

Normalization: Normalizing image data is an important step to ensure that values across all images (after embedding operations) are on a common scale. This can prevent the dominance of certain features due to differences in their ranges of values. In general, when dealing with numerical features, centering the data by mean or median can help to shift the data so that the central value is closer to zero. Scaling the data by standard deviation can help to further adjust the data to a suitable range for analysis.

Randomization: Randomizing instances and classes is a useful step for reducing bias and ensuring that the future classification model is robust to different orders of presentation of data. This can help to ensure that the final model(s) is (are) not biased towards any particular pattern or structure in the data.

Removing sparse features: This is another useful step that can help to simplify the data and remove noise. Features with a high percentage of missing or zero values may not contribute much to the classification task and can be safely removed.

PCA: Principal component analysis is a common technique used for dimensionality reduction. It can help to identify the most important features in the data and reduce the number of features, while still retaining much of the original information.

CUR matrix decomposition: This is another technique for dimensionality reduction that can help to reduce the computational complexity of our analysis. Similar to PCA, it identifies the most important features in the data and reduces the number of features without sacrificing too much information. We tested it, but finally, we only applied PCA.

3.1.4. Fully Connected Neural Network (FCNN) Hyper-Parameters

After data preparation, features' extraction, and pre-processing, the next crucial step was to optimize the hyper-parameters of our FCNN. There are many parameters to play with in order to make a deep neural network effective. Adjusting these parameters in an optimal way can be a difficult and subjective task. However, there are automatic approaches for that purpose, such as using specific reinforcement learning methods that help define the optimal parameters of a neural network for reproducing a desired output. Many combinations of hyper-parameters can be automatically tested, and the effectiveness of each combination is verified through cross-validation tests, as explained in the following. The crucial network parameters are:

- N-hl and N-Neurons: These represent, respectively, the number of hidden layers and the number of neurons populating each hidden layer. We tested neural networks with a minimum of 1 up to a maximum of 10 hidden layers, using a number of neurons ranging from 100 to 300 for each hidden layer. Among the possible choices, after many tests, we selected a quite simple architecture with three hidden layers and 200 neurons for each layer.
- Activation function for the hidden layers: In deep neural networks, activation functions are mathematical functions that are applied to the output of each neuron in the network. These functions introduce non-linearity to the output of each neuron,

allowing the network to learn complex patterns in the input data. There are several types of activation functions, including:

Logistic Sigmoid Function: This function takes an input value and returns a value between 0 and 1, which can be interpreted as a probability. It is defined as:

$\sigma(\mathbf{x}) = 1/(1 + \exp(-\mathbf{x}))$, where $\mathbf{x}$ represents the input vector.

Hyperbolic Tangent Function: This function takes a vector of input values $\mathbf{x}$ and returns values between -1 and 1. It is defined as:

$$\tanh(\mathbf{x}) = (\exp(\mathbf{x}) - \exp(-\mathbf{x}))/(\exp(\mathbf{x}) + \exp(-\mathbf{x}))$$

Rectified Linear Unit (ReLU) Function: This function returns the input value if it is positive, and 0 otherwise. It is defined as:

$$\text{ReLU}(\mathbf{x}) = \max(0, \mathbf{x})$$

Leaky ReLU Function: This function is similar to ReLU but introduces a small slope for negative values, preventing the "dying ReLU" problem that can occur when the gradient of the function becomes 0. It is defined as:

$$\text{Leaky ReLU}(\mathbf{x}) = \max(0.01\mathbf{x}, \mathbf{x})$$

Exponential Linear Unit (ELU) Function: This function is similar to leaky ReLU but uses an exponential function for negative values, allowing it to take negative values. It is defined as: $\text{ELU}(\mathbf{x}) = \{\mathbf{x} \text{ if } \mathbf{x} >= 0, \text{alpha} * (\exp(\mathbf{x}) - 1) \text{ if } \mathbf{x} < 0\}$, where alpha is a small constant.

Choosing the right activation function is important for the performance of a neural network. Some functions work better than others depending on the type of problem being solved and the architecture of the network. For example, ReLU and its variants are widely used in deep learning because of their simplicity and effectiveness, while sigmoid and tanh functions are less popular due to their saturation and vanishing gradient issues. In fact, we used ReLU in our final setting.

- Solver for weight optimization: A solver for weight optimization in a neural network is a method used to find the set of weights that minimize the loss function of the network. The loss function measures the difference between the predicted output of the network and the actual output, and the goal of the solver is to find the set of weights that minimize this difference. There are various types of solvers used for weight optimization in neural networks, each with their own strengths and weaknesses. Outlined below are three commonly used types:

L-BFGS-B: This is an optimizer in the family of quasi-Newton methods, which are used for unconstrained optimization problems. It is a popular choice for optimizing the weights in a neural network because it is fast and efficient and can handle a large number of variables.

SGD: Stochastic gradient descent is a popular optimization algorithm that works by iteratively updating the weights in the network based on the gradient of the loss function with respect to the weights. It is simple to implement and computationally efficient, but it can be sensitive to the choice of the learning rate and can get stuck in local minima.

Adam: This is a stochastic gradient-based optimizer that is a modification of SGD. It uses an adaptive learning rate that adjusts over time based on the past gradients and includes momentum to prevent oscillations. Adam is often preferred over traditional SGD because it is less sensitive to the choice of the learning rate and can converge faster.

The choice of solver depends on the specific problem being solved, the size and complexity of the network, and the available computational resources. Finally, we selected the Adam solver for our network.

- Alpha: L2 penalty (regularization term) parameter. In a neural network, the alpha parameter is a regularization term used to control the amount of L2 regularization applied to the weights of the network during training. L2 regularization is a technique used to prevent overfitting, which occurs when the network learns the training data too well and performs poorly on new, unseen data. The alpha parameter is multiplied by the sum of squares of all the weights in the network and added to the loss function during training. This penalty term encourages the network to learn smaller weights, which helps to reduce overfitting. Increasing the value of alpha increases the amount of regularization applied to the weights, which can help to reduce overfitting but may also result in underfitting if the regularization is too strong. On the other hand, decreasing the value of alpha reduces the amount of regularization applied to the weights, which can lead to overfitting. The optimal value of alpha depends on the specific problem being solved and the complexity of the network. It can be determined using techniques such as grid search or cross-validation. Regularization is an important technique for improving the performance of neural networks and should be considered when building a network. We ran many tests in a wide range of values for this parameter, from relatively small values (0.0005) in order to avoid underfitting, up to high values (100), in order to exclude the possibility of the opposite problem (overfitting). Finally, we selected an average value (0.1).

- Max iterations: Maximum number of iterations. The maximum number of iterations in a deep neural network is the maximum number of times the training algorithm updates the weights of the network during training. The training process in a neural network involves feeding the input data into the network, computing the output, comparing it to the actual output, and updating the weights to minimize the difference between them. The number of iterations needed to train a deep neural network depends on various factors, such as the size and complexity of the network, the amount and complexity of the input data, the choice of activation functions, the optimization algorithm used, and the convergence criteria. To prevent overfitting, it is common to monitor the performance of the network on a validation set during training and stop the training when the performance on the validation set starts to degrade. This can help to avoid training the network for too many iterations, which can lead to overfitting. For our tests, we used a Max iterations number ranging from 100 to 200.

3.1.5. FCNN Training and Cross-Validation Tests

After setting the parameters of our FCNN, as explained above, we performed an automatic sequence of cross-validation tests. These represent a common technique used in machine learning, including deep neural networks, to evaluate the performance of a model on unseen data. It involves partitioning the available dataset into several subsets, or "folds", where one fold is used as the validation set, while the remaining folds are used for training. This process is repeated several times, with each fold taking turns as the validation set.

In the context of deep neural networks, we applied cross-validation to assess how well our FCNN model generalizes to new data, as well as to tune hyper-parameters such as the learning rate, the number of layers, and the number of neurons in each layer. The goal was to find a model that performs well on the validation sets across all folds, without overfitting to the training data.

There are several types of cross-validation tests, including k-fold cross-validation and stratified k-fold cross-validation. In k-fold cross-validation, the dataset is divided into k equally sized folds, and the model is trained and evaluated k times, with each fold serving as the validation set once. In stratified k-fold cross-validation, the dataset is divided into k-folds that preserve the proportion of samples for each class, which can be especially useful when dealing with imbalanced datasets. In our case, we performed mainly k-fold tests using a number of folds ranging between 3 and 8.

3.1.6. FCNN Performance Evaluation

Once the cross-validation process was complete, the performance metrics for each fold were averaged to provide an estimate of the model's performance on unseen data. These metrics can include accuracy, precision, recall, F1 score, and "area under the receiver operating characteristic curve" (AUC-ROC), depending on the specific problem being addressed. The following is a brief explanation of these metric indexes.

Accuracy: This metric measures the proportion of correct predictions made by the model. It is calculated by dividing the number of correct predictions by the total number of predictions. Accuracy is a straightforward metric, but it can be misleading if the dataset is imbalanced, meaning that one class has significantly more samples than the other.

Precision: This metric measures the proportion of true positive predictions out of all positive predictions made by the model. It is calculated by dividing the number of true positive predictions by the sum of true positive and false positive predictions. Precision is useful when the cost of false positives is high.

Recall: This metric measures the proportion of true positive predictions out of all actual positive samples in the dataset. It is calculated by dividing the number of true positive predictions by the sum of true positive and false negative predictions. Recall is useful when the cost of false negatives is high.

F1 Score: This metric is the harmonic mean of precision and recall and provides a single score that balances both metrics. It is calculated as: 2 × (precision × recall)/(precision + recall). The F1 score is useful when both precision and recall are important.

AUC-ROC: This metric measures the model's ability to distinguish between positive and negative samples. It is calculated by plotting the true positive rate (TPR) against the false positive rate (FPR) at various threshold values and calculating the area under the curve of this plot. A perfect classifier will have an AUC-ROC score of 1, while a random classifier will have an AUC-ROC score of 0.5. AUC-ROC is useful when the cost of false positives and false negatives is similar.

Table 1 shows an example of performance evaluation through a k-fold test on our image data. In this specific case, we applied a fully connected deep neural network consisting of 5 hidden layers with 200 neurons each. The network uses a Rectified Linear Unit activation function, and an Adam solver running for a maximum of 200 iterations. We can see that all the indexes showed relatively high values, indicating that our FCNN model has good generalization performance on unseen data.

Table 1. An example of performance indexes for a fully connected neural network after one of many cross-validation tests (see the text for a detailed explanation of each index).

Model	Area under Curve (AUC)	Classification Accuracy (CA)	F1	Precision	Recall
FCNN (5 hidden layers)	0.947	0.796	0.791	0.794	0.797

3.1.7. Classification

After optimizing the network parameters and after the cross-validation tests, we applied our "best" FCNN model for classifying the unlabeled ("unseen") mineralogical thin sections not included in the training dataset. To be more precise, we selected a set of FCNN models with good performances. Finally, we created an "average model" from that set, simply by averaging the optimal hyper-parameters determined through the cross-validation tests (see FCNN parameters related to Table 1). Figure 3 shows one illustrative case of a classification result using the above neural network architecture. The classification performance was good, even though one image of Biotite was misclassified as Olivine (the top-right image).

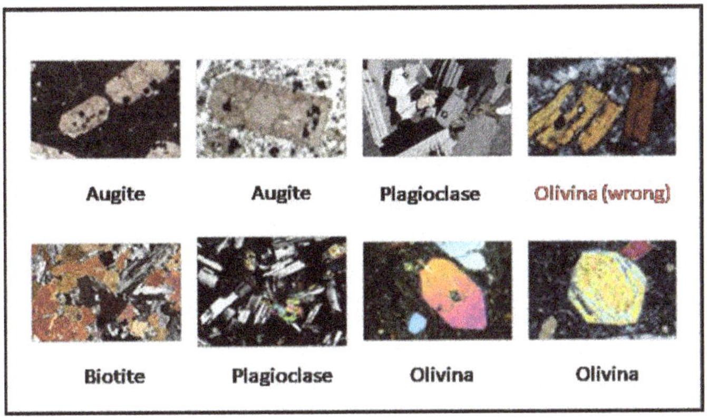

Figure 3. Example of classification result using FCNN. Despite misclassifying one Biotite image as Olivine (in the top-right image), the classification performance can still be considered good.

3.2. Classification of Mineralogical Thin Sections Using ResNet

In order to improve the classification performances, we applied the same workflow shown in Figure 2, but this time using deep convolutional residual neural networks (ResNets) rather than FCNNs. The number of hidden layers in convolutional residual networks can be parameterized, allowing for the testing of their effectiveness with quantitative performance indicators. In our experiments, we tested ResNet_18, ResNet_34, ResNet_50, and ResNet_152, which have 18, 34, 50, and 152 deep layers, respectively. To accomplish this, we created a Jupyter notebook (Python) that utilizes the residual networks open-source code available for download from the "TORCHVISION.MODELS.RESNET" website: https://pytorch.org/vision/0.8/_modules/torchvision/models/resnet.html (accessed on 10 January 2023).

For each network architecture, we conducted cross-validation tests to assess its performance and plotted the accuracy and loss function against the iteration number. Here, accuracy represents the ratio of correct predictions to total input samples, and it increased over time, while the loss function decreased, theoretically converging towards zero. In particular, for ResNet_152, the accuracy reached almost the ideal value of 1 after a few iterations. Figure 4 illustrates the graph of both the training and validation loss functions, plotted together with accuracy versus the iteration number.

Figure 4 is interesting because it shows both benefits and limitations of our classification approach based on ResNet. We remarked that training loss and validation loss are both measures of how well the model is performing on a given dataset, but they have different purposes. Training loss is the error metric used during the training phase to optimize the network parameters. It is calculated as the difference between the predicted output of the network and the true output of the training set. The objective during training is to minimize the training loss by adjusting the weights and biases of the network. The training loss is typically calculated after each batch or epoch of training and is used to update the model parameters. Instead, validation loss is a measure of how well the model generalizes to new, unseen data. It is calculated using the validation set, which is a subset of the data that is not used during training. The validation loss is a metric used to evaluate the performance of the model during training and to prevent overfitting. Overfitting occurs when the model performs well on the training data but poorly on the validation or test data. The goal during training is to minimize the validation loss, which indicates that the model is generalizing well to new data.

It is clear that in our test with ResNet_152, there was some overfitting, because the validation loss started increasing after seven iterations. In general, a good model should show decreasing values versus iterations for both training and validation losses. Sometimes, choosing a residual network with a high number of hidden layers (>100) could be the reason for overfitting effects, as occurred in our case with ResNet_152. For that reason, we tried to classify our thin-section images using a ResNet_50 (with 50 hidden layers). Figure 5 shows the trend of the training and validation loss versus iterations for ResNet_50. This time, both curves decreased with regularity, showing much less overfitting problems than ResNet_152. Unfortunately, the accuracy was not as good as in the previous case. In other words, there was a trade-off between accuracy and overfitting. In conclusion, this tutorial test shows that reliable classification results should derive from a balanced compromise between an accurate training and limited overfitting effects.

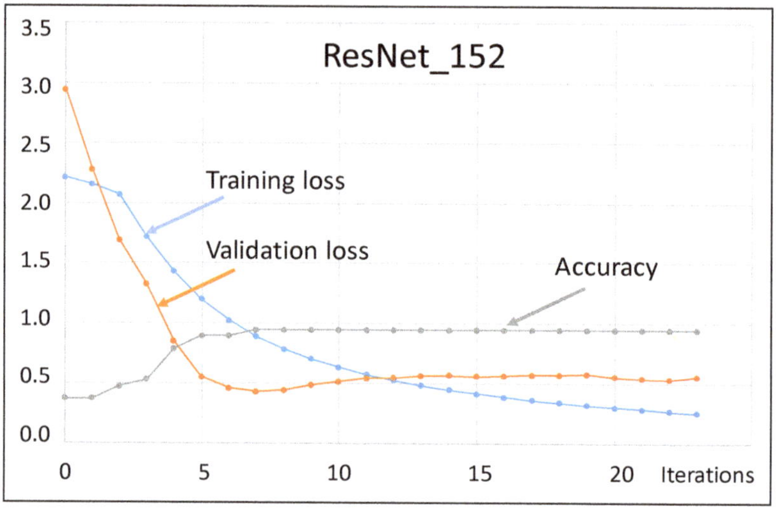

Figure 4. Training and validation loss functions, and accuracy versus iteration number related to ResNet_152. Training loss is the error metric used during the training phase to optimize the network parameters. The validation loss quantifies how effective a model is at extrapolating to new and unseen data. This measure is obtained by utilizing the validation set, which is an independent subset of data not used for training the model. In this specific test, there is some overfitting, because the validation loss starts increasing after a few iterations.

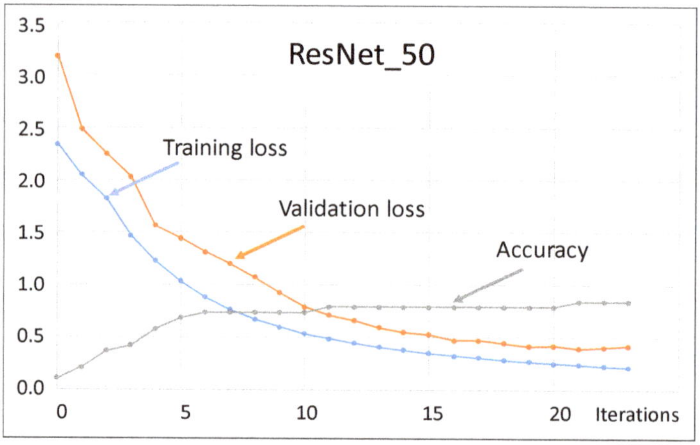

Figure 5. Training and validation loss functions, and accuracy versus iteration number related to ResNet_50. In this test, both curves decrease with regularity, showing much less overfitting problems than the test with Res-Net_152.

Despite the different overfitting problems, in our specific case, both ResNet_152 and ResNet_50 performed well when we applied them for classifying our set of unlabeled thin mineralogical sections. Figure 6 shows an example of classification obtained using ResNet_50. All the test images were properly classified, with variable values of probability (see Table 2). For each test image, the value of the loss function was generally around 0.1–0.2 and the accuracy was generally around 0.8–0.9. This occurred after a few (20–25) iterations, indicating that the network converged quickly towards the correct predictions. Results with ResNet_152 were not very different (with similar probabilities of classification).

Figure 6. Example of classification results with ResNet_50. Here, only 9 samples are considered for illustrative purposes. The main part of the dataset was properly classified using this ResNet_50 architecture, with a low percentage of mistakes. For each classified thin section, the value of the loss function is generally close to 0.1–0.2 and the accuracy is generally greater than 0.8–0.9, after around 20–25 iterations.

Table 2. Classification probabilities for each mineralogical class (for the test images of Figure 6). It can be observed that the classification probabilities for correctly identified mineral species are very high, almost nearing 100%.

	Classification Probability [%] for Each Class			
Sample ID	**Augite**	**Biotite**	**Olivine**	**Plagioclase**
1	98	0	2	0
2	95	3	2	0
3	0	100	0	0
4	1	98	1	0
5	1	0	99	0
6	0	1	99	0
7	1	0	99	0
8	0	1	0	99
9	0	0	0	100

3.3. Comparison with Other Classification Approaches

The deep learning (DL) methods considered in the previous applications (FCNN, ConvNet, and ResNet) use learned features automatically extracted from the raw image data. In order to compare DL methods with some other completely data-driven techniques, or based on handcrafted features, we performed an independent classification of the same dataset using different machine learning methods, not based on neural networks, including decision tree, support vector machine, random forest, naive Bayes, and adaptive boosting. We applied these methods to features of the thin sections manually designed by human experts, including specific patterns, textures, or shapes in the image. Table 3 below shows the comparison of classification performances between deep learning (FCNN, in this example) and the suite of alternative classifiers. Although techniques such as random forest or support vector machine seem to produce satisfactory classification results, as we can notice, a simple FCNN with just five hidden layers showed better performance indexes (highlighted in bold in Table 3) than the other methods. However, none of the indexes were equal to one, indicating that the classification performance was good, but not perfect. Indeed, there were still misclassification cases, even when using FCNNs. We can expect that by increasing the size of the training dataset (that in our tests was relatively small), it will be possible to improve the classification results.

Table 3. Comparison of classification performances between deep learning and a suite of different classifiers. The indexes of FCNN are generally higher than the values for the other machine learning methods.

Classifiers	Area under Curve (AUC)	Classifiaction Accuracy (CA)	F1	Precision	Recall
FCNN (5 hidden layers)	**0.947**	**0.796**	**0.791**	**0.794**	**0.797**
Decision Tree	0.761	0.632	0.624	0.646	0.632
Support Vector Machine	0.832	0.779	0.772	0.712	0.779
Random Forest	0.887	0.716	0.702	0.712	0.716
Naive Bayes	0.854	0.516	0.498	0.721	0.516
Adaptive Boosting	0.671	0.526	0.525	0.524	0.526

As an additional test, we expanded the dataset, including thin sections obtained from sedimentary rock samples. Figure 7 shows some examples of thin sections used for training all the classification methods mentioned above, in order to perform an additional performance comparison with DL methods.

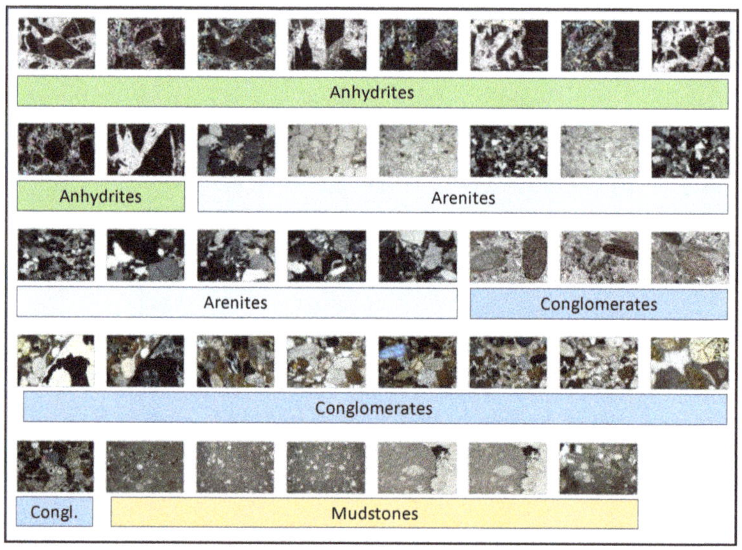

Figure 7. Examples of training images of thin sections obtained from four classes of sedimentary rocks. For this test, we trained the neural network(s) and the other classifiers on a limited set of mineral images, in order to perform the same type of classification test on sedimentary rocks, too.

Figure 8 shows an example of a performance comparison on the training dataset. In this case, instead of using performance indexes as in Table 3, we used a visual evaluation approach based on a confusion matrix. Note that a confusion matrix in machine learning is a table that summarizes the performance of a model on a set of test data by comparing the actual target values with the predicted values. It is a way to evaluate the accuracy of a classification model and helps to identify where errors in the model were made. The matrix typically has N rows and N columns, representing the actual and predicted classes, respectively. True positives, false positives, true negatives, and false negatives are the four types of values in the matrix, and they are used to calculate metrics such as accuracy, precision, recall, and F1 score. Figure 8 shows an example of a confusion matrix for a neural network (FCNN) and random forest, obtained through a cross-validation test applied to the sedimentary thin-section images shown in Figure 7. We can notice that the values on the principal diagonal (percentage of correct predicted versus actual results) for the neural network were significantly higher than for random forest, indicating that higher classification accuracy was obtained through the deep learning approach. In conclusion, the confusion matrix technique also showed that the classification performance of deep neural networks was good, although not perfect.

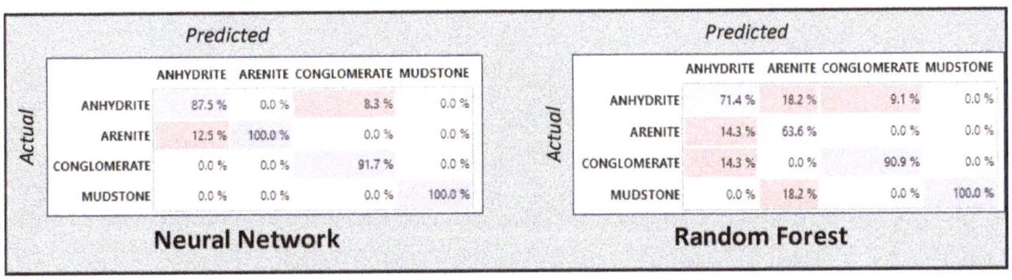

Figure 8. Comparison of two confusion matrices for neural network and random forest applied to the same image dataset (see Figure 7). The values on the principal diagonal (percentage of correct predicted versus actual results, in blue) for the neural network are significantly higher than for random forest, indicating a higher classification accuracy. The percentages of misclassifications are in pink).

4. Extension of the Workflow to Other Geo-Data

The same deep learning approach can be generalized to a multiscale/multipurpose methodology that is addressed to the analysis and automatic classification of multidisciplinary information (Figure 9).

This can include paleontological thin sections, composite well-logs, geophysical models, and so forth. The structure of the workflow is substantially the same as that shown in Figure 2. The deep learning algorithms are the same, even though the hyper-parameters need to be optimized in relation to the different types of images/data to classify. Furthermore, as we have seen in the examples discussed above, the different types of deep learning methods can be supported by additional machine learning techniques, such as adaptive boosting, decision trees, random forest, support vector machine, Bayesian methods, and so forth. In its complete implementation, our integrated machine learning framework includes a suite of all these algorithms working in parallel. The performances of all these algorithms are quantitatively estimated and compared. Finally, all the classification results are compared, too. We have discussed such a comparative approach in previous papers, for a specific application to composite well-log analysis and litho-fluid facies' classification [4].

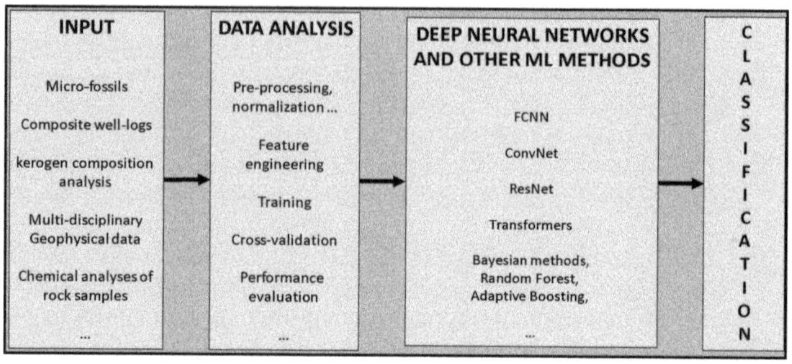

Figure 9. Scheme of the integrated DL/ML platform for automatic analysis and classification of multidisciplinary geo-data. The structure of the workflow largely remains unchanged with respect to Figure 2. Of course, the hyper-parameters must be optimized with respect to the different types of images/data being classified. Furthermore, additional deep learning methods can be complemented by further machine learning techniques, such as adaptive boosting, decision trees, random forest, support vector machine, Bayesian methods, etc. In our holistic machine learning framework, all these algorithms work concurrently. Their individual performances are quantitatively estimated and compared.

An additional useful application of our integrated machine learning platform is aimed at the integration of multiple geophysical models retrieved from multidisciplinary geophysical measurements. In that case, we create a multi-physics attribute matrix that comprises characteristics extracted from both data and model space. Geophysical measurements or observations of any type (such as seismic travel times, EM, DC, gravity data, etc.) can be used as data, while spatial distributions of geophysical parameters (such as seismic velocity, electric resistivity, density, etc.) can be used as models. In our approach, these models are progressively generated using an iterative multi-domain process that includes constrained, cooperative, and joint inversion of multi-physics data. Both data and models are calibrated at well locations to create a robust labeled data/model set for training the suite of automatic learners mentioned above. If well data are unavailable, the training dataset is created using multi-domain forward modeling in realistic scenarios. The effectiveness of each automatic learner is evaluated using cross-validation techniques, performance indices, and confusion matrices. The final step of the workflow involves classifying/predicting the remaining part of the data and models (located away from the calibration points). Ultimately, the results are presented in the form of a probabilistic spatial distribution of classes, such as "Brine", "Oil", "Gas", etc. This workflow is particularly useful (but not exclusively) in areas where drilling results are available and where there is a desire to expand our knowledge of probabilistic multi-physics models over large distances from the wells. A case history is discussed by Dell'Aversana [16].

5. Discussion

Despite its high complexity, we remark that the most time-consuming part of the workflow described in the previous sections consists in the preparation of a pre-labeled dataset for training the network models. Human experts perform that part in advance, whereas the part of the job performed in the field/lab consists in running an automatic chain of steps under human supervision. As described above, these steps include feature engineering (embedding and extraction), pre-processing, model optimization, training, cross-validation, performance evaluation, model selection, and the final automatic classification of new unlabeled images/data. All these steps require just a few seconds of computation using a standard PC (for instance, we used a System with a Dual-core Intel processor, 2.5 GHz, RAM 12.0 GB, Windows 10, 64 bit). Obviously, the larger the training dataset and the more balanced it is with respect to the various classes, the better the performance of the adopted neural network models will be. However, the size of the training dataset has a relatively low impact on the computation times of the automatic classification workflow in the field/lab.

Based on the classification tests discussed here, the main deep learning techniques (fully connected, convolutional, and residual neural networks) proved to be effective in recognizing and classifying mineral images (microscope thin sections), as well as other types of geological–geophysical data. The FCNN showed some limitations in classification accuracy, although the algorithm's performance remained generally high. This performance can be quantified using appropriate indices (as well as a confusion matrix) that, in the test we conducted, showed moderately high values: accuracy and precision were around 0.8, with 1.0 being the ideal value (corresponding to the correct classification of all images included in the dataset). These limitations of accuracy and precision were partially resolved using ResNets. In this case, accuracy generally exceeded 0.9 for ResNet_50, while it approached 1.0 for ResNet_152 (after 20–25 iterations). In ResNet_152, the values of the validation loss function tended to increase rather than decrease after only 7–8 iterations. This is clear evidence of overfitting problems that can increase with the number of hidden layers. Therefore, a general rule is that it is important to test ResNets with a variable number of hidden neural layers, in order to find the right balance between classification accuracy and the overfitting risk. In our test, ResNet_50 showed a good performance, reaching a satisfactory balance between accuracy and generalization on unseen data. In summary, ResNets guarantee good, although not perfect, classification performances. The reason for some inaccurate results can be the relatively small size of the training dataset that we used in our tests, and it is highly probable that the accuracy could improve with an expanded labeled dataset.

We recognize that there are additional deep neural network architectures that can work properly for image classification tasks that have not been applied in this paper. For example, long short-term memory (LSTM) deep networks are capable of processing both individual data points and entire sequences of data, making them well-suited for image classification tasks that require temporal information. For instance, there are interesting applications based on the integration of a block-chain layer with an LSTM architecture [17]. This is particularly useful when dealing with videos or sequential data. In fact, because LSTMs have feedback connections, they can remember information over longer periods, allowing them to maintain a more accurate representation of the input data. This memory retention also enables them to identify subtle patterns or similarities in images that may be challenging for other models to detect. However, using LSTMs for image classification also has some limitations. One of the main challenges is the training time required, which can be significantly longer than that for other models. Additionally, training an LSTM network requires a larger dataset, especially when compared to traditional feedforward neural networks. This is because an LSTM network needs to consider and integrate data over several time steps, so it needs more data to learn the temporal dependencies.

However, as discussed in the previous section, deep learning methods represent just one among a suite of machine learning techniques that can be applied for automatic analysis and classification of multidisciplinary geo-data. Many algorithms can run in parallel on the same data, in order to perform a sort of cooperative and comparative automatic interpretation of complementary big datasets. An integrated system of machine learning and deep learning, as schematically shown in Figure 9, represents a tool of fundamental importance to support the decision-making process of geologists, geophysicists, engineers, and managers. In fact, it allows for the rapid and reliable integration of a large amount of heterogeneous information at an extremely variable scale. Finally, the integrated models can be passed as inputs to another system of automatic analysis, this time based on reinforcement learning techniques [18]. These latter techniques allow for the optimization of decision-making policies in highly complex and dynamic environments, based on input variables that vary over time. In this way, our integrated framework of machine learning, deep learning, and reinforcement learning becomes an agile and robust tool at the same time, supporting the operational work of geoscientists and managers [19,20].

6. Conclusions

In this article, we have attempted to provide the main criteria and methods for developing a generalized machine learning and deep learning approach aimed at fast automatic classification of mineral image thin sections. The main conclusion is that fully connected (FCNNs), convolutional, and residual neural networks (ResNets) are effective techniques in recognizing and classifying mineral images directly in the field, with ResNets outperforming the other techniques in terms of accuracy and precision. Using appropriate hyper-parameters, ResNets demonstrated good (but improvable) performances in all the classification tests performed in this work. The presence of few unavoidable inaccuracies can be easily explained by the fact that we used a training dataset of a limited size. We expect to improve these results using a larger labeled dataset for training our networks.

We remark that accurate classification is never the result of the application of a single algorithm, but rather the result of a complex workflow of analysis, training, optimization of neural network parameters, the selection and choice of the type of classifier algorithm, possible retraining of the selected algorithm, cross-validation tests, and final verification by a human expert. The advantage offered by a deep learning approach is that the main part of this workflow can be automatized and requires very short computation times.

Finally, the same workflow discussed here, with some appropriate variations, can be used to classify other types of geologically relevant images. For instance, we have already applied it to analyze and classify microscope images of microfossils, as well as thin sections of rocks with different types of kerogens. Furthermore, we have integrated this deep learning methodology into a broader machine learning context, in order to create a general methodology for integrated analysis and classification of multidisciplinary/multiscale information (chemical analysis of rock samples, composite well-logs, geophysical data, and so forth).

Funding: This research received no external funding.

Data Availability Statement: All the images (microscope mineral thin sections) discussed and shown in this paper have been obtained courtesy of Dr. Alessandro Da Mommio. Link: http://www.alexstrekeisen.it/index.php (accessed on 20 February 2023). The specific jpeg files used in this paper can be obtained upon request by writing an email to dellavers@tiscali.it. An excellent Python tutorial about a complete workflow for image classification can be found and tested directly on Google Colab, at the following link: https://colab.research.google.com/github/tensorflow/docs/blob/master/site/en/tutorials/images/classification.ipynb#scrollTo=5fWToCqYMErH (accessed on 10 April 2023).

Acknowledgments: The author would like to thank Alessandro Da Mommio for providing permission to use the thin-section images for scientific purposes.

Conflicts of Interest: The author declares no conflict of interest.

References

1. Aminzadeh, F.; de Groot, P. *Neural Networks and Other Soft Computing Techniques with Applications in the Oil Industry*; EAGE Publications: Houten, The Netherlands, 2006; Volume 129, p. 161.
2. Barnes, A.E.; Laughlin, K.J. Investigation of methods for unsupervised classification of seismic data. In *Expanded Abstracts*; SEG Technical Program: Salt Lake City, UT, USA, 2002; pp. 2221–2224. [CrossRef]
3. Bestagini, P.; Lipari, V.; Tubaro, S. A machine learning approach to facies classification using well logs. In *Expanded Abstracts*; SEG Technical Program: Houston, TX, USA, 2017; pp. 2137–2142. [CrossRef]
4. Dell'Aversana, P. Comparison of different Machine Learning algorithms for lithofacies classification from well logs. *Bull. Geophys. Oceanogr.* **2017**, *60*, 69–80. [CrossRef]
5. Dell'Aversana, P. Deep Learning for automatic classification of mineralogical thin sections. *Bull. Geophys. Oceanogr.* **2021**, *62*, 455–466. [CrossRef]
6. Hall, B. Facies classification using machine learning. *Lead. Edge* **2016**, *35*, 906–909. [CrossRef]
7. She, Y.; Wang, H.; Zhang, X.; Qian, W. Mineral identification based on machine learning for mineral resources exploration. *J. Appl. Geophys.* **2019**, *168*, 68–77.
8. Liu, K.; Liu, J.; Wang, K.; Wang, Y.; Ma, Y. Deep learning-based mineral classification in thin sections using convolutional neural network. *Minerals* **2020**, *10*, 1096.

9. Raschka, S.; Mirjalili, V. *Python Machine Learning: Machine Learning and Deep Learning with Python, Scikit-Learn, and TensorFlow*, 2nd ed.; Packt Publishing Ltd.: Birmingham, UK, 2017.
10. Rosenblatt, F. *The Perceptron, a Perceiving and Recognizing Automaton*; Cornell Aeronautical Laboratory: New York, NY, USA, 1957.
11. Minsky, M.; Papert, S. Perceptrons. In *An Introduction to Computational Geometry*; M.I.T. Press: Cambridge, UK, 1969.
12. LeCun, Y.; Bottou, L.; Bengio, Y.; Haffner, P. Gradient-based Learning Applied to Document Recognition. *Proc. IEEE* **1998**, *86*, 2278–2324. [CrossRef]
13. Simard, P.Y.; Steinkraus, D.; Platt, J.C. Best Practices for Convolutional Neural Networks Applied to Visual Document Analysis. In *Icdar*; IEEE: Piscataway, NJ, USA, 2003; p. 958.
14. He, K.; Zhang, X.; Ren, S.; Sun, J. Deep Residual Learning for Image Recognition. In Proceedings of the IEEE Conference on Computer Vision and Pattern Recognition (CVPR), Las Vegas, NV, USA, 20–22 June 2016; pp. 770–778. [CrossRef]
15. Dell'Aversana, P. Artificial Neural Networks and Deep Learning: A Simple Overview. In *A Global Approach to Data Value Maximization. Integration, Machine Learning and Multimodal Analysis*; Cambridge Scholars Publishing: Newcastle upon Tyne, UK, 2019.
16. Dell'Aversana, P. An integrated multi-physics Machine Learning approach for exploration risk mitigation. *Bull. Geophys. Oceanogr.* **2020**, *61*, 517–538.
17. Mendi, A.F. A Sentiment Analysis Method Based on a Blockchain-Supported Long Short-Term Memory Deep Network. *Sensors* **2022**, *22*, 4419. [CrossRef] [PubMed]
18. Mendi, A.F.; Doğan, D.; Erol, T.; Topaloğlu, T.; Kalfaoğlu, E.; Altun, H.O. Applications of Reinforcement Learning and its Extension to Tactical Simulation, November 2021. *Int. J. Simul. Syst. Sci. Technol.* **2021**, *22*, 14–15. [CrossRef]
19. Ravichandiran, S. *Deep Reinforcement Learning with Python*; Packt Publishing: Birmingham, UK, 2020.
20. Dell'Aversana, P. Reservoir prescriptive management combining electric resistivity tomography and machine learning. *AIMS Geosci.* **2021**, *7*, 138–161. [CrossRef]

Disclaimer/Publisher's Note: The statements, opinions and data contained in all publications are solely those of the individual author(s) and contributor(s) and not of MDPI and/or the editor(s). MDPI and/or the editor(s) disclaim responsibility for any injury to people or property resulting from any ideas, methods, instructions or products referred to in the content.

minerals

MDPI

Review

Twenty Years of CSEM Exploration in the Brazilian Continental Margin

Paulo T. L. Menezes [1,2,*], Sergio M. Ferreira [1], Jorlivan L. Correa [1,3] and Everton N. Menor [1]

1 Petrobras-EXP/PEN/AB, Av. Henrique Valadares 28, 4 andar, Rio de Janeiro 20231-030, Brazil
2 Departamento de Geologia Aplicada, Faculdade de Geologia/UERJ, Rua São Francisco Xavier, 524, sala 2009 Bloco A, Rio de Janeiro 20550-900, Brazil
3 Department of Geology and Geophysics, University of Wyoming, Laramie, WY 82072, USA
* Correspondence: ptarsomenezes@petrobras.com.br or paulo.menezes@uerj.br

Abstract: The controlled source electromagnetic (CSEM) method is frequently used as a risk reduction tool in hydrocarbon exploration. This paper aims to provide a comprehensive historical review of the CSEM method's twenty-year history in the Brazilian continental margin. Since 2003, we have significantly improved our understanding of CSEM resistivity data across various geological scenarios. This review presents a roadmap of the technical advancements in acquisition design and interpretation techniques. As a result, our understanding of the methodology has broadened from traditional to more general use, such as salt imaging, gas hydrates, geohazard mapping, and reservoir characterization. Finally, we indicate the potential upcoming CSEM applications in new energy resources and carbon capture and storage.

Keywords: CSEM exploration; reservoir monitoring; geohazard; energy resources

Citation: Menezes, P.T.L.; Ferreira, S.M.; Correa, J.L.; Menor, E.N. Twenty Years of CSEM Exploration in the Brazilian Continental Margin. *Minerals* **2023**, *13*, 870. https://doi.org/10.3390/min13070870

Academic Editor: Michael S. Zhdanov

Received: 30 May 2023
Revised: 22 June 2023
Accepted: 23 June 2023
Published: 28 June 2023

Copyright: © 2023 by the authors. Licensee MDPI, Basel, Switzerland. This article is an open access article distributed under the terms and conditions of the Creative Commons Attribution (CC BY) license (https://creativecommons.org/licenses/by/4.0/).

1. Introduction

The marine controlled source electromagnetic (CSEM) method is a risk reduction tool used in hydrocarbon exploration [1]. In its most frequently practiced setup, the CSEM operates a towed horizontal electric dipole (HED) source that transmits a low-frequency EM pulse, commonly in the 0.01–10 Hz range, and free-fall ocean bottom node receivers are set on the seafloor (Figure 1). The receivers usually register the two horizontal electric fields and two horizontal magnetic components. During the transmitter's off-time periods, the nodes still register the natural electromagnetic fields, making magnetotelluric (MT) data available as a byproduct of the acquisition.

The marine CSEM has been applied in de-risking deep-water high-cost drilling decisions in many basins worldwide since the beginning of the 2000s, when the first survey was performed in October 2000 at the Girassol Prospect, offshore Angola [2]. Since the early stages, CSEM has been expanded to a broader range of geographic areas, geological environments, and application scenarios.

In Brazil, CSEM's history started 2003, when Petrobras planned and contracted the first multiclient survey in the Brazilian offshore margin in April 2004. Shell and Exxon also acquired these data. Since then, we have achieved impressive numbers, only 20 years after the beginning of CSEM's usage in Brazil.

Forty-seven surveys have been performed along the Brazilian coast, from north to south (Figure 2). These surveys span 19,350 line kilometers of deep towed CSEM data and 5410 receiver deployments. Petrobras plays a leading role in CSEM usage, acquiring almost all of the data shown in Figure 2. Most datasets were acquired by EMGS (80%), while Schlumberger (SLB) was responsible for 20%.

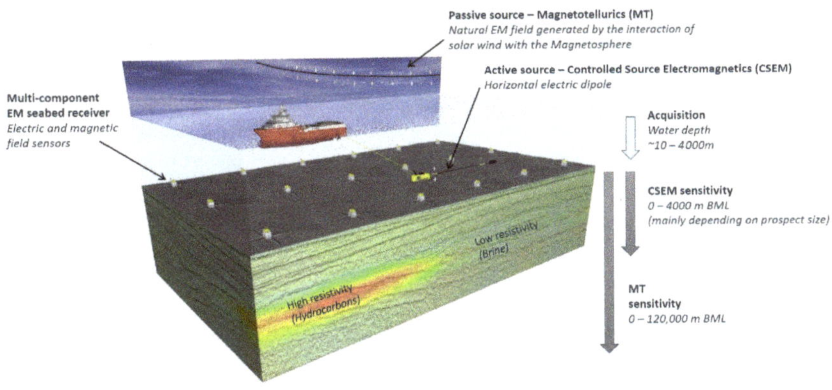

Figure 1. CSEM acquisition scheme and the state-of-art technology sensitivity below mud line (BLM) of the MT (up to 120 km depths) and CSEM (reservoirs up to 4 to 5 km burial) methods. Courtesy of EMGS.

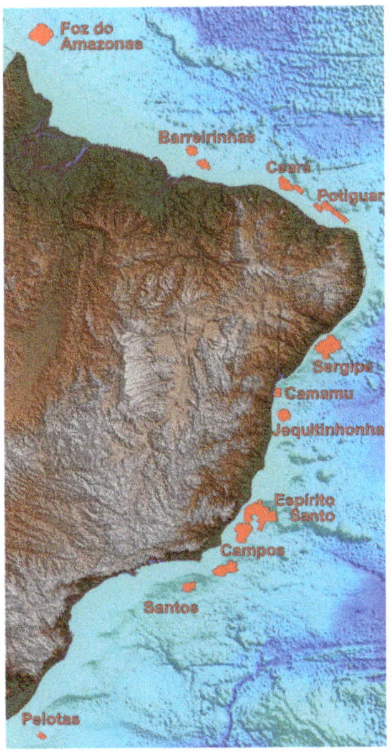

Figure 2. CSEM schematic acquisition map at Brazilian continental margin at several offshore basins (labeled in red). Red polygons indicate the position of the CSEM surveys.

This paper aims to provide a historical review, which is accessible to non-experts, of the CSEM development in the Brazilian continental margin, from the initial period, with the inherent distrust of our asset teams about the usability of the method, until the current period, where CSEM has consolidated itself as a reliable prospective tool for hydrocarbons. We can divide this timeline into three important periods:

- 2003–2010—first steps in CSEM and the consolidation of the methodology;
- 2011–2020—this phase includes the expansion of commercial surveys for exploration and appraisal purposes;

- 2021 and beyond—this includes important information about where we are now. Is CSEM worthwhile? What results have we obtained after 20 years of usage? Finally, we have the opportunity to plan the future, moving from exploration towards monitoring and energetic transition applications.

This review addresses the developments in the acquisition's survey design—most importantly, in the CSEM processing and interpretation schemes, beyond its earlier usage for reservoir de-risking. The idea is to provide a roadmap of the technical advances in acquisition design (Figure 3) and interpretation techniques (Figure 4) over the past twenty years, mainly in understanding how the CSEM resistivity data can effectively contribute to the overall geological knowledge of a given area.

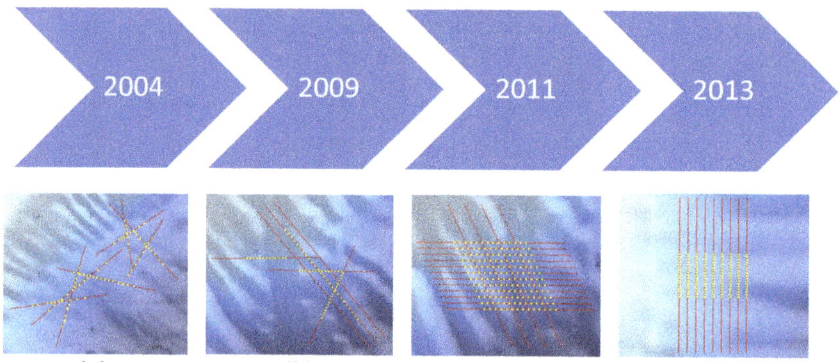

Figure 3. Schematic cartoons (maps out of scale) to illustrate the evolution of the CSEM acquisitions in the Brazilian continental margin. Yellow dots represent the CSEM receivers and red lines indicate the source tow lines. (**a**) 2004 acquisition. 2D regional acquisition with uneven +2 km receiver spacing. (**b**) 2D acquisition with regular 1.5 km receiver spacing; additional source tow lines were collected, simulating a pseudo-3D acquisition. (**c**) Full-azimuth 3D acquisition at the same spot as (**c**), regular grid with 1.5 km receiver spacing. (**d**) High-resolution 3D acquisition (1 × 0.5 km receiver grid).

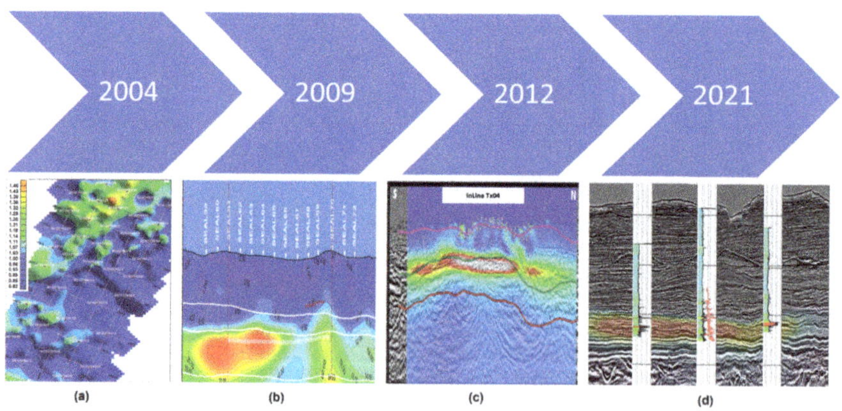

Figure 4. Schematic cartoons (figures out of scale) to illustrate the evolution of CSEM interpretation in Petrobras. (**a**) Normalization map calculated in the data domain. Background values are close to 1 (blue portions in the map); green-to-orange colors represent anomalous spots (areas where the electric field is 15% to 45% higher than the background). (**b**) Output resistivity model from an unconstrained 2D inversion; anomalies are represented by yellow to reddish colors. Three main regional stratigraphic horizons are superimposed on the model. The red dashed line represents a

volcanic layer, and the studied reservoir is shown by the white dashed line. (**c**) Output resistivity model from an unconstrained 3D inversion superimposed on the seismic amplitude attribute; anomalies are represented by yellow to reddish colors. (**d**) Seismic, constrained 3D CSEM inversion results and well-log data integrated into a single interpretation platform.

2. The 2003–2010 Period, Consolidation of the Methodology

2.1. First Steps in CSEM

CSEM's history in Petrobras started curiously. By the end of 2002, Ricardo Catellani, a senior consultant with expertise in seismic amplitude versus offset (AVO) analysis and reservoir modeling [3], had read the paper presenting the results of the first CSEM experiment in Angola acquired for Equinor (former Statoil) [2]. Initially skeptical but interested in this new technology, he challenged his colleague, Marco Polo Buonora, a potential-field methods expert, regarding whether this new method was trustworthy and could be applied to the exploration of the Brazilian offshore turbidites.

To respond to this challenge, Buonora carried out research, aimed at comprehending the theoretical principles of electromagnetic methods and reviewing case studies of their use in marine exploration. Then, Buonora scheduled a series of technical visits to the headquarters of the three commercial CSEM service contractors available at that time [4]: EMGS (a descendent of Statoil); OHM Surveys, arising from Southampton University; and Arnold Orange Associates (AOA) Geomarine Operations (AGO, later acquired by Schlumberger [4]).

By July 2003, a group of geophysicists from AOA had visited the Petrobras office in Rio de Janeiro (RJ) to present the CSEM methodology and discuss the parameters of a multiclient survey. To these meetings, Petrobras invited EM experts from Brazilian universities to collaborate in the project and compose a scientific advisory board to help Petrobras staff. A 700 MB CD-ROM-R containing the presentations on CSEM applied to hydrocarbon mapping remains in the Petrobras collection as a register of the first meeting in the Petrobras headquarters, dated 3 July 2003 (Figure 5a).

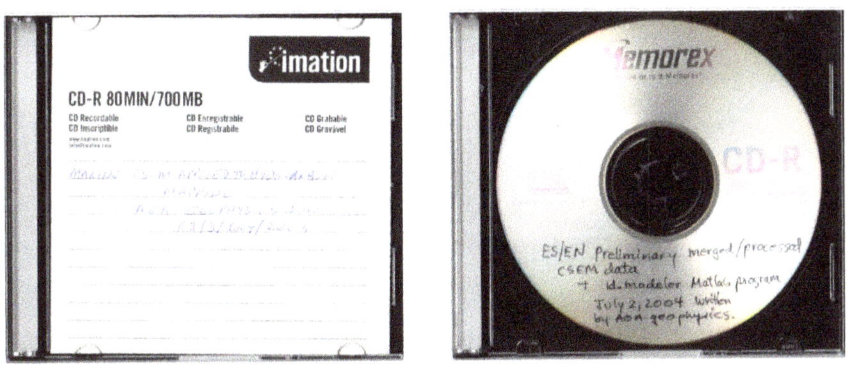

(a) **(b)**

Figure 5. Digital media used at the beginning of the 2000s to exchange information. (**a**) 700 MB CD-ROM-R (read-only) with CSEM material provided by AOA Geophysics (2003). (**b**) 700 MB CD-ROM-R with the first CSEM data delivered for Petrobras (2004). Courtesy of Petrobras.

As a result, Petrobras agreed to undertake the first multiclient survey in Brazil. The survey comprised approximately 1600 line kilometers acquired along 36 towed lines [5]. The survey was organized into three major areas in the Santos, Campos, and Espirito Santo basins (Figure 2). The CSEM data were acquired using two distinct patterns, two surveys using a sparse 2D layout (receivers with +2 km spacing), in a star-like shape (Figure 3a), and the first regional 3D one employing a 5 km rectangular grid [5].

These three areas were chosen because of the proven reservoirs that could be used to validate and calibrate the responses of the methodology over known targets [5] and undrilled prospects identified by seismic interpretation to be tested [6].

The preliminary CSEM dataset of the multiclient was delivered in CD-ROM-R form by July 2004, together with some Matlab scripts for 1D modeling and data plotting tools (Figure 5b).

At that time, there was no commercial software available, and 2D and 3D imaging processing was rather primitive when compared to seismic [6]. Therefore, the interpretation in the industry was limited in identifying and correlating the CSEM anomalies in the data domain (Figure 4a) with the outline of prospects [5,6].

Buonora et al. (2005) [5] showed that the 2004 CSEM 3D grid data were sensitive to known oil reservoirs in selected areas of the Campos basin. They could identify electric/magnetic field anomalies as small as 20% above the background response using state-of-the-art processing and modeling tools. Finally, they forewarned about the need to push the CSEM technology forward through the development of interpretation workflows based on a novel suite of multidimensional software tools.

Smit et al. (2006) [6] interpreted the 2004 dataset to downgrade an undrilled prospect where seismic amplitude studies were inconclusive. The CSEM receiver collected along the prospect displayed no anomaly in the normalized electric field, indicating that the selected area probably had resistivity similar to the surrounding conductive shales (see Figure 3 of [6]).

Newman et al. (2010) [7] used the regional grid dataset to validate their 3D imaging algorithm. They showed that incorporating electrical anisotropy into 3D inversion produces reliable models with a superior data fit compared to isotropic models.

2.2. Building the House

With increasing confidence in the CSEM method, Petrobras decided, in 2006, to build a strong internal group dedicated to multiphysics methods under Marco Polo Buonora's leadership. To this end, Petrobras hired some EM experts, combined them with new, talented young geologists and geophysicists, and then engaged in intensive investment in the continuous education of all its personnel. Petrobras created in-house EM courses to train the multiphysics group and spread awareness of the EM methods' applicability to the exploration asset teams.

Following this guideline, Petrobras joined some academic EM consortia in the USA, such as Scripps (University of California San Diego) and the Consortium for Electromagnetic Modeling and Inversion (University of Utah), and, more recently, the Electromagnetic Methods Research Consortium (Lamont-Doherty Earth Observatory, Columbia University). In Brazil, Petrobras stimulated the development of EM groups at Universidade Federal do Pará (UFPA), Universidade do Estado do Rio de Janeiro (UERJ), and Observatório Nacional (MCTI-ON).

In 2007, Petrobras established a three-year joint industry project (JIP) with the Schlumberger Brazilian Research Geoscience Center in Rio de Janeiro, Brazil. The main goal of the JIP was to develop an EM interpretation workflow and integrate deep-reading EM tools into the full cycle of hydrocarbon reservoir exploration. The JIP was responsible, among other items, for the first marine magnetotelluric [8,9] and full-azimuth CSEM [10,11] surveys acquired for offshore Brazil.

Other important products of the JIP were the development of a commercial implementation of 2.5D anisotropic inversion as a user-friendly graphical user interface (GUI) [12] and the first results from 3D inversions [11].

The availability of 2.5D inversion as an easy-to-use GUI changed how CSEM was interpreted from the data to the model (Earth) domain. These inversions could be more easily executed by a larger group of non-expert geophysicists and produce more meaningful images to be shared with the asset teams. This approach's major benefit is exporting the recovered resistivity models in the SEG-Y seismic standard format, which can be easily uploaded into any seismic interpretation software. Consequently, it allows qualitative correlations by co-rendering the inverted CSEM 2D Earth models with the depth-converted/migrated seismic data (Figure 4b).

From 2005 to 2009, six commercial surveys were acquired for Petrobras in several basins, all using a 2D layout to investigate leads as imaged by seismic interpretation.

Figure 4b shows the CSEM interpretation of one of these surveys. Acquired in 2009 by SLB in the Sergipe-Alagoas basin, this survey is considered the turning point in Petrobras' high management's acceptance of the CSEM method. To interpret the data, we used the 2.5D inversion to show a resistive anomaly response coinciding with the lateral boundaries of the investigated lead. Nonetheless, the center of the resistive anomaly was above the expected depths given by the seismic AVO anomaly.

Drilled in 2010, the successful wildcat well confirmed the CSEM and seismic forecasts, with the reservoir being hit shallower than expected by the seismic interpretation. The velocity field was then updated using the wildcat well information. Figure 4b displays the co-rendering of the 2009 CSEM anomaly with the 2010 updated depth-converted seismic line, where the low-resolution resistivity anomaly is centered over the top of the reservoir.

3. The 2011–2020 Period, Expansion of the Commercial Surveys

3.1. CSEM for Exploration

The year 2010 ended with a great deal of in-house work for the multiphysics group, performing a huge feasibility study over the Brazilian shelf. This task was named the Varredura (sweep) project. The main idea behind Varredura was to bring a dedicated ship to execute a large multiclient survey over prospects at several basins (Figure 2).

We conducted 2.5D feasibility studies to assess the CSEM effectiveness in distinguishing oil-filled turbiditic reservoirs from the background geology. Our colleagues from the asset teams provided information about the leads: the main stratigraphic horizons, the outline of the leads with their top and base, and resistivity well logs of the nearest well, if available. We analyzed 115 leads; 79 were detectable, and 36 were non-detectable. After a round of analyses based on economic criteria, 56 leads were approved for acquisition.

The Varredura studies were responsible for selecting the target areas and defining all relevant survey parameters, such as the receiver's positioning, sampling frequencies, and source towing directions [13].

EMGS was responsible for Varredura's acquisitions [14]. The surveys started in 2011 in the Barreirinhas basin (Brazilian equatorial margin), followed by the Ceará, Potiguar, Sergipe, Jequitinhonha, Espírito Santo, and Campos basins (Figure 2). The survey designs moved from a 2D to a 3D layout in a similar movement to that which occurred with the seismic method, moving from 2D to 3D seismic acquisitions. At the beginning of the Varredura acquisition, the 3D grids were planned with 1.5 to 2 km receiver spacing (Figure 3c). Nonetheless, with increased confidence in the CSEM method, high-resolution surveys were executed with 500 m receiver spacing, aiming for detailed prospect characterization (Figure 3d) [13].

One of the key lessons that we learned after the start of the CSEM acquisitions was to shift towards using 3D surveys. This type of design provides more information and reduces ambiguity in interpretation. The CSEM vertical resolution is much poorer than the horizontal resolution [15]. To increase the likelihood of a successful drill, we propose imaging the shapes of anomalies along horizons or depth slices and correlating them with the lead's outline [15].

Varredura's CSEM data spanned more than 5000 km^2 coverage and 3103 deployed receivers [14]. Interpreting this massive quantity of data at many distinct basins and leads with different drilling time schedules, and aiming to provide quick answers to allow the asset teams to make drilling decisions, was a great challenge. At this time, 3D inversions were already available [16], but the task was time-consuming, with a single inversion run usually taking several weeks.

To provide quick responses to the asset teams, Petrobras developed a fast interpretation workflow [13] comprising the identification of anomalies via frequency ratio normalization of the CSEM data [17] and then performing constrained 1D CMP inversions [18], followed by high-resolution 2.5D anisotropic polygonal inversions [12]. We applied the workflow in a complex geological setting where the reservoir dipped toward salt domes. An asset team colleague joined the multiphysics group, providing important geological information from the seismic interpretation that helped us to obtain useful information about the local geology. Two successfully drilled wells corroborated the integrated seismic–CSEM interpretation [13].

The multiclient contract included providing the output of unconstrained 3D anisotropic inversions based on a quasi-Newton BFGS optimization algorithm [19] as a by-product (Figure 4c). Petrobras then acquired two commercial 3D inversion suites, Sblwiz (EMGS) and the Petrel™ EM plugin (SLB), and implemented them in a dedicated cluster for the multiphysics group. Since then, Petrobras has been continuously upgrading the capacities of its high-performance computing (HPC) park. Today, we use cluster- and cloud-based solutions to perform our 3D inversions.

Outside the Varredura project, Petrobras acquired, in 2011, the CSEM multiclient dataset collected by SLB [20]. The dataset included more than 250 km of towlines recorded by 136 receivers at five different spots in the Ceará and Potiguar basins (Figure 2). A package with the 3D inversion results of the studied areas (resistivity models, data fits, etc.) was also made available by SLB in their EM plugin at Petrel™.

3.2. *CSEM for "Non-Standard" Exploration Applications*

3.2.1. Sub-Salt Imaging

Beyond the standard use of CSEM in de-risking prospects, Petrobras understood that the CSEM data could provide information about the subsurface geology and be used for other purposes within the oil industry. Thus, Petrobras signed a second JIP cooperation agreement with SLB, aiming at several goals, such as sub-salt imaging, reservoir characterization, and the first steps in reservoir monitoring.

One of the aims of the JIP was seismic sub-salt imaging, usually a difficult task, mainly due to complex bodies with steeply dipping flanks of salt domes and large acoustic impedance and velocity contrasts between the high-velocity salt and the lower-velocity sediments beneath. On the other hand, EM fields easily penetrate through resistive (high-velocity) salt with little attenuation, thus providing information on the underlying conductive (low-velocity) sediments [21].

Among the EM methods, MT is the classical method used for sub-salt imaging [22] but also for sub-carbonate [23] and sub-basalt imaging [24,25], which face similar challenges presented by autochthonous and allochthonous salt body zones.

The highly conductive seawater layer attenuates the high-frequency MT fields in deep to ultra-deep waters. Frequencies higher than 0.1 Hz are commonly unusable [21]. A possible approach to overcome this issue is to combine CSEM and MT to interpret using joint-inversion, clearer subsurface images [26,27].

Zerilli et al. (2016) [21] developed an integrated seismic–CSEM two-step interpretation workflow and applied it to a broadband ultra-long offset CSEM research survey acquired over a selected ultra-deepwater area of the Espirito Santo basin [28] (Figure 2).

In the first step, [21] conducted a 3D pixel-based inversion [29] to obtain the first estimates of the geometry and resistivity of an allochthonous salt body, as indicated by previous seismic interpretation conducted by the asset team, and the background resistivity. A priori information about the sedimentary background resistivity was provided by available nearby wells.

In the second step, [21] ran structure-based MT–CSEM joint-inversions [30,31] using the recovered model of the previous step as input. Then, [21] used a start model based on the previous seismic interpretation and constrained the inversion domain to an area enclosing the top of salt and the base of salt horizons. The background resistivity and the top of salt horizon were fixed during the inversion process. This strategy allowed the better recovery of the base of salt horizon, 300 to 700 m shallower than previously interpreted by the asset team.

Interpreting the same broadband dataset, [32] established a resistivity-to-velocity model by calculating the Hacikoylu petrophysical cross-property relationship [33]. They calibrated the Hacikoylu coefficients (a,c) at a well-log scale. Then, they applied them to the regional resistivity models derived from the 3D CSEM inversions to obtain P-wave velocity (Vp) models.

Zerilli et al. (2017) [34] then used the Vp models provided by [32] as a cost-effective starting model for full waveform inversion (FWI), to provide better seismic images in the complex salt environment of the Espírito Santo basin. Moreover, [34] showed that their seismic data alone had insufficient information on the salt geometries and the salt, around-salt, and sub-salt velocity distributions. They concluded that including the knowledge from the CSEM allowed an unambiguous and robust strategy to improve the seismic images, allowing the interpreters to achieve a more accurate seismic interpretation.

3.2.2. Gas Hydrates and Geohazards

Gas hydrates can be an unconventional energy resource [35,36], a potential climate forcer once methane from hydrate deposits is freed into the atmosphere [37]. Nonetheless, they are also a problem as they represent a drilling geohazard [38].

Petrobras supported a CSEM research survey in the Pelotas basin, Southern Brazil (Figure 2), to investigate the origin and distribution of gas hydrate deposits in the basin. The dense 3D survey layout included 132 receivers spaced 1000 m along the source tow and 500 m across the source tow direction. With a 1 to 19 Hz range, the transmitted frequencies were higher than usually acquired for deeper targets.

Tharimela et al. (2019) [39] applied an unconstrained 3D inversion [19] to define the location and extent of the saturated gas hydrates and free gas in the shallow subsurface. The CSEM results were integrated with other near-surface geophysical data, including 2D seismic, sub-bottom profiler, and multibeam bathymetry data, identifying faults, chimneys, and seeps conducting to pockmarks in the seafloor.

As pointed out by [39], some resistivity anomalies revealed an excellent spatial correlation with some of these features. Thus, some seismic features were filled in with free gas and other potential geohazards impacting drilling operations, while others were not. This is a noteworthy contribution of CSEM in solving ambiguities in seismic interpretation.

Additionally, features previously mapped as gas-hydrate-bearing were reinterpreted as residual or low-saturated gas features due to the lack of a significant resistivity response associated with them. Moreover, using Archie's equation, Ref. [39] used the inverted resistivity volume to calculate the saturation volume of the subsurface.

3.2.3. Reservoir Characterization

Other objectives of the second Petrobras–SLB JIP cooperation agreement included developing new techniques for reservoir characterization, aiming to comprehend the reservoir rocks and fluids through accurate multiphysics measurements, to help the asset teams to develop optimal appraisal and production/monitoring plans.

Following this guideline, Miotti et al. (2018) [40] developed a new workflow to perform a petrophysical joint inversion (PJI) of seismic and CSEM data to determine important reservoir properties. The workflow uses seismic, CSEM, and well-log data information to enhance the reservoir's characterization.

The PJI workflow was applied to a deepwater oil field in offshore Brazil in the Sergipe-Alagoas basin, where we had available CSEM data. This approach successfully retrieved an accurate estimate of the reservoir's porosity and saturation from the electric and seismic domains.

The Jubarte experiment (JE) [41–43] was the first attempt to develop an understanding and assess the sensitivity of the CSEM method to water flooding associated with oil production in a complex and heterogeneous deep-water Brazilian turbiditic reservoir. The Jubarte Reservoir was chosen because it hosted the first fully optical deep-water permanent reservoir monitoring (PRM) seismic system installed in Brazil [44].

The JE studies indicate that production effects and associated variations in saturation produce changes in the reservoir's resistivity structure over time. Instead of trying to identify the associated changes in the time-lapse data domain, it was demonstrated that working in the model domain through inversions on each 3D dataset can retrieve reservoir-production-related resistivity differences. The advantage of the latter procedure is to avoid several issues related to receiver positioning and repeatability when working in the data domain, permitting easier and cheaper CSEM time-lapse monitoring.

In 2017, Petrobras embraced and supported the Marlim R3D (MR3D) project [45–47], which produced an open-source realistic Earth modeling project for electromagnetic simulations of the post-salt turbiditic reservoirs of the Brazilian offshore margin.

MR3D provides a realistic anisotropic geoelectric model aiming to be a standard for CSEM studies of the turbiditic reservoirs of the Brazilian continental margin. The MR3D model includes fine-scale stratigraphy and fluid-filled reservoirs whose geometries are based on a detailed 3D seismic interpretation [48]. MR3D also includes a CSEM [46] and an MT [47] dataset.

By using MR3D as a testing scenario in developing its interpretation workflows, Petrobras avoids confidentiality issues for external communications such as papers and conference presentations.

4. 2021 and Beyond

4.1. CSEM for Exploration and Appraisal

By the end of 2020, Petrobras had acquired a new, optimized Gauss–Newton 3D anisotropic inversion commercial code [49], which has several advantages compared to the quasi-Newton (BFGS) inversion code, as it runs faster, provides a higher resolution in defining much-improved and more stable anomaly images positioned at the correct depths, and yields much more accurate resistivity values [50].

Besides better inversion algorithms, by 2021, advanced interpretation software suites were also available, allowing more precise interpretations via correlation with seismic and well-log data, such as the one shown in Figure 4d, where three wells, the wildcat and two appraisals, were successfully drilled along a CSEM anomaly, positioned correctly over the reservoir.

Figure 6 shows the high success rate of the CSEM interpretation at Petrobras, calculated at 44 drilled wells. In the proposed classification, hits indicate true positives and true negatives, while misses indicate false positives and false negatives.

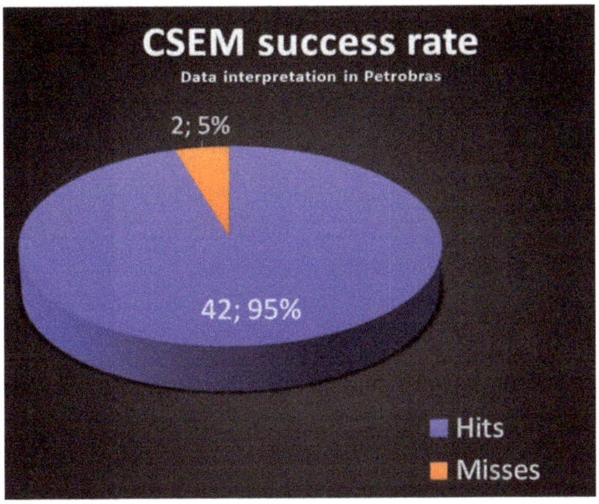

Figure 6. Petrobras CSEM cases at 44 wells and a 95% success rate. Hits (true positives/negatives), misses (false positives/negatives). Results from integrated CSEM–seismic interpretations.

Almost all wells (39) were drilled using the results of the Varredura project, which required the cost of one well to acquire a huge amount of CSEM data over several basins.

Petrobras' success rate of 95% exceeds the published 80% rate [51]. What are the main reasons for this high success rate?

Firstly, we attribute these numbers to the extensive and rigorous feasibility studies described previously in Section 3.1. Non-detectable targets or those with very ambiguous CSEM responses were discarded at once. Some of them occurred because they were too deeply buried to be imaged by the 2010 technology. With upgrades in the acquisition technology, more accurate receivers, more potent sources, and new 3D algorithms, a new 3D feasibility study is planned to reevaluate them.

Secondly, we believe that the in-house training, not only for the geophysicists of the multiphysics group but mainly in exporting the EM culture to the explorationists of the asset teams by allowing them to better understand the physics, advantages, and limitations of the CSEM methodology, allowed a fruitful collaboration between the two groups. As a result, it facilitated an integrated seismic–CSEM interpretation by incorporating all a priori knowledge in every studied area.

Finally, the scientific and technological advances developed in the integrated interpretation workflows must be acknowledged. One of the most recent is the Multiphysics Anomaly Map (MAM) [52], a data fusion solution consisting of a spatial representation of the correlations between anomalies from distinct geophysical methods.

The MAM was applied to CSEM and seismic inversion results from the offshore Sergipe-Alagoas basin. The MAM helped to differentiate between a dry and an oil-bearing channel previously outlined in seismic data. Both channels had the same seismically anomalous response. By applying the MAM, it was possible to resolve the seismic ambiguity. Our results were confirmed by drilling.

Moreover, we may consider the two failures from the drilled wells. The first (Case 1) was a false positive, and the other (Case 2) was a false negative.

Case 1 was an interpretation of the early period, with a 2D star-like shape of the 2007 survey design. CSEM data were interpreted by associating anomalous responses in the data domain with the interpreted reservoir outline and posterior 2D modeling. Nonetheless, we identified a slight anomaly at depth. The well was drilled near the lines, at the best position defined by seismic interpretation, and hit a reservoir with a gas show and filled in with moderately resistive (10–15 ohm.m) freshwater.

Case 2 was a 2012 survey with a small 3D design, with three parallel lines in a complex area with allochthonous salt. The CSEM data were inverted using the BFGS 3D algorithm (unconstrained inversion), which revealed an anomalous body. However, this body was found to be located far above the hydrocarbon reservoir that was discovered in the drilled well. When we performed the statistical analysis, we pinpointed it as a miss in the interpretation. Case 2 is a classic example of a lesson learned. With our state-of-the-art interpretation tools, we could probably position the anomaly correctly at the proper depth [50] and correlate it with seismic anomalies [52].

4.2. Planning for the Future

Figure 7 summarizes Petrobras' thoughts about the usability of the CSEM method through the various phases of a hydrocarbon reservoir's life cycle, from the early stages of exploration to time-lapse studies, including the data acquired for reservoir characterization, which can be used to support geohazard determinations and drilling operations.

CSEM usage for exploration is still very active at Petrobras. We have recently acquired a huge CSEM dataset in the Foz do Amazonas basin (Figure 2) [53]. This dataset is currently being interpreted not only in search for new opportunities but to understand the whole petroleum system in the area, as pointed out by [54].

Beyond exploration and appraisal, we are moving towards reservoir monitoring. Menezes et al. (2021) [55] have shown the ability of CSEM to produce reliable estimates of the SoPhiH maps [56] at a given reservoir. These maps provide knowledge of the remaining oil thickness in the studied reservoir. The reservoir teams frequently use the SoPhiH map as a subsidiary tool to define the best drilling locations for production or injection wells.

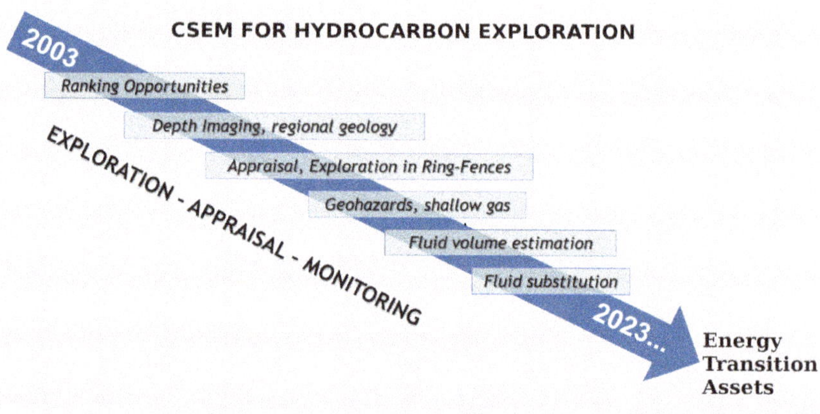

Figure 7. Petrobras' view of CSEM's applicability in the energy market.

Finally, we propose a new concept of ocean-bottom multiphysics nodes (OBMP) designed for reservoir monitoring purposes [15]. The idea is to find a feasible means to increase the demand for CSEM for 4D monitoring programs by increasing the value of information and reducing survey costs by performing joint operations, where seismic and CSEM data are acquired during the same survey at equivalent spatial densities. An important by-product of such joint acquisition is the reduction of the carbon footprint, a challenge for the oil industry. The global deep-water industry is being strengthened and is preparing for the next phase of upgrades. This time, it may be possible for a JIP to build an OBMP reservoir monitoring system that will add substantial value to reservoir management decisions, leading to greater oil recovery, reduced expenses, and improved sustainability for the oil industry.

Regarding the future, the deepening of the climate crisis is driving the world to transition to low-carbon energy sources such as native hydrogen and geothermal resources [57–59]. Another option in reducing the CO_2 concentration in the atmosphere is injecting and storing it in saline aquifers [60], a technique popularly known as carbon capture usage and storage (CCUS).

It is well known that the CSEM method is very sensitive to fluid saturation in rocks. Therefore, CSEM and other EM methods are expected to be prominent in exploring all transition energy assets depicted above. The same is expected in the monitoring phases, as EM acquisitions tend to be cheaper than seismic ones [61].

Similarly to the circle of life, Petrobras has returned 2023 to the beginning. Now, however, the multiphysics group has gained considerable knowledge in EM applications. We expect to achieve the same success with energy transition assets as in hydrocarbon exploration.

Author Contributions: Conceptualization, P.T.L.M. and E.N.M.; methodology, P.T.L.M., J.L.C. and E.N.M.; validation, P.T.L.M., J.L.C. and E.N.M.; writing—original draft preparation, P.T.L.M.; writing—review and editing, P.T.L.M., E.N.M., S.M.F. and E.N.M.; supervision, S.M.F.; project administration, S.M.F. All authors have read and agreed to the published version of the manuscript.

Funding: This research was funded by Petrobras.

Data Availability Statement: Data associated with this research are confidential and cannot be released.

Acknowledgments: We are grateful to the two pioneers, Ricardo Castellani and Marco Polo Buonora, for the initiation of the CSEM method in Petrobras. We acknowledge the continuous support of our colleagues from the multiphysics group. We thank the Petrobras/PEN/AB management for their support and authorization to publish the present work. We also acknowledge our colleagues in the asset teams for the discussions and collaborations that helped to promote the methodology within the company. We are grateful to Petrobras management for their continued support over these past 20 years and for allowing the publication of the present paper. We are grateful to the assigned editor Leo Du for the support and to two anonymous reviewers for their suggestions that helped us to improve our paper.

Conflicts of Interest: The authors declare no conflict of interest.

Abbreviations

The following abbreviations are used in this manuscript:

CSEM	Controlled source electromagnetic method
HED	Horizontal electric dipole
EM	Electromagnetic
MT	Magnetotelluric
AVO	Amplitude versus offset
VP	Compressional velocity
FWI	Full waveform inversion
BLM	Below mud line
1/2/2.5/3/4D	One/two/three/four-dimensional
JIP	Joint industry project
GUI	Graphical user interface
CMP	Common midpoint
PJI	Petrophysical joint inversion
JE	Jubarte experiment
MR3D	Marlim Resistivity 3D
BFGS	Broyden–Fletcher–Goldfarb–Shanno algorithm
MAM	Multiphysics anomaly map
PRM	Permanent reservoir monitoring
OBMP	Ocean-bottom multiphysics nodes
CCUS	Carbon capture usage and storage

TLA	Three-letter acronym
LD	Linear dichroism
SoPhiH	Oil saturation (So), porosity (Phi), and thickness (H)
Petrobras	Petroleo Brasileiro S.A
EMGS	Electromagnetic Geoservices
SLB	Schlumberger
AOA	Arnold Orange Associates
AGO	Geomarine Operations
MCTI-ON	Ministério de Ciência, Tecnológia e Inovações—Observatório Nacional
UFPA	Universidade Federal do Pará
UERJ	Universidade do Estado do Rio de Janeiro

References

1. Constable, S.; Srnka, L.J. An introduction to marine controlled-source electromagnetic methods for hydrocarbon exploration. *Geophysics* **2007**, *72*, WA3–WA12. [CrossRef]
2. Ellingsrud, S.; Eidesmo, T.; Johansen, S.; Sinha, M.C.; MacGregor, L.M.; Constable, S. Remote sensing of hydrocarbon layers by seabed logging (SBL): Results from a cruise offshore Angola. *Lead. Edge* **2002**, *21*, 972–982. [CrossRef]
3. Castellani, R.T. Incorporation of seismic intermediate-scale data for improving reservoir description. In Proceedings of the 7th International Congress of the Brazilian Geophysical Society, Salvador, Brazil, 28–31 October 2001; European Association of Geoscientists & Engineers: Utrecht, The Netherlands, 2001; p. cp-217-00215.
4. Constable, S. Perspectives on Marine Electromagnetic Methods. *Perspect. Earth Space Sci.* **2020**, *1*, e2019CN000123. [CrossRef]
5. Buonora, M.P.P.; Reddig, R.; Heelan, W.; Schofield, J.D.; Zerilli, A.; Labruzzo, T. Some preliminary interpretation on the marine controlled source electromagnetic (MCSEM) data acquired on Campos Basin, Brazil. In Proceedings of the 9th International Congress of the Brazilian Geophysical Society & EXPOGEF, Salvador, Bahia, Brazil, 11–14 September 2005; Society of Exploration Geophysicists and Brazilian Geophysical Society: Rio de Janeiro, Brazil, 2005; pp. 147–152.
6. Smit, D.; Saleh, S.; Voon, J.; Costello, M.; Moser, J. Recent Controlled Source EM results show positive impact on exploration at Shell. In *SEG Technical Program Expanded Abstracts 2006*; Society of Exploration Geophysicists: Houston, TX, USA, 2006; pp. 3536–3541.
7. Newman, G.A.; Commer, M.; Carazzone, J.J. Imaging CSEM data in the presence of electrical anisotropy. *Geophysics* **2010**, *75*, F51–F61. [CrossRef]
8. de Lugao, P.P.; Fontes, S.L.; La Terra, E.F.; Zerilli, A.; Labruzzo, T.; Buonora, M.P. First application of marine magnetotellurics improves depth imaging in the Santos Basin, Brazil. In Proceedings of the 70th EAGE Conference and Exhibition incorporating SPE EUROPEC 2008, Rome, Italy, 9–12 June 2008; European Association of Geoscientists & Engineers: Utrecht, The Netherlands, 2008; p. cp-40-00421.
9. Gallardo, L.; Fontes, S.; Meju, M.; Buonora, M.; De Lugao, P. Robust geophysical integration through structure-coupled joint inversion and multispectral fusion of seismic reflection, magnetotelluric, magnetic, and gravity images: Example from Santos Basin, offshore Brazil. *Geophysics* **2012**, *77*, B237–B251. [CrossRef]
10. Buonora, M.P.; Zerilli, A.; Labruzzo, T.; Rodrigues, L.F. Detecting Hydrocarbon Reservoirs from Marine CSEM in the Santos Basin, Brazil. *Search Discov.* **2009**, 40402.
11. Zerilli, A.; Labruzzo, T.; Buonora, M.P.; de Tarso Luiz Menezes, P.; Rodrigues, L.F.; Lovatini, A. 3D inversion of total field mCSEM data: The Santos Basin case study. In *SEG Technical Program Expanded Abstracts 2010*; Society of Exploration Geophysicists: Houston, TX, USA, 2010; pp. 629–633.
12. Zerilli, A.; Buonora, M.P.; Abubakar, A.; Labruzzo, T. Inversion of marine Controlled Source Electromagnetic data using a "structure"-based approach. In Proceedings of the 12th International Congress of the Brazilian Geophysical Society, Rio de Janeiro, Brazil, 15–18 August 2011; European Association of Geoscientists & Engineers: Utrecht, The Netherlands, 2011; p. cp-264-00039.
13. Buonora, M.P.P.; Correa, J.L.; Martins, L.S.; Menezes, P.T.L.; Pinho, E.J.C.; Silva Crepaldi, J.L.; Ribas, M.P.P.; Ferreira, S.M.; Freitas, R.C. mCSEM data interpretation for hydrocarbon exploration: A fast interpretation workflow for drilling decision. *Interpretation* **2014**, *2*, SH1–SH11. [CrossRef]
14. Lorenz, L.; Pedersen, H.T.; Buonora, M.P. First results from a Brazilian mCSEM calibration campaign. In Proceedings of the 13th International Congress of the Brazilian Geophysical Society & EXPOGEF, Rio de Janeiro, Brazil, 26–29 August 2013; Society of Exploration Geophysicists and Brazilian Geophysical Society: Rio de Janeiro, Brazil, 2013; pp. 131–136.
15. Menezes, P.T.L.; Zerilli, A.; Correa, J.L.; Menor, E.N.; Ferreira, S.M.; Labruzzo, T. New Concept Ocean-Bottom Multiphysics (OBMP) Nodes for Reservoir Monitoring. *Minerals* **2023**, *13*, 602. [CrossRef]
16. Zerilli, A.; Buonora, M.P.; Menezes, P.D.L.; Crepaldi, J.L.; Labruzzo, T. Seismic-EM Integration Tackles Deep Water E&P Challenges. In Proceedings of the 76th EAGE Conference and Exhibition Workshops, Amsterdam, The Netherlands, 16–19 June 2014; EAGE Publications BV: Utrecht, The Netherlands, 2014; p. cp-401-00077.
17. Buonora, M.P.; Zerilli, A.; Labruzzo, T. Marine controlled source electromagnetic: New insights in data interpretation. In *SEG/SBGf D&P Forum Deep Water Challenges in Exploration, Development, and Production*; Society of Exploration Geoscientists & Sociedade Brasileira de Geofísica: Rio de Janeiro, Brazil, 2006; p. cp–1.

18. Silva Crepaldi, J.L.; Pereira Buonora, M.P.; Figueiredo, I. Fast marine CSEM inversion in the CMP domain using analytical derivatives. *Geophysics* **2011**, *76*, F303–F313. [CrossRef]
19. Maaø, F.A. Fast finite-difference time-domain modeling for marine-subsurface electromagnetic problems. *Geophysics* **2007**, *72*, A19–A23. [CrossRef]
20. Lovatini, A.; Myers, K.; Watterson, P.; Campbell, T. The Potiguar integrated exploration project: CSEM prospectivity assessment offshore Brazil. *Lead. Edge* **2010**, *29*, 848–851. [CrossRef]
21. Zerilli, A.; Buonora, M.P.; Menezes, P.T.; Labruzzo, T.; Marçal, A.J.; Silva Crepaldi, J.L. Broadband marine controlled-source electromagnetic for subsalt and around salt exploration. *Interpretation* **2016**, *4*, T521–T531. [CrossRef]
22. Hoversten, G.M.; Constable, S.C.; Morrison, H.F. Marine magnetotellurics for base-of-salt mapping: Gulf of Mexico field test at the Gemini structure. *Geophysics* **2000**, *65*, 1476–1488. [CrossRef]
23. Travassos, J.; Menezes, P. Geoelectric structure beneath limestones of the Sao Francisco Basin, Brazil. *Earth Planets Space* **1999**, *51*, 1047–1058. [CrossRef]
24. Paulo de Tarso, L.M.; Travassos, J.M. EM modeling of the central–northern portion of Ponta Grossa Arch, Paraná Basin, Brazil. *Phys. Earth Planet. Inter.* **2005**, *150*, 145–158.
25. Menezes, P.T.; Travassos, J.M. Magnetotellurics as a modeling tool in the extensive magmatic context of Paraná Basin, Brazil. *Lead. Edge* **2010**, *29*, 832–840. [CrossRef]
26. Colombo, D.; McNeice, G.; Curiel, E.S.; Fox, A. Full tensor CSEM and MT for subsalt structural imaging in the Red Sea: Implications for seismic and electromagnetic integration. *Lead. Edge* **2013**, *32*, 436–449. [CrossRef]
27. Mantovani, M.; Hokstad, K.; Trezzi, S.; Navaza, M.; Duffaut, K.; Wiik, T.; Lovatini, A.; Guerra, I. Cooperative, structural and grid-based simultaneous joint inversion for salt base and presalt imaging. In Proceedings of the 76th EAGE Conference and Exhibition 2014, Amsterdam, The Netherlands, 16–19 June 2014; European Association of Geoscientists & Engineers: Utrecht, The Netherlands, 2014; pp. 1–5.
28. Zerilli, A.; Labruzzo, T.; Marçal, A.A.; Buonora, M.P.; Crepaldi, J.L.; Menezes, P.T. Broadband and ultra-long offset mCSEM—A recipe for complex subsalt imaging. In Proceedings of the 14th International Congress of the Brazilian Geophysical Society & EXPOGEF, Rio de Janeiro, Brazil, 3–6 August 2015; Brazilian Geophysical Society: Rio de Janeiro, Brazil, 2015; pp. 261–265.
29. Mackie, R.; Watts, M.D.; Rodi, W. Joint 3D inversion of marine CSEM and MT data. In Proceedings of the 2007 SEG Annual Meeting, San Antonio, TX, USA, 23–26 September 2007; OnePetro: Richardson, TX, USA, 2007.
30. Habashy, T.; Abubakar, A. A general framework for constraint minimization for the inversion of electromagnetic measurements. *Prog. Electromagn. Res.* **2004**, *46*, 265–312. [CrossRef]
31. Zerilli, A.; Labruzzo, T.; Buonora, M.P.; Abubakar, A. Joint inversion of marine CSEM and MT data using a "structure"-based approach. In *SEG Technical Program Expanded Abstracts 2011*; Society of Exploration Geophysicists: Houston, TX, USA, 2011; pp. 604–608.
32. Crepaldi, J.L.; Zerilli, A.; Labruzzo, T.; dos Santos, G.B.; Buonora, M.P. Inversão conjunta sísmica/em para sal alóctone. In Proceedings of the 14th International Congress of the Brazilian Geophysical Society & EXPOGEF, Rio de Janeiro, Brazil, 3–6 August 2015; Brazilian Geophysical Society: Rio de Janeiro, Brazil, 2015; pp. 190–193.
33. Hacikoylu, P.; Dvorkin, J.; Mavko, G. Resistivity-velocity transforms revisited. *Lead. Edge* **2006**, *25*, 1006–1009. [CrossRef]
34. Zerilli, A.; Miotti, F.; Mantovani, M.; Menezes, P.T.; Crepaldi, J.L.S. Broadband CSEM-guided starting model for full waveform inversion–Impact in the Brazilian deep offshore. In Proceedings of the 15th International Congress of the Brazilian Geophysical Society & EXPOGEF, Rio de Janeiro, Brazil, 31 July–3 August 2017; Brazilian Geophysical Society: Rio de Janeiro, Brazil, 2017; pp. 138–142.
35. Boswell, R.; Collett, T.S. Current perspectives on gas hydrate resources. *Energy Environ. Sci.* **2011**, *4*, 1206–1215. [CrossRef]
36. Jong, J.; Goh, H.S.; McGiveron, S.; Fitton, J. A Case Study Of Natural Gas Hydrates (NGH) In Offshore NW Sabah: Identification, Shallow Geohazard Implication For Exploration Drilling, Extraction Challenges And Potential Energy Resource Estimation. *Bull. Geol. Soc. Malays.* **2020**, *70*, 57–75. [CrossRef]
37. Hunter, S.; Goldobin, D.; Haywood, A.; Ridgwell, A.; Rees, J. Sensitivity of the global submarine hydrate inventory to scenarios of future climate change. *Earth Planet. Sci. Lett.* **2013**, *367*, 105–115. [CrossRef]
38. McConnell, D.R.; Zhang, Z.; Boswell, R. Review of progress in evaluating gas hydrate drilling hazards. *Mar. Pet. Geol.* **2012**, *34*, 209–223. [CrossRef]
39. Tharimela, R.; Augustin, A.; Ketzer, M.; Cupertino, J.; Miller, D.; Viana, A.; Senger, K. 3D controlled-source electromagnetic imaging of gas hydrates: Insights from the Pelotas Basin offshore Brazil. *Interpretation* **2019**, *7*, SH111–SH131. [CrossRef]
40. Miotti, F.; Zerilli, A.; Menezes, P.T.L.; Crepaldi, J.L.; Viana, A.R. A new petrophysical joint inversion workflow: Advancing on reservoir's characterization challenges. *Interpretation* **2018**, *6*, SG33–SG39. [CrossRef]
41. Zerilli, A.; Labruzzo, T.; Andreasi, F.G.; Menezes, P.; Crepaldi, J.; Alvim, L. Realizing 4D CSEM Value on Deep Water Reservoirs – The Jubarte Case Study. In Proceedings of the 80th EAGE Conference and Exhibition 2018, Copenhagen , Denmark, 10 June–15 July 2018; EAGE Publications BV: Utrecht, The Netherlands, 2018. [CrossRef]
42. de Tarso Menezes, P.; Crepaldi, J.L.; Zerilli, A.; Labruzzo, T.; Alvim, L.; Correa, J.; Pinho, E.; Lyrio, J.C.S.; Viana, A.R. 4-D CSEM, A Cost-Effective Tool for Deep-Water Clastic Reservoir Monitoring. In Proceedings of the 2018 AAPG International Conference and Exhibition, Salt Lake City, UT, USA, 20–23 May 2018; American Association of Petroleum Geologists: Tulsa, OK, USA, 2018.

43. Zerilli, A.; Viana, A.; Menezes, P.; Correa, J. Ocean Bottom Multi-Physics Nodes—A Long Way to Next Reservoir Monitoring Wave? In Proceedings of the 81st EAGE Conference and Exhibition 2019, London, UK, 3–6 June 2019; European Association of Geoscientists & Engineers: Utrecht, The Netherlands, 2019; pp. 1–5.
44. Ramos Filho, W.L.; Dariva, P.; Born, C.C.; Zorzanelli, I.B.; Goertz, A.; Smith, A. Permanent Reservoir Monitoring at Jubarte Field-4D Results and Reservoir Characterization. In Proceedings of the First EAGE Workshop on Practical Reservoir Monitoring, Rio de Janeiro, Brazil, 21–22 November 2017; European Association of Geoscientists & Engineers: Utrecht, The Netherlands, 2017; p. cp–505.
45. Carvalho, B.R.; Menezes, P.T.L. Marlim R3D: A realistic model for CSEM simulations-phase I: model building. *Braz. J. Geol.* **2017**, *47*, 633–644. [CrossRef]
46. Correa, J.L.; Menezes, P.T. Marlim R3D: A realistic model for controlled-source electromagnetic simulations—Phase 2: The controlled-source electromagnetic data set. *Geophysics* **2019**, *84*, E293–E299. [CrossRef]
47. Correa, J.L.; Menezes, P.T. Marlim R3D phase 3: The marine magnetotelluric regional model and associated data set. *Lead. Edge* **2021**, *40*, 686–692. [CrossRef]
48. Nascimento, T.M.; Menezes, P.T.L.; Braga, I.L. High-resolution acoustic impedance inversion to characterize turbidites at Marlim Field, Campos Basin, Brazil. *Interpretation* **2014**, *2*, T143–T153. [CrossRef]
49. Nguyen, A.K.; Nordskag, J.I.; Wiik, T.; Bjørke, A.K.; Boman, L.; Pedersen, O.M.; Ribaudo, J.; Mittet, R. Comparing large-scale 3D Gauss–Newton and BFGS CSEM inversions. In Proceedings of the 2016 SEG International Exposition and Annual Meeting, Dallas, TX, USA, 16–21 October 2016; OnePetro: Richardson, TX, USA, 2016.
50. Streich, R.; Súilleabháin, L.; Frantzen, P. Ambiguity in EM Images Reduced by Gaussnewton Inversion: A Case Study of a Basement High. In Proceedings of the 82nd EAGE Annual Conference & Exhibition, Online, 18–21 October 2021; EAGE Publications BV: Utrecht, The Netherlands, 2021; pp. 1–5.
51. Price, A. What's the problem with CSEM? In *Second International Meeting for Applied Geoscience & Energy*; Society of Exploration Geophysicists and American Association of Petroleum: Houston, TX, USA, 2022; pp. 667–671.
52. Lyrio, J.C.S.O.; Menezes, P.T.L.; Correa, J.L.; Viana, A.R. Multiphysics anomaly map: A new data fusion workflow for geophysical interpretation. *Interpretation* **2020**, *8*, B35–B43. [CrossRef]
53. Rodriguez, K.; Pedersen, H.; Svendsen, R.; Negri, D. CSEM Anomalies co-rendered with Modern Seismic to De-risk Exploration Leads–Foz do Amazonas Brazil. In Proceedings of the 77th EAGE Conference and Exhibition 2015, Madrid, Spain, 1–4 June 2015; European Association of Geoscientists & Engineers: Utrecht, The Netherlands, 2015; pp. 1–5.
54. Rodriguez, K.; Berryman, J.; Kearns, H.; Saunders, M. Frontier Basin Exploration Through Integration of BSR-Derived Geothermal Gradient and CSEM Anomaly-Seismic Co-Rendering—Foz do Amazonas Basin, Brazil. In Proceedings of the AAPG/SEG International Conference & Exhibition, Cancun, Mexico, 6–9 September 2016.
55. Menezes, P.T.L.; Correa, J.L.; Alvim, L.M.; Viana, A.R.; Sansonowski, R.C. Time-Lapse CSEM Monitoring: Correlating the Anomalous Transverse Resistance with SoPhiH Maps. *Energies* **2021**, *14*, 7159. [CrossRef]
56. Brinkerhoff, R. Pitfalls in Geological Mapping Within Unconventional Plays. In Proceedings of the AAPG Annual Convention and Exhibition, Calgary, AB, Canada, 19–22 June 2016.
57. Zgonnik, V. The occurrence and geoscience of natural hydrogen: A comprehensive review. *Earth-Sci. Rev.* **2020**, *203*, 103140. [CrossRef]
58. Tester, J.W.; Beckers, K.F.; Hawkins, A.J.; Lukawski, M.Z. The evolving role of geothermal energy for decarbonizing the United States. *Energy Environ. Sci.* **2021**, *14*, 6211–6241. [CrossRef]
59. Meju, M.A.; Saleh, A.S. Using Large-Size Three-Dimensional Marine Electromagnetic Data for the Efficient Combined Investigation of Natural Hydrogen and Hydrocarbon Gas Reservoirs: A Geologically Consistent and Process-Oriented Approach with Implications for Carbon Footprint Reduction. *Minerals* **2023**, *13*, 745. [CrossRef]
60. Ayani, M.; Grana, D.; Liu, M. Stochastic inversion method of time-lapse controlled source electromagnetic data for CO_2 plume monitoring. *Int. J. Greenh. Gas Control* **2020**, *100*, 103098. [CrossRef]
61. Huang, L.; Yang, X. Geophysical Monitoring Techniques. *AGU J.* **2022**, 439–440. [CrossRef]

Disclaimer/Publisher's Note: The statements, opinions and data contained in all publications are solely those of the individual author(s) and contributor(s) and not of MDPI and/or the editor(s). MDPI and/or the editor(s) disclaim responsibility for any injury to people or property resulting from any ideas, methods, instructions or products referred to in the content.

minerals

MDPI

Article

Geophysical Characterization and Attenuation Correction Applied for Hydrate Bearing Sediments in Japan Sea

Luiz Alberto Santos [1,2,*], Ryo Matsumoto [3], Fernanda Darcle Silva Freitas [1,2] and Marco Antonio Cetale Santos [2]

1 Petróleo Brasileiro S.A. (Petrobras), Av. Henrique Valadares 28, Centro, Rio de Janeiro 20231-170, RJ, Brazil
2 Departamento de Geologia e Geofísica (GGO), Instituto de Geociências (EGG), Universidade Federal Fluminense (UFF), Campus da Praia Vermelha, Av. Gen. Milton Tavares de Souza, s/n, bloco P, Boa Viagem, Niterói 24210-346, RJ, Brazil
3 Department of Earth and Planetary Science, University of Tokyo, Hongo 7-3-1, Bunkyo-ku, Tokyo 113-0033, Japan
* Correspondence: luiz_alberto@petrobras.com.br or luizalb1@gmail.com

Abstract: Estimation of rock properties from seismic data is important for exploration and production activities in the petroleum industry. Considering the compressional velocity—the speed of propagating body waves in formations—and the quality factor (Q)—a measure of the frequency-selective energy losses of waves propagating through formations—both properties are usually estimated from multichannel seismic data. Velocity is estimated during multichannel processing of seismic reflection data in either the time or depth domain. In marine seismic acquisition, Q can be estimated from the following sources: Vertical Seismic Profile (VSP) surveys, where sources are located near the sea surface and geophones are distributed at depth along a borehole; and multichannel reflection data, where sources are also located near the sea surface and receivers are distributed either at the sea surface (conventional seismic survey with streamers) or on the sea floor (use of nodes or Ocean Bottom Cables (OBC)). The aforementioned acquisition devices, VSP, conventional streamers, nodes, and OBCs are much more expensive than single-channel acquisition with one receiver per shot due to the cost of operation. There are numerous old and new datasets from academia and the oil industry that have been acquired with single-channel acquisition devices. However, there is a paucity of work addressing the estimation of velocity and Q from this type of equipment. We investigate the estimation of Q and velocity from single-channel seismic data. Using the windowed discrete Fourier transform for a single seismic trace, we calculate the peak and dominant frequency that changes with time. In the geologic environment, higher frequencies are attenuated at shallow depths (time), while lower frequencies remain at deeper positions. From the rate at which higher frequencies are attenuated with time, we estimate the effective quality factor (Q_{eff}). However, when using Kirchhoff migration to process single-channel seismic data, events far from the vertical projection of the receiver contribute to the trace at a given time. Then, an underestimation of the effective quality factor occurs. To compensate for the effects of more distant events with lower-frequency content contaminating the shorter events, we propose a linear equation to correct the effective quality factor estimated from migrated seismic data. Effective Q and its correction are estimated in five single-channel seismic lines surveyed along the Joetsu Knoll, a SW-NE anticline structure on the eastern margin of the Sea of Japan. These results are linked to geomorphological and geological features and the velocity field. Joetsu Knoll is a known site of massive gas hydrates (GH), which occur in the first hundred metres of Neogene sediments and, together with gas chimneys, play an important role in seismic wave absorption. Q_{eff} estimated from migrated seismic data maintains the spatial relationship between high and low Q regions. The region of low Q, which is below 124 and has an average value of 57, occurs near the anticlinal hinge and tends to coincide with the region in which the Bottom Simulating Reflector (BSR) resides. The coexistence of GH and free gas coincides with the very low P velocity gradient of 0.225 s^{-1}. BSR occurrence, Q_{eff} and the geometry of the Joetsu anticline testify to progressive gas hydrate depletion northward along the dome.

Keywords: absorption; gas hydrates; single channel seismic; time migration; Japan Sea

Citation: Santos, L.A.; Matsumoto, R.; Freitas, F.D.S.; Santos, M.A.C. Geophysical Characterization and Attenuation Correction Applied for Hydrate Bearing Sediments in Japan Sea. *Minerals* **2023**, *13*, 655. https://doi.org/10.3390/min13050655

Academic Editor: Neil M. Ribe

Received: 7 February 2023
Revised: 10 March 2023
Accepted: 3 April 2023
Published: 10 May 2023

Copyright: © 2023 by the authors. Licensee MDPI, Basel, Switzerland. This article is an open access article distributed under the terms and conditions of the Creative Commons Attribution (CC BY) license (https://creativecommons.org/licenses/by/4.0/).

1. Introduction

The porous filling of sedimentary formations can have a strong influence on the absorption behavior of seismic signals. The use of absorption in seismic data to study and understand the subsurface has been the subject of studies worldwide. Several studies have been conducted on vertical seismic profiles (VSP), direct waves, and seismic reflection data, which provide important formation and fluid properties and are sometimes able to infer fluid properties, which is a valuable requirement for hydrocarbon prospecting.

In the paper [1], the author states that attenuation by absorption is always observed in seismic data and adds that absorption has a significant effect on recorded waveforms and amplitudes due to the dispersive nature of the phenomenon. In reference [2], the Q factor is estimated using VSP data from the Campos Basin (off the Brazilian coast) using spectral ratio and frequency shift methods. In paper [3], the concept of spectral ratio is applied to the common midpoint (CMP) gathering of conventional seismic surface data. The authors successfully estimated Q-factors for the following three case studies: (i) comparison with VSP data at a site in the North Sea; (ii) use of the Q-factor to distinguish between sedimentary and crystalline basement, with the latter having higher Q-values; and (iii) evaluation of the gas effect on saturated reservoirs in a field in the UK North Sea.

In the paper [4] a stabilized inverse Q filter is used for a land VSP dataset and the author concludes that the method promotes a robust estimation of Q values. The subsequent processing flow flattens the amplitude spectrum, strengthens the time-variant amplitude, increases the spectral bandwidth, and improves the signal-to-noise ratio (S/N). In reference [5], the authors point out three direct indicators of gas hydrate (GH) at Blake Ridge, in deep waters of the eastern coast of United States: a paleo bottom simulating reflector (BSR); stratigraphic intervals of high velocity and low amplitude (blanking); and bright spots within the hydrate stability zone due to upward gas.

Based on observations in different depositional environments and frequency ranges at the Outer Blake Ridge (U.S. coast), Malik (Canada), and Nankai Trough (offshore Japan), [6] find that hydrated sediments always provide very high attenuation of seismic waves, despite their stiffness and higher velocities. The results of [7,8] confirm the observation that attenuation increases with hydrate concentration.

In the paper [9], the authors state that wave velocities and attenuation are two important properties used to estimate the lithology, saturation and in situ conditions of rocks. The authors' model predicts that, in general, velocity increases while attenuation decreases with increasing gas hydrate concentration. References [7,10,11] support the model of decreasing attenuation with increasing gas-hydrate concentration.

An equation for estimating gas hydrate and free gas from seismic data is derived in [12], emphasizing the need for high-resolution geophysical data. The authors of [13] use the concept of attenuated travel time tomography to estimate the interval Q and successfully delineate a known zone of low Q along the 3D survey offshore Australia. In [14], the high-resolution Q tomography technique is used to automatically detect shallow gas pockets in Brunei. In reference [15], the authors apply an improved peak frequency shift to estimate the Q factor using time migration data in a 2D seismic line in the Sichuan Basin (China) and conclude that the quality factor can be used to detect hydrocarbons. The authors of [16] discuss stabilized inverse Q filtering and apply this technique to improve 3D seismic stacking data in the Blake Ridge.

The above bibliographies deal with Q estimation using VSP, OBN, land and offshore seismic reflection, direct and refraction data in different environments, some of them dealing with sites with free gas and gas hydrate. Characterization and better geometric and structural delineation of gas hydrate deposits is becoming increasingly important for subsequent, and perhaps future, methane production. Energy-dependent countries such as Japan and Ukraine have large gas hydrate deposits. Modeling of gas production from hydrates in the Black Sea using artificial [17] or natural energy from mud volcanoes [18] demonstrates the feasibility of significant amounts of methane meeting local energy needs.

Although there are a large number of scientific and industrial, recent and older, single-channel seismic data (SCS) studies around the world, there is very little work on estimating Q from SCS data—zero-offset studies—that could contribute to energy supply from gas hydrates through better geophysical characterization.

Our goal is the geophysical characterization of zero-offset seismic data (in this work, we consider SCS a zero-offset acquisition), focusing on the estimation of the quality factor (Q) and the P velocity. To achieve the proposed goals, we use the frequency peak method of [19] to estimate the effective Q. Thus, we derive the time migration equations in non-conservative media, we show the difference between the frequency content between pre- and post-migrated data, and then we estimate a linear relationship between the effective Q (Q_{eff}) and the Q estimated from the post-migrated data (Q_{mig}), which is proposed to correct Q_{mig}.

In the application section, the derivations for Q estimation are applied to an integrative study of the single-channel seismic dataset (SCS) in the Joetsu Basin on the Sea of Japan, where new properties are estimated at this site. Finally, we conclude that the Q_{eff} estimated from migrated seismic data preserves the spatial relationships between regions of high and low Q. We quantify the Q_{eff} and the gradient of P velocity along the Joetsu Knoll and estimate the geometry of the northward narrowing of the GH.

2. Materials and Methods

2.1. Theoretical Aspects of Spectral Characterization

When viewing a seismic section in the time domain, the observer often notes a loss of resolution with increasing time. There is a time-dependent frequency variation. Consider $U(x_{src}, x_{rec}, \omega)$ the Fourier transform of each trace of a recorded wavefield, where ω is the angular frequency; x_{src} is the seismic source and x_{rec} is the location of the receiver, both in the horizontal. The amplitude spectrum of the seismic data, calculated from $U(x_{src}, x_{rec}, \omega)$, shows the general behavior of each trace and is often used to estimate the peak frequency and lateral energy variation. However, the time-dependent spectrum variation cannot be localized and evaluated using the discrete Fourier Transform (DFT). In [20], several spectral decomposition methods are presented, including: windowed discrete Fourier transform (WDFT), maximum entropy method (MEM), continuous wavelet transform (CWT), matching pursuit decomposition (MPD), and exponential pursuit decomposition (EPD). Even though the authors state that there is no right or wrong method for time–frequency analysis, they highlight the advantages of EPD.

In this work, the WDFT is used because it directly relates the amplitude spectrum to a specific time window. Thus, the window used is a Gaussian function $g(t)$, as shown in Equation (1).

$$g(t) = A\, e^{\frac{-(t-\tau)^2}{2\sigma^2}} \quad (1)$$

where A is a scalar, usually 1; σ determines the function width; and τ specifies the position t at which the apex of the function is centered. Equation (2) can then be used to evaluate the time-dependent spectrum.

$$U(x_{src}, x_{rec}, \tau, \omega) = \int_{-\infty}^{\infty} u(x_{src}, x_{rec}, t) g(t, \tau) e^{-i\omega t} dt \quad (2)$$

$U(x_{src}, x_{rec}, \tau, \omega)$ is the windowed Fourier transform of the recorded wavefield. Equations (1) and (2) are described for the general 2D seismic survey. For 2D zero offset or migrated data, instead of $u(x_{src}, x_{rec}, \tau)$, $U(x_{src}, x_{rec}, \omega)$, $U(x_{src}, x_{rec}, \tau, \omega)$ we can briefly consider $u(x_{rec}, t)$, $U(x_{rec}, \omega)$, and $U(x_{rec}, \tau, \omega)$, respectively. From the amplitude spectrum of $U(x_{rec}, \tau, \omega)$, we extract the peak frequency at each time using Equation (3):

$$F(x_{rec}, \tau) = arg_{\omega} max\{U(x_{rec}, \tau, \omega)\} \quad (3)$$

We apply $F(x_{rec}, \tau)$ directly to estimate Q_{eff} using the method of Ref. [19]), where we consider $f_m = F(x_{rec}, \tau_1)$ as the dominant frequency at a reference time (τ_1); $f_p = F(x_{rec}, \tau_2)$ as the peak frequency at another time (τ_2) higher than the reference time; and $\Delta t = \tau_2 - \tau_1$. Q_{eff} is thus calculated as:

$$Q_{eff} = \frac{\pi \Delta t f_p f_m^2}{2\left(f_m^2 - f_p^2\right)} \tag{4}$$

The calculation of the effective Q by the centroid shift of the frequency peak. Ref. [21] states that Q_{eff} is the result of the path integral along the ray path, which cumulates the individual absorption effects of each unit or layer. In this work, we restrict our analysis only to the effective Q.

2.2. Effects of Time Migration in the Estimation of Attenuation

Consider a point source at a point A on the subsurface where the wavelet is in frequency domain $w(\omega)$, and the wavefield $U(x_j, \omega)$ is recorded along a perfectly horizontal and infinite surface at z_0. If the medium is conservative, the time migration can be calculated according to [22]:

$$U_{mig}(x_{rec}, \omega) = \frac{1}{2\pi} \frac{\partial}{\partial z} \int_{-\infty}^{\infty} \int_{-\infty}^{\infty} dx\, dy\, U(x_j, \omega)\, G(x_j, r_S, \omega) \tag{5}$$

where $G(x_j, r_S, \omega)$ is the Green function in the frequency domain from a receiver at x_j on the surface to the point S; dx and dy are infinitesimal horizontal segments of a reference system.

Consider the argument of integration in Equation (5):

$$I(x_{rec}, \omega) = U(x_j, \omega)\, G(x_j, r_S, \omega) \tag{6}$$

This operation corresponds to a phase shift scaled by the reciprocal of the distance between x_j and r_S, say Δr_j. The frequency content of $I(x_j, \omega)$ does not change compared to $U(x_j, \omega)$. Since

$$G(x_j, r_S, \omega) = \frac{e^{i\omega . t_j}}{\Delta r_j} \tag{7}$$

the integral of Equation (6) may be approximated by

$$S(x_j, \omega) = \Delta x \Delta y \sum_{j=1}^{n} U(x_j, \omega) \frac{e^{i\omega . t_j}}{\Delta r_j} \tag{8}$$

In a conservative medium, after the phase shift operation, all contributions $U(x_j, \omega) . e^{i\omega . t_j}$ are exactly the same for the wavelet $w(\omega)$, but scaled by the geometric spreading effect, which we call K_j. Thus, the summation of Equation (8) is:

$$S(x_j, \omega) = \Delta x \Delta y\, w(\omega) \sum_{j=1}^{n} \frac{K_j}{\Delta r_j} \tag{9}$$

Again, the frequency content of $S(x_j, \omega)$ does not change compared to $U(x_j, \omega)$.

However, if the medium is not conservative, the attenuated wavefield, $S_{atten}(x_j, \omega)$, can be calculated by replacing $U(x_j, \omega)$ in Equation (9) with the attenuated signal $U_{atten}(x_j, \omega)$:

$$S_{atten}(x_j, \omega) = \Delta x \Delta y \sum_{j=1}^{n} U_{atten}(x_j, \omega) \frac{e^{i\omega . t_j}}{\Delta r_j} \tag{10}$$

We can consider $U_{atten}(x_j, \omega) = U(x_j, \omega)e^{\frac{-\omega.t_j}{2Q}}$, whose attenuation is given by the term $e^{\frac{-\omega.t_j}{2Q}}$, where Q is the quality factor and t_j is the time of the event. For a model with constant Q, the function $U_{atten}(x_j, \omega)$ has lower amplitude and energy at higher frequencies ω compared to $U(x_j, \omega)$.

$$S_{atten}(x_j, \omega) = \Delta x \Delta y \sum_{j=1}^{n} u(x_j, \omega)e^{\frac{-\omega.t_j}{2Q}} \frac{e^{i\omega.t_j}}{\Delta r_j} \tag{11}$$

Considering $K_j w(\omega) = U(x_j, \omega)e^{i\omega.t_j}$:

$$S_{atten}(x_j, \omega) = \Delta x \Delta y\ w(\omega) \sum_{j=1}^{n} \frac{K_j}{\Delta r_j} e^{\frac{-\omega.t_j}{2Q}} \tag{12}$$

Equation (12) shows that the summation $S_{atten}(x_j, \omega)$ has much smaller amplitudes and energies than $S(x_j, \omega)$ at higher frequencies. In a non-conservative medium, the migrated seismic data then have lower-frequency content at the event position because the term $e^{\frac{-\omega.t_j}{2Q}}$ contributes to a larger shift. The natural consequence of the above derivation is that using Equation (4) to calculate the effective quality factor for migrated data, Q_{mig}, leads to lower results than the real $S_{atten}(x_j, \omega) < S(x_j, \omega)$. This effect is mainly larger when the reference frequency (f_m) is taken at the source position of real surveys or at the seafloor. The above derivation can be extended to the reflection data. Then, we expect an underestimation of Q in time migrated data.

2.3. Q_{mig} and Q_{eff} Relation

The derivation of the previous topic shows that the quality factor in time-migrated data, Q_{mig}, is underestimated compared to non-migrated data. Compensation can be made to obtain a more realistic effective quality factor. We can estimate the relationship between Q_{mig} and Q_{eff} as a function of the acquisition device. In this work, we are interested in zero-offset seismic surveys.

Consider the experiment shown in Figure 1, where a point source is located at S 1000 m depth. Two hundred receivers are evenly spaced at 25 m intervals along a horizontal surface at z = 0 on the surface. The source is located just below receiver station 100, and we use a Ricker wavelet with a peak frequency of 25 Hz. The model has a constant velocity of 2000 m/s. Three tests are performed, A, B and C, with a constant Q of 50, 100 and 200, respectively. For each test, we calculate the corresponding seismogram in the time domain, $u_{atten}(x_j, t)$—and $U_{atten}(x_j, \omega)$ is the seismogram in the frequency domain.

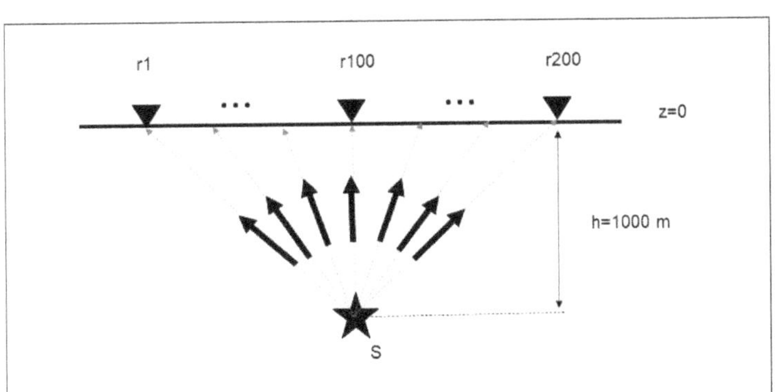

Figure 1. Forward modeling from a point source with the wavefield recorded at a horizontal surface z0 with 200 receivers, from r1 to r200 evenly spaced at 25 m.

The seismogram of experiment A, with Q equal to 50 (Figure 2a), is flattened at 0.5 s, the apex of the hyperbola (Figure 2b). The amplitude spectrum of $U_{atten}(x_j, \omega)$ for Q equals 50 is shown in Figure 2c, where the frequency peak changes toward lower frequencies, away from the central trace x_j = 100.

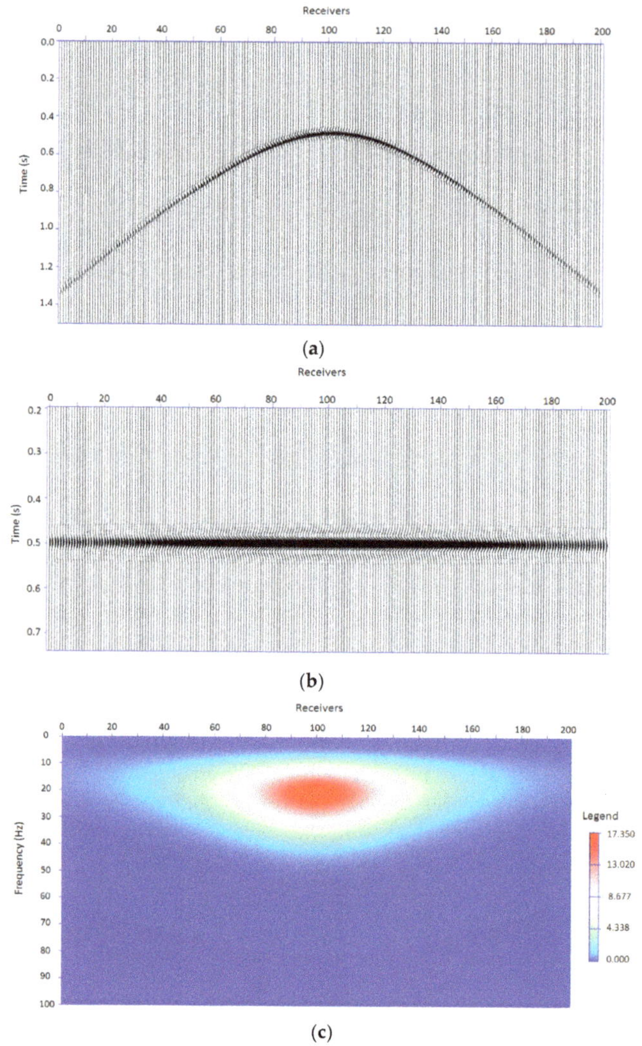

Figure 2. (**a**): Seismogram of the experiment designed at Figure 1. (**b**): Flattened seismogram of 2a. (**c**): The amplitude spectra of $U_{atten}(x_j, \omega)$ for Q equal to 50 (experiment A).

The summation of Equation (10) corresponds to the stacking of all 200 traces of $u_{atten}(x_j, t)$ in the time domain, with a radius of aperture equal to 2.5 km. The resulting trace $S_{atten}(x_j, \omega)$ has a peak frequency f_p of 19.5 Hz for experiment A, with Q = 50. Using Equation (5) for t = 0.5 s, $f_m = 25$ Hz, $f_p = 19.5$ Hz, the effective Q for migrated data, Q_{mig} is 39.

Additionally, in experiment A, if we assume a migration aperture equal to zero, we cannot use the stacked trace, but only the central one, $u_{atten}(x_j = 100, \omega)$, for Figure 2b, whose peak frequency f_p is 20.6 Hz. The calculation of the quality factor for a migration aperture of zero Q_{mig} is equal to 50, the expected result.

The procedure described in the previous sections is repeated in experiments B and C, where the models have Q of 100 and 200, respectively. The results are shown in Figure 3, where the horizontal axis represents the effective quality factor of the migrated data (Q_{mig}) and the vertical axis represents the expected or actual effective Q. The orange squares and blue circles correspond to the migration apertures of zero and 2.5 km. For all experiments, the migrated data with zero aperture Q are equal to the expected one, $Q_{mig} = Q_{eff}$. For a migration aperture of 2.5 km, the relationship between effective Q and Q_{mig} is given by Equation (13).

$$Q_{eff} = 1.3402\, Q_{mig} - 9.6892 \tag{13}$$

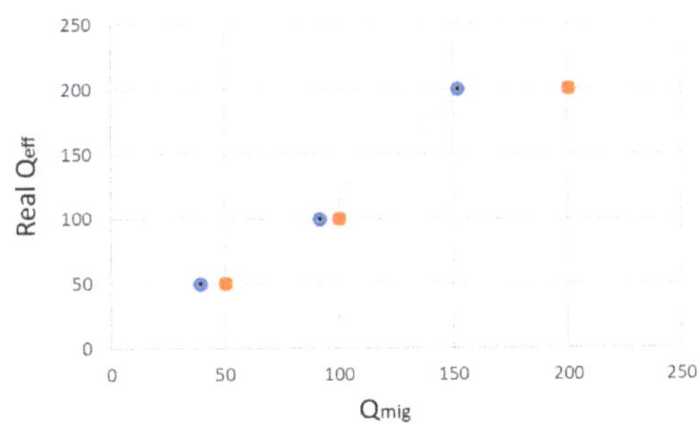

Figure 3. Q_{mig} vs. Q_{eff} plot for migration apertures of zero and 2.5 km, respectively orange squares and blue circles. As expected, $Q_{mig} = Q_{eff}$ for aperture equal to zero (orange squares). The regression of the line joining the blue circles leads to Equation (13).

The relationship between Q_{eff} and Q_{mig} derived above is based on a single diffraction curve for each experiment. However, in real data, there are multiple events that overlap. The migration process may use these unwanted overlapping signals to compose a particular diffraction pattern, which actually reduces the accuracy of the Q estimate. Despite this loss of accuracy, migration preserves the $Q_{eff} > Q_{mig}$ characteristics and the relative and lateral contrast of high and low Q regions, providing semiquantitative subsurface information.

3. Application

The estimation of effective Q and its correction (Equation (13)) was performed on a migrated single-channel seismic dataset from the Joetsu Basin in the Japan Sea.

3.1. Joetsu Basin Geology and Characteristics

The Joetsu Basin is a known site of massive gas hydrates. Geological evolution and current tectonic activity favor the existence of upwelling gas plumes that cut through soft sediments and solid gas hydrate in the first hundred meters below the seafloor. The difference in the porous content of sediments results in unequal physical properties for the same lithotype. Heterogeneous velocity field and attenuation patterns are thus expected due to energy losses of the propagating wave front.

Gas Hydrate in the Joetsu Knoll, in the homonymous basin, lay along the first hundred meters of Neogene sediments that are mainly composed of clay and silt [23,24]. Upwelling gas locally cut the sediment pile through chimneys that reaches the seafloor and continues upward along the water column. Such a geologically active environment results in heterogeneous physical properties of sediments and rocks and their fillings (gas, water, or solid GH), which are sensitive to seismic signals.

Despite the available seismic data and geologic features, little research has been conducted on the Joetsu Basin to understand Q distribution and its relationship to gas hydrate and compression velocity (V_p). In reference [23], an average V_p between 1618 and

1659 m/s is reported for the Haizume Formation based on two wells drilled in Umitaka Spur, a structure south of the Joetsu anticline. The authors in reference [24] correlate the heat flow data, pressure–temperature curve of gas hydrate stability, and seismic two-way time of BSR and estimate the velocity of host sediments between 1000 m/s and 1150 m/s along the crest and between 1170 and 1230 m/s in the troughs. In paper [25], the authors use diffraction velocity analysis (DVA) to estimate a rms velocity between 1500 and 1580 m/s in the Haizume Formation.

3.2. Joetsu Knoll Shallow Sediments and Dynamics

The Joetsu Basin was formed 25 million years ago [26] as a segment of the Back-Arc system west of the Japan island arc. Its geologic and tectonic evolution can be found in [27–29] and are briefly summarized in [25]. Our analyses are conducted over the clay-rich sediments of the Haizume Formation, which have been deposited from the Late Pleistocene to the present.

The Joetsu Basin is a very active geological system, as folding of the very young Haizume Formation provides evidence of extremely active tectonics at the present time. In the paper [30], the authors report that the muddy seafloor of the Joetsu Basin currently has an average heat flow of 98 ± 13 mW/m^2, which exceeds 150 mW/m^2 at the methane venting and mounds along the Joetsu Knoll line. Reference [24] reports that mounds and pockmarks are 50 to 500 m wide and 10 to 50 m deep, which is related to gas chimneys below. The existing source rock of the Miocene units, the Nanatani and Teradomari Formations [28], the thermal history of the Joetsu Basin, and the vertical fault system at the apex of the Joetsu anticline provide conditions for the precipitation of hydrates in the Haizume Formation. According to [24], bottom simulating reflectors (BSRs) are often parallel to stratification and, based on gas hydrate stability conditions and heat flow information, BSRs are estimated to be 115 m below seafloor (bsf) at the crest of the Joetsu Knoll and 135 m in the trough.

3.3. Description of Geophysical Data

Seismic data were acquired in 2007 by the Japan Agency for Marine-Earth Science and Technology (JAMSTEC). The source, an air-gun cluster, was towed 30 m from the ship and depths ranged from 1.5 to 7.4 m; the receiver device was a streamer with 48 channels evenly spaced at 1 m intervals with a minimum offset of 136.5 m from the source. More details about the acquisition device can be found in [25].

After time shift correction, all recorded channels were stacked to create a single station per shot. The data were recorded at a sampling rate of 0.001 s. In this study, the five seismic lines used (Figure 4) are referred to as JKn, where n indicates the line number: 109, 114, 117, 124, and 153. JK153 is a strike line parallel to the Joetsu Knoll hinge and has a northeast direction. From southwest to northeast (Figure 4), observed dip lines JK109, JK114, JK117, and JK124 are orthogonal to JK153, and in a northwest direction. These lines are selected because of their extents and their long linear segments, which favor the processing flow and analysis.

In the following topics, sections with five attributes or properties are described: seismic amplitude, EHA, peak frequency, effective Q and V_p, showing the BSR and intersections with the crossing lines. The strike line JK153 always shows four vertical lines corresponding to the intersection with the dip lines JK109, JK114, JK117, and JK124. The dip lines are detailed in the discussion section and are shown with the interpreted BSR and with a vertical line representing the intersection with JK153. The JK-153 strike line is shown from left (SW) to right (NE), while the dip lines are all shown from left (SE) to right (NW).

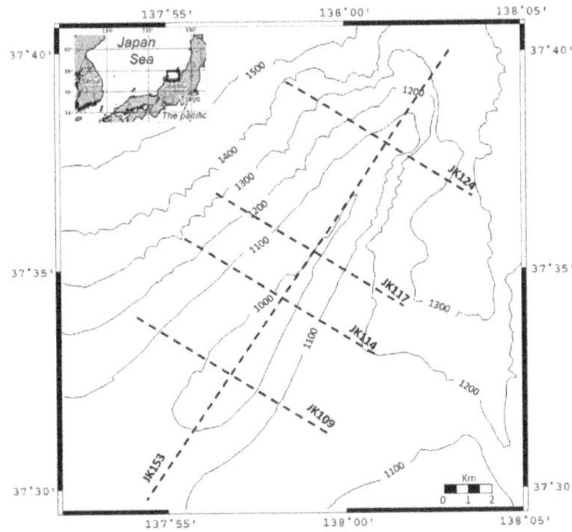

Figure 4. Location map of the Joetsu Knoll and bathymetry showing the thick dotted lines used in this study.

3.4. Processing Flow and Property Estimation

The processing flow described in [25] was repeated individually for seismic lines JK109, JK114, JK117, JK124, and JK153. It included trace edition, static correction, bandpass filtering, predictive deconvolution, diffraction velocity analysis (DVA), Kirchhoff time migration, and muting above sea level. A trapezoidal filter with a frequency of 5–15–180–200 Hz was used in the filtering step. After applying the described processing procedure, the resulting strike section JK153 is shown in Figure 5.

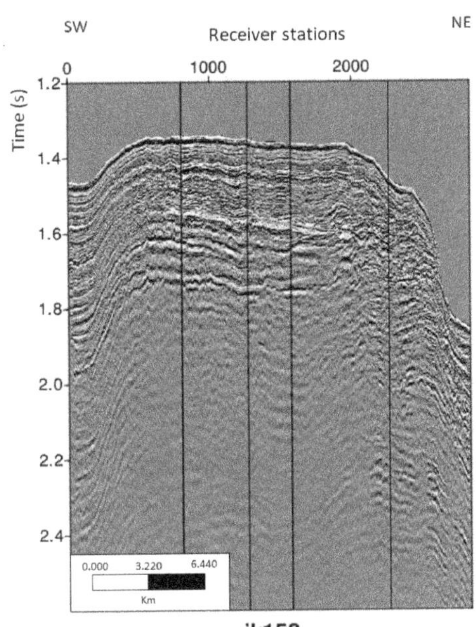

Figure 5. Amplitude section of the JK153 line in the time domain. The vertical straight black lines represent the dip lines crossing JK153 from left (SW) to right (NE): JK109, JK114, JK117 and JK124. The BSR is mapped with the cyan horizon.

3.5. Estimation of Velocity along Haizume Formation

The equivalent normal moveout velocity was estimated for each line using the DVA method described in [25]. The interval velocity V_p was estimated from the rms velocity [31] and used for imaging. The resulting V_p allowed the identification of the main trends, although it was too smooth and lacked the appropriate resolution to capture thin events and gas chimneys. Figure 6 shows the resulting interval velocity for the JK153 strike line in the time domain, indicating a general trend that increases from 1500 m/s at seafloor to 1750 m/s at 2.4 s. Smooth lateral velocity changes are observed along the section in the shallow region while, one of them, near station 2000, is close to a mound and the steeper slope of the Joetsu Knoll NE side (right side of Figure 6).

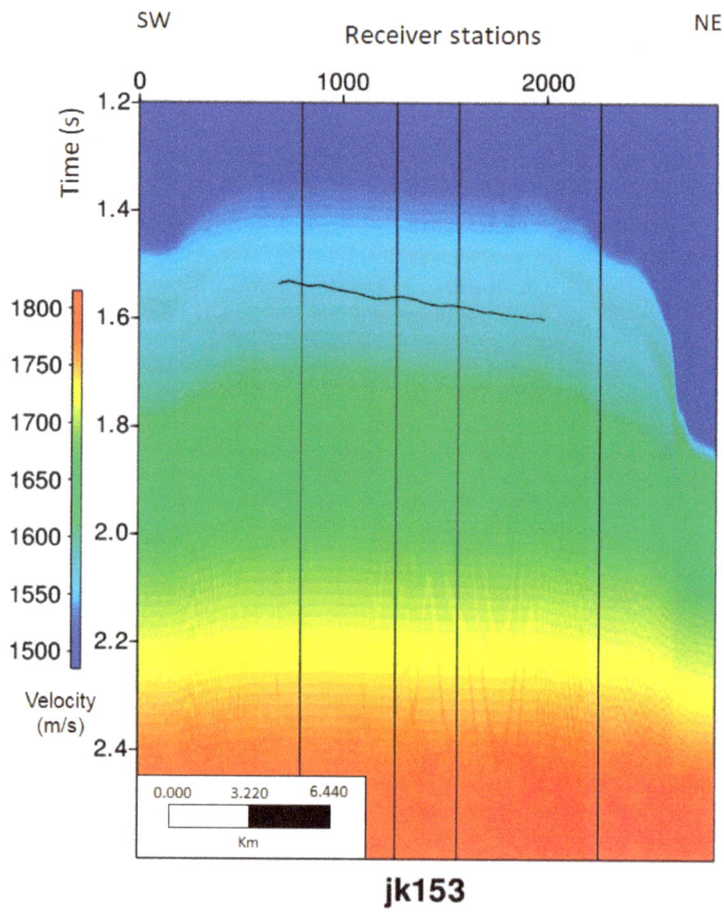

Figure 6. Interval velocity in the JK153 line in the time domain. The vertical straight black lines represent the JK109, JK114, JK117 and JK124 dip lines, from left (SW) to right (NE). Horizontal axes represent station numbers. The BSR is mapped with the horizon in black.

Each interval velocity section was time-to-depth converted. Then, we calculated the vertical gradient of V_p from the smooth velocity field in depth of Figure 6. The vertical velocity gradient provides information on how velocity increases with depth and is also a guide to the behavior of the compaction curve. A normal clastic sequence in a marine environment typically has vertical velocity gradients of about 0.600 s^{-1} [32].

By counting the velocity gradients and distributing them at regular intervals, we constructed a histogram, which is shown in Figure 7. The highest value of the vertical axes (called density) shows the dominant trend of the velocity gradient in the horizontal axes, that is 0.225 s^{-1} for the Haizume Formation. This value is much lower than the expected value of 0.600 s^{-1} for a normal clastic compaction curve. The negative gradients present are mapped by DVA because they would be useful for mapping gas clouds, but higher resolution is required to use this information. Gradients of lower magnitude near to and higher than 0.600 s^{-1} are also found in the dataset.

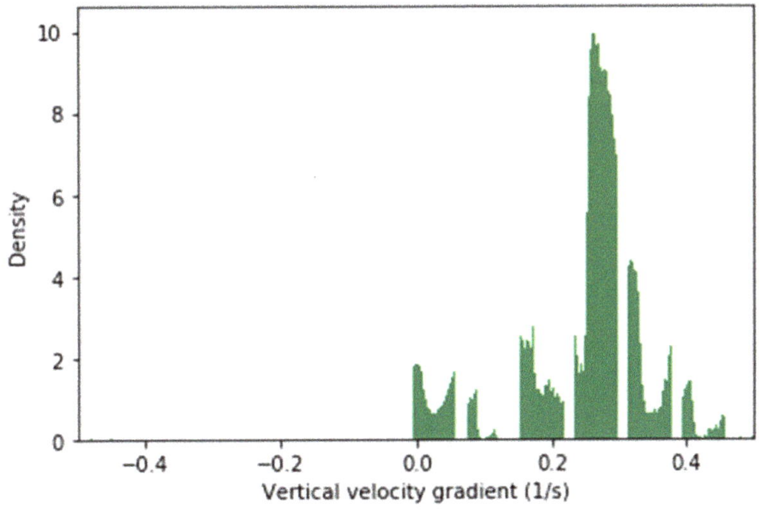

Figure 7. Histogram of P-velocity vertical gradient. The vertical axis represents the number of samples for a specific velocity gradient.

3.6. Enhancement of BSR

BSRs are generally not difficult to map. However, int the Joetsu Basin, however, they often occur parallel to stratification [23,24], making BSR mapping a difficult task. Under constant pressure, the bottom simulating reflector is an isothermal boundary that defines a phase transition (solid hydrate on top and gas on bottom) with high reflectivity that causes high amplitude in seismic records.

After normalizing the amplitudes, when powering the entire seismic data, using a value p higher than 1, the contrast between high and low amplitudes, respectively $u_{high}(x_{rec}, t)^p$ and $u_{low}(x_{rec}, t)^p$, increases. If p is a very high positive odd value, high-amplitude derived events ($u_{high}(x_{rec}, t)^p$) are preserved with their signal polarity, and ($u_{low}(x_{rec}, t)^p$) tends to zero. If p is a very high positive even value, we obtain abs($u_{high}(x_{rec}, t)^p$) and, again, ($u_{low}(x_{rec}, t)^p$) tends to zero. Thus, with suitable high p, the resulting section with enhanced high amplitudes (EHA) $u_{enhance}(x_{rec}, t)$ is calculated as follows

$$u_{enhance}(x_{rec}, t) = u(x_{rec}, t)^p \tag{14}$$

We power the processed seismic lines using p equal to 10 (Equation (14)). This procedure enhances high-amplitude events and highlights the BSR along its expected position in the EHA lines. Figure 8 shows the EHA section for the JK153 strike line.

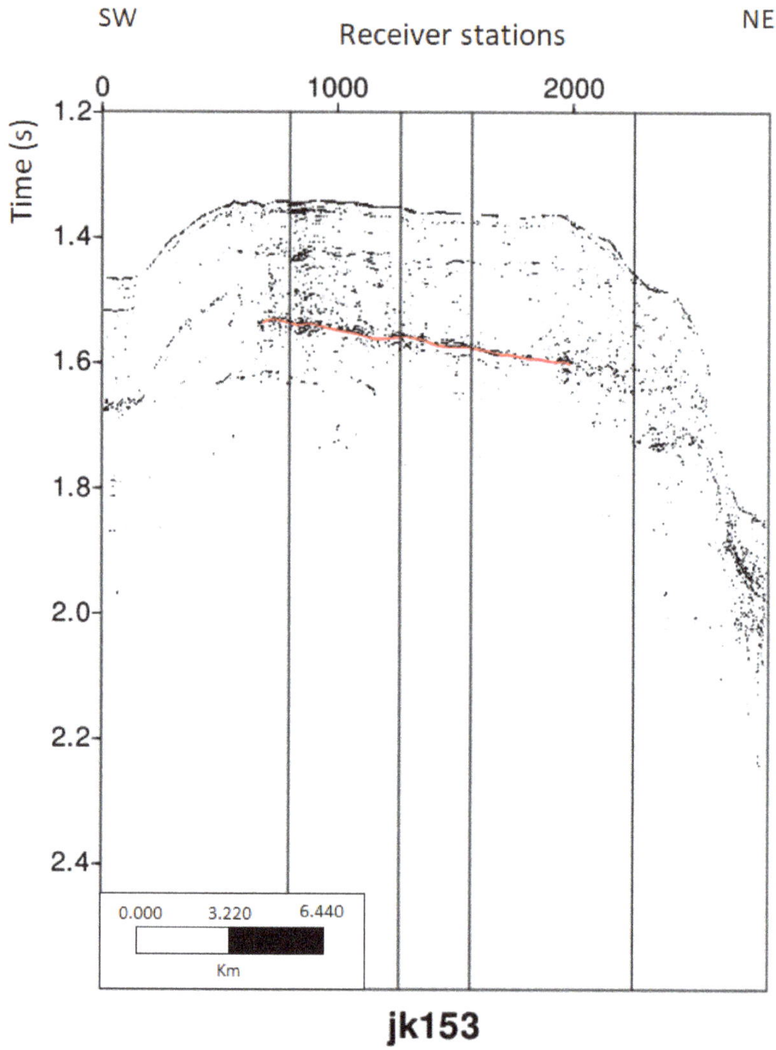

Figure 8. Seismic section enhancing high amplitude events in the JK153 line in the time domain. The vertical straight black lines represent the JK109, JK114, JK117 and JK124 dip lines, from left (SW) to right (NE). The BSR is mapped with the horizon in red.

3.7. Q Estimation along Haizume Formation

The application of the windowed Fourier Transform, Equation (3), requires calibration of the σ parameter of the Gaussian function $g(t)$ to determine an appropriate time window size, since a low σ focuses the spectral response, while a high σ causes smearing with neighboring events. Since we performed trapezoidal filtering (5–15–180–200 Hz) in the processing flow, the data did not have a period above 0.2 s. We apply a σ that imposes a time window of 0.4 s, twice the maximum period, to allow proper evaluation of periods equal to or less than 0.2 s in the central region of the Gaussian curve, thus accounting for a low relevant frequency near 5 Hz. The existing smearing due to adjacent events is attenuated by the Gaussian tails.

From the amplitude spectrum of $U(x_{rec}, \tau, \omega)$, the peak frequency at each position and time is calculated using Equation (4). Figure 9 shows the spatial behavior of the peak frequency in section JK153. The attribute changes laterally, but it clearly shows the decreasing values from the seafloor to greater depths or higher times.

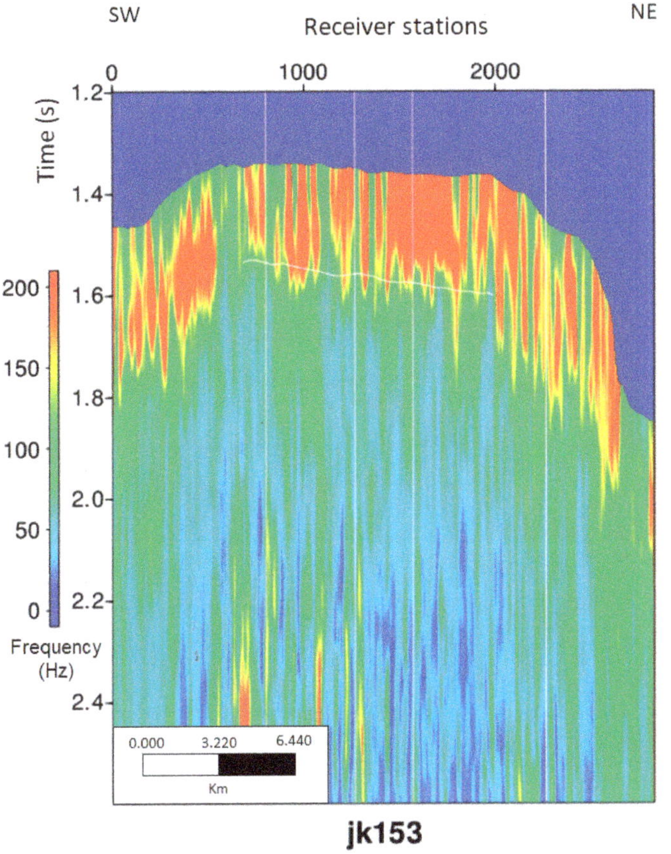

Figure 9. Peak frequency in the JK153 line. The vertical straight white lines represent the JK109, JK114, JK117 and JK124 dip lines, from left (SW) to right (NE). The BSR is mapped with the horizon in white.

The effective Q (Q_{eff}) is calculated for each line using Equation (5). Equation (13) is used to compensate for the Q estimate because the aperture from which it is derived, 2.5 km, is the same size as that used for the migration of the Joetsu seismic lines.

Due to the lateral variations of the peak frequencies and their vertical gradients, the effective Q changes considerably. In the strike section JK153, Figure 10, a region with a mean Q_{eff} of 50 is strongly confined to the occurrence of BSR. The NE limit of BSR, near station 2000, coincides with a remarkably low effective Q penetrating shallow sediments and also coincides with a known gas chimney, expressed by a mound on the seafloor (Figures 5 and 9).The NE limit of BSR, near station 2000, coincides with a remarkably low effective Q penetrating shallow sediments, and also coincides with a known gas chimney, expressed by a mound on the seafloor (Figures 5 and 9).

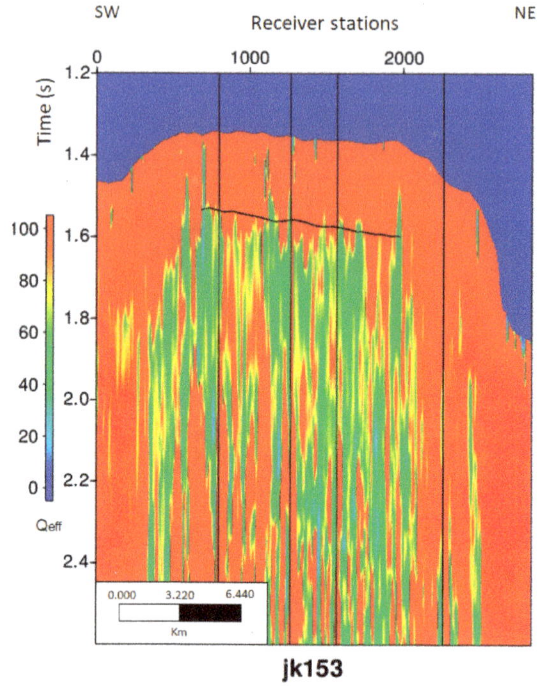

Figure 10. Effective Q factor in the JK153 strike line. The vertical straight black lines represent the JK109, JK114, JK117 and JK124 dip lines, from left (SW) to right (NE). The BSR is mapped with the horizon in white.

4. Discussion

We integrate the sections with enhanced high amplitude (EHA), peak frequency, effective Q, and velocity of each dip section, JK109, JK114, JK117, and JK124, as shown in Figures 11–14, respectively.

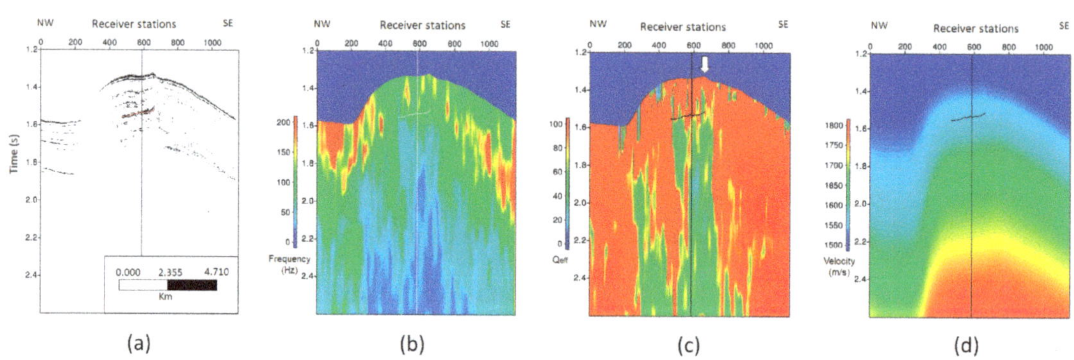

Figure 11. Section JK109 with the different attributes. All sections have the same vertical and horizontal scale in time and space, respectively. They show the intersection with the JK153 strike line represented by the vertical lines close to station 600. Finally, all lines display BSR. (**a**) Enhanced high amplitude; (**b**) peak frequency—the white arrow highlights a low peak frequency just below a mound; (**c**) effective Q; and (**d**) interval velocity.

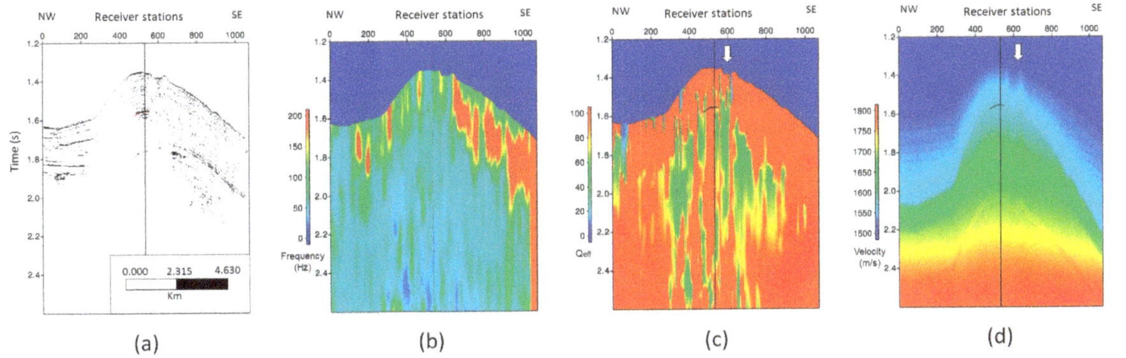

Figure 12. Section JK114 with different attributes. All sections have the same vertical and horizontal scale in time and space, respectively. They show the intersection with the JK153 strike line represented by the vertical lines close to station 500. (**a**) Enhanced high amplitude and BSR drawn in red; (**b**) peak frequency with BSR drawn in black; (**c**) effective Q and BSR mapped in white—the black arrow highlights two mounds and low-Q zones reaching the sea-floor just below the mounds; and, (**d**) interval velocity and BSR mapped in black—the black arrow indicates the lateral velocity change related to the region below the two mounds.

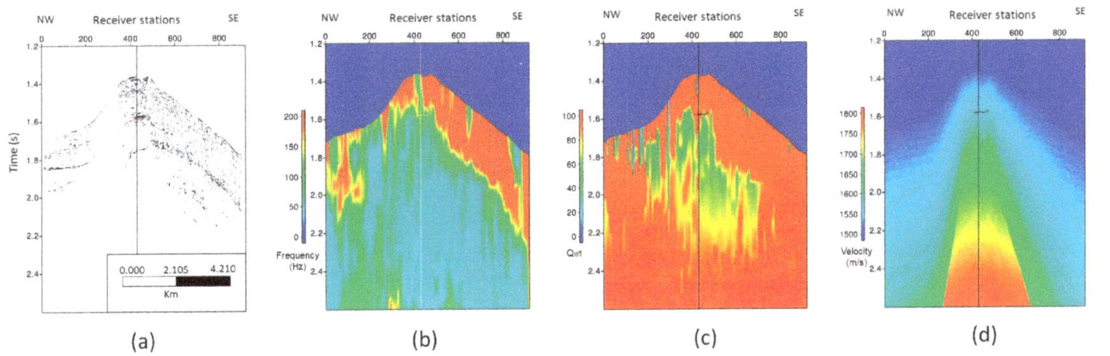

Figure 13. Section JK117 with different attributes. All sections have the same vertical and horizontal scale in time and space, respectively. They show the intersection with the JK153 strike line represented by the vertical lines close to station 400. (**a**) Enhanced high amplitude and BSR drawn in red; (**b**) peak frequency with BSR drawn in black; (**c**) effective Q and BSR mapped in white; and (**d**) interval velocity and BSR mapped in black.

The EHA sections, calculated by Equation (15), are very useful to map BSRs and confirm their existence, mainly along dip lines where this feature is harder to track. The EHA sections of lines JK109, JK114 and JK117 shown in Figures 11a, 12a and 13a, respectively, indicate a continuous BSR event that correlates perfectly with the strike section JK153 (Figure 5). The BSR is longer in JK109 than in JK114 and JK117. Comparison between the amplitude and EHA sections of the JK153 line in Figures 5 and 8, respectively, shows that BSR mapping in the studied Joetsu area is facilitated by an enhanced high-amplitude attribute. In JK124, Figure 14, BSR is hardly observed with the EHA. Solid hydrate may disappear or become too sparse somewhere between lines JK117 and JK124.

The peak frequency values show a decreasing trend from the seafloor to higher times, or depths, along all seismic sections analyzed. However, the observed lateral changes impose vertical frequency gradient variation along each line. In the JK153 line, frequencies are higher than 120 Hz and reach 200 Hz near the seafloor (Figure 9). The gradient to low frequencies, below 90 Hz, are approximately higher along the region where the BSR is interpreted, indicating higher absorption in this region. Along lines JK109, JK114, and JK117, the higher gradient to lower peak frequencies is also consistent with the mapped BSR (Figures 11b, 12b and 13b) and coincides with the Joetsu anticline hinge. In section JK124, despite the absence of the mapped BSR, the higher gradient from high to low frequencies also occurs along the anticline hinge (Figure 14b), suggesting that this region is a site without massive hydrate, but with upward gas migration adjacent to a giant gas chimney, as also observed at station 2000 of the JK153 strike line.

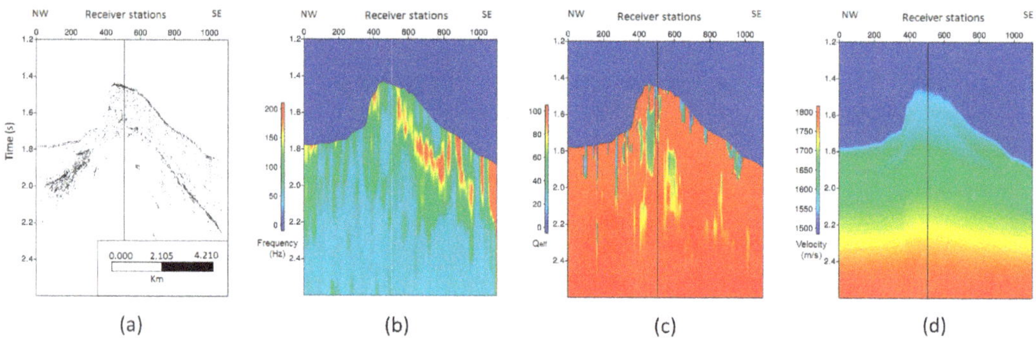

Figure 14. Section JK124 with different attributes. All sections have the same vertical and horizontal scale in time and space, respectively. They show the intersection with the JK153 strike line represented by vertical line close to station 500. Differing from the other seismic lines, BSR is not observed along the JK124. (**a**) enhanced high amplitude; (**b**) peak frequency; (**c**) effective Q; and (**d**) interval velocity.

A pattern change in the lateral distribution of peak frequencies is observed alongside the anticline structure. In JK109, frequencies are higher on the SE flank than on the NW flank (Figure 11b). All other dip sections, JK114, JK117, and JK124, show an opposite behavior, as the peak frequencies are lower in their SE flank than in the NW flank (Figures 12b, 13b and 14b). The reason for this behavior is not fully understood.

Effective Q values less than 100 and averaging 50 occur along a region larger than the BSR boundary, as observed in the JK153 strike line (Figure 10). These low Q_{eff} values are found at the SE flank and the anticline hinge of JK109 (Figure 11c), while the average value of 50 is observed at the hinge and at both structure flanks on the JK114 (Figure 12c). The line pattern of JK117 is similar to that of JK109, but narrower than that of JK114 (Figure 13c). In addition, line JK124 shows the narrowest region with Q_{eff} averaging 50 (Figure 14c). This northeastern narrowing and low Q_{eff} value are consistent with northeastern narrowing of the anticline structure and the progressive decrease in hydrate concentration in the shallow region of the Haizume Formation. The Q factor is lower in hydrate-prone sediments than in sediments without hydrates [6]. The resulting geometry of low Q_{eff} in the studied sections of the Joetsu anticline allows us to map the gas hydrate zone near the seafloor and the gas chimneys at depth.

In addition, hydrates do not occur as a continuous layer, but filling centimeter- to meter-thick pores and voids in the sediments of the Haizume Formation (Figure 15), due to the pressure and temperature near the surface, which are 9.0 MPa and 2 C, respectively. The contrast between the soft sediments and the distributed stiff solid hydrate results in scattering of the propagating wave field, which enhances attenuation and resembles apparent absorption. The combined effect of gas and seismic signal scattering by hydrate increases the attenuation, but it is not possible to distinguish the contribution of each one.

Figure 15. Hydrate sampled in a piston core immerse in silty and clay sediments of Haizume Formation. The rule shows centimeter scale in the picture.

All analyzed lines have frequency peaks above 100 Hz, sometimes reaching 200 Hz, near the sea floor. Additionally, below a few seconds from the seafloor, all lines have frequencies close to 50 Hz. Therefore, seismic data with high-frequency content are needed for the proposed study, otherwise the absorption effect cannot be detected.

All estimated velocities show rather monotonous behavior, as can be seen in Figures 11d, 12d, 13d and 14d. Despite their low spatial frequency, the DVA-derived interval velocity shows some upward gas chimneys directly beneath the mapped mounds. An example of this is at station 600 of line JK114 (Figure 12d), where the lateral velocity changes due to a gas chimney are also expressed by two mounds on the seafloor. DVA allows extraction of geologically possible features and provides normal move-out velocities for time migration. However, it lacks high-frequency information [12] for further estimation and interpretation. A detailed V_p field can be obtained with high-resolution tomography over multichannel seismic data, which were not accessed in this study. The most often vertical V_p gradient of 0.225 s^{-1} is, indeed, too low for a shallow clastic sequence. The high heat flow [24,30,33], hydrocarbon (gas) generation, upward fluid migration, gas chimneys and local fluid entrapment increase the pore pressure, reduce the effective stress, and thus reduce the compression velocity and its vertical gradient.

5. Conclusions

In this study, we show that velocity gradient, peak frequency, effective Q, and EHA can be estimated from SCS datasets for subsurface geophysical characterization. In addition, we propose a linear correction for effective Q estimated from migrated datasets. Application of the techniques presented to the Joetsu Knoll SCS dataset leads to a number of new estimates and interpretations of Q and velocity at this site.

The DVA in the SCS does not allow estimation of detailed velocities, but provides a useful, smooth V_p field that can estimate the general trend. In the Haizume Formation along the Joetsu Knoll, velocities range from 1500 to 1750 m/s with a smooth trend of 0.225 s^{-1}, suggesting under-compaction of sediments.

The estimate of effective Q from migrated seismic data (Q_{mig}) preserves the relationship between areas of high and low Q. Q_{mig} is always smaller than the effective Q estimated from non-migrated data (Q_{eff}). A linear correction with respect to the acquisition aperture (Equation (13)) can be applied to improve the Q_{eff} estimate.

The area of low effective Q averages 50 and encloses gas hydrate. In addition to mapping the BSR and GH occurrence areas, it is also useful for locating gas chimneys. Gas hydrate geometry and width are due to anticline geometry, while northeastward narrowing is observed within the low-Q_{eff} area. In this geologic context, sediment composition and granulometry, effective Q section or volume are important and useful seismically derived properties that can be used to automatically locate gas exudation and GH to support integration and interpretation tasks.

The gas hydrate and gas exudation region has a lower Q value than the sediment without free gas and GH. The tools presented in this research are useful for applications with similar datasets, SCS, in basins where there are hydrates and gas chimneys that affect seismic signals with strong absorption effects.

Author Contributions: L.A.S., R.M., F.D.S.F. and M.A.C.S. equally contributed to the conceptualization, methodology development, validation, investigation, writing and reviewing the original draft. All authors have read and agreed to the published version of the manuscript.

Funding: This research received no external funding.

Data Availability Statement: The seismic data used in this research are not available.

Acknowledgments: We would like to express our gratitude for Petrobras for the support and encouragement regarding the publication of this research.

Conflicts of Interest: The authors declare no conflict of interest.

References

1. Futterman, W.I. Dispersive body waves. *J. Geophys. Res.* **1962**, *67*, 5279–5291. [CrossRef]
2. Santos, R.A. Perfil Sísmico Vertical: Separação de Campos de Onda e Obtenção do Fator Q. Master Thesis, Universidade Federal da Bahia, Bahia, Brazil, 1986.
3. Dasgupta, R.; Clark, R.A. Estimation of Q from surface seismic reflection data. *Geophysics* **1998**, *63*, 2120–2128. [CrossRef]
4. Wang, Y. Quantifying the effectiveness of stabilized inverse Q filtering. *Geophysics* **2003**, *68*, 337–345. [CrossRef]
5. Hornbach, M.J.; Holbrook, W.S.; Gorman, A.R.; Hackwith, K.L.; Lizarralde, D.; Pecher, I. Direct seismic detection of methane hydrate on the Blake Ridge. *Geophysics* **2003**, *68*, 92–100. [CrossRef]
6. Dvorkin, J.; Uden, R. Seismic wave attenuation in a methane hydrate reservoir. *Lead. Edge* **2004**, *23*, 732–739. [CrossRef]
7. Rossi, G.; Gei, D.; Bohm, G.; Madrussani, G.; Carcione, J.M. Attenuation tomography: An application to gas-hydrate and free-gas detection. *Geophys. Prospect.* **2007**, *55*, 655–669. [CrossRef]
8. Guerin, G. Goldberg, Sonic wave attenuation in gas hydrate-bearing sediments from the Mallik 2L-38 research well, Mackenzie Delta, Canada. *J. Geophys. Res.* **2002**, *107*, B5.
9. Gei, D.; Carcione, J. Acoustic properties of sediments saturated with gas hydrate, free gas and water. *Geophys. Prospect.* **2003**, *51*, 141–157. [CrossRef]
10. Tinivella, U.; Carcione, J.M. Estimation of gas-hydrate concentration and free-gas saturation from log and seismic data. *Lead. Edge* **2001**, *20*, 200–203. [CrossRef]
11. Carcione, J.M.; Gei Davide Rossi, G.; Madrussani, G. Estimation of gas-hydrate concentration and free-gas saturation at the Norwegian-Savalbard continental margin. *Geophys. Prospect.* **2005**, *53*, 803–810. [CrossRef]
12. Zillmer, M. A method for determining gas-hydrate or free-gas saturation of porous media from seismic measurements. *Geophysics* **2006**, *71*, N21–N32. [CrossRef]
13. Cavalca, M.; Moore, I.; Zhang, L.; Leng, S.; Fletcher, R.; Bayly, M. Ray-based tomography for Q estimation and Q compensation in complex media. In Proceedings of the 81st Annual International Meeting, SEG, San Antonio, TX, USA, 18 September 2011.
14. Gamar, F.; Carotti, D.; Guillaume, P.; Gacha, A.; Lopes, L. Success of high-resolution volumetric Q-tomography in the automatic detection of gas anomalies on offshore Brunei data. In Proceedings of the 85th Annual International Meeting, New Orleans, LA, USA, 18 October 2015.
15. Li, S.; Gao, J.; Zhang, B.; Wang, S. Q estimation method and its application on the middle of Sichuan basin. In Proceedings of the 88th Annual International Meeting, SEG, Anaheim, CA, USA, 14 October 2018.
16. Ramos, E.C.; Costa, F.T.; Santos, M.A.C.; Santos, L.A.; Santos, R.A.; Soares Filho, D.M. Frequency Domain Q Analysis and Stabilized Inverse Q Filtering: Blake Ridge data application. In Proceedings of the 16th International Congress of the Brazilian Geophysical Society, Rio de Janeiro, Brazil, 20 August 2019.
17. Bazaluk, O.; Sai, K.; Lozynskyi, V.; Petlovanyi, M.; Saik, P. Research into Dissociation Zones of Gas Hydrate Deposits with a Heterogeneous Structure in the Black Sea. *Energies* **2021**, *14*, 1345. [CrossRef]
18. Klymenko, V.; Ovestskyi, S.; Martynenki, V.; Vytyaz, O.; Uhrynovskyi, A. An alternative method of methane production from deposits of subaquatic gas hydrates. *Min. Miner. Depos.* **2022**, *3*, 11–17. [CrossRef]

19. Zhang, C.; Ulrych, T.J. Estimation of quality factors from CMP records. *Geophysics* **2002**, *67*, 1542–1547. [CrossRef]
20. Castagna, J.P.; Sun, S. Comparison of spectral decomposition methods. *First Break* **2006**, *24*, 75–79. [CrossRef]
21. Quan, Y.; Harris, J.M. Seismic attenuation tomography using the frequency shift method. *Geophysics* **1997**, *14*, 905–985. [CrossRef]
22. Gisolf, D.; Verschuur, E. *The Principles of Quantitative Acoustical Imaging*, 1st ed.; EAGE Publications: DB Houten, The Netherlands, 2010; p. 271.
23. Freire, A.F.M.; Matsumoto, R.; Santos, L.A. Structural-stratigraphic control on the Umitaka Spur gas hydrates of Joetsu Basin in the eastern margin of Japan Sea. *Mar. Pet. Geol.* **2011**, *28*, 1967–1978. [CrossRef]
24. Matsumoto, R.; Hiromatsu, M.; Sato, M. Fluid flow and evolution of gas hydrate mound of Joetsu Basin, Eastern margin of Japan Sea: Constraints from high-resolution geophysical survey by AUV. In Proceedings of the 7th International Conference of gas hydrates, Edimburgh, UK, 17–21 July 2011.
25. Santos, L.A.; Neves, E.H.P.; Freire, A.F.M.; Santos, M.A.C.; Matsumoto, R.; Ajus, C.M.I. Diffraction velocity analysis in a single-channel seismic survey in the Joetsu Basin. *Geophysics* **2020**, *85*, U47–U53. [CrossRef]
26. Tamaki, K.; Isezaki, N. Tectonic Synthesis of the Japan Sea Based on the Collaboration of the Japan-URSS Monograph Project. In *Geology and Geophysics of the Japan Sea*; Isezaki, N., Ed.; Japan-Russia Monograph Series 1; Terra Scientific Publishing Company: Tokyo, Japan, 1996; pp. 483–487.
27. Tamaki, K. Geological structure of the Japan Sea and its tectonic Implications. *Bull. Geol. Surv. Jpn.* **1988**, *39*, 269–365.
28. Okui, A.; Kaneko, M.; Nakanishi, S.; Monzawa, N.; Yamamoto, H. An Integrated Approach to Understanding the petroleum system of a Frontier deep-water area, offshore Japan. *Pet. Geosci.* **2008**, *14*, 223–233. [CrossRef]
29. Son, B.K.; Yoshimura, T.; Fukazawa, H. Diagenesis of Dioctahedral and Trioctahedral Smectites from Alternating beds in Miocene to Pleistocene rocks of the Niigata basin, Japan. *Clays Clay Miner.* **2001**, *49*, 333–346. [CrossRef]
30. Machiyama, H.; Kinoshita, M.; Takeuchi, R.; Matsumoto, R.; Yamano, M.; Hamamoto, H.; Hiromatsu, M.; Satoh, M.; Komatsubara, J. Heat flow distribution around the Joetsu Gas hydrate field, Western Joetsu Basin, Eastern Margin of the Japan Sea. *J. Geogr.* **2009**, *118*, 986–1007. [CrossRef]
31. Dix, C.H. Seismic velocities from surface measurements. *Geophysics* **1955**, *20*, 68–86. [CrossRef]
32. Rosa, A.L.R. *Análise do Sinal Sísmico*, 1st ed.; Sociedade Brasileira de Geofísica: Rio de Janeiro, Brasil, 2010; p. 668.
33. Yasui, M.; Kishii, T.; Watanabe, T. Heat flow in the Sea of Japan. In *The Crust and Upper Mantle of the Pacific Area*, 1st ed.; Knopoff, L., Drake, C.L., Hart, P.M., Eds.; AGU: Washington, DC, USA, 2012; Volume 12, pp. 3–16.

Disclaimer/Publisher's Note: The statements, opinions and data contained in all publications are solely those of the individual author(s) and contributor(s) and not of MDPI and/or the editor(s). MDPI and/or the editor(s) disclaim responsibility for any injury to people or property resulting from any ideas, methods, instructions or products referred to in the content.

We first present our inversion method and illustrate it using synthetic gravity data, simulating a portion of a sedimentary basin. We then apply the method to a set of field gravity data acquired from the Recôncavo Basin, Brazil.

2. Methodology

The goal of our inversion is to find a reliable model that approximates the interface separating the sediments and the basement. The interface is assumed to represent the geometry of the basement in a portion of a sedimentary basin. The importance of defining this interface lies in the fact that in some sedimentary basins, especially rift-related ones, the basement geometry controls the distribution of potential oil fields. We restrict ourselves to working with only a portion of a basin since the assumptions involving the physical characteristics of the media, such as constant density, for instance, are more likely to be valid in smaller areas. In addition, this approach seeks to broaden the contribution of the gravity method in oil exploration because it focuses the work at an oil-field scale rather than at a basin scale for study.

To solve the problem numerically, we discretize the basement depth into a set of rectangular patches of a constant size and, therefore, represent the 3D sedimentary basin with a set of contiguous rectangular prisms of a constant density contrast (Figure 1). The tops of the prisms are at the surface, and their thicknesses (or heights) are to be determined from observed gravity data. To allow for flexibility in the model in terms of representing varied basement structures, we required the number of prisms in the model, M, to be always greater than the number of gravity observations, N. This approach allows for a higher resolution in the recovered models because, in contrast to other inversion methodologies that require the number of observations and prisms to be the same, here, we can have a large number of prisms even when only a small number of field observations are available.

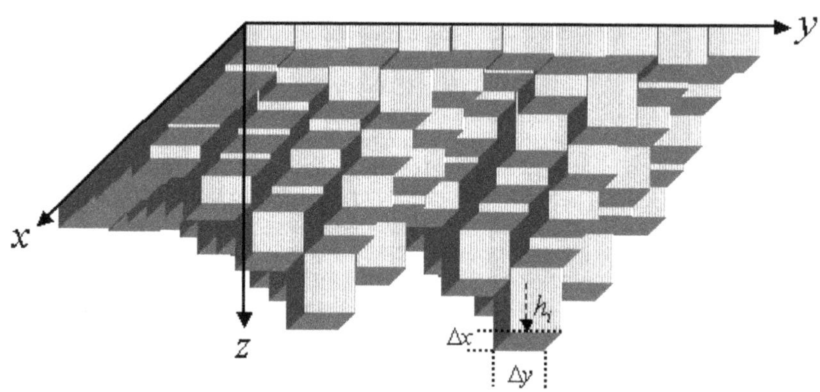

Figure 1. Sketch representing the discretization of a sedimentary basin as a set of rectangular prisms having fixed horizontal dimensions dx and dy . The heights of the different prisms, h_i, are the parameters to be inverted.

The general relationship between gravity anomaly and its sources is given by (e.g., [14]):

$$g(r) = \int_V \rho(r')\psi(r,r')dv, \tag{1}$$

where $g(r)$ is the gravity field at the observation position r outside of the volume V that is occupied by the source; $\rho(r')$ is the source density at location r', and $\psi(r,r')$ is a function that depends on the geometric relations between positions r and r'. Gravity inversion makes use of field measurements to find the main characteristics of either density ρ (linear problem) or some aspects of ψ, such as the region of the source. The former is a linear problem, whereas the latter is nonlinear since it intends to recover a geometric aspect of the problem. The problem of recovering the basement depth falls into the latter category.

The relationship between the gravity field at the origin and a single prism with a constant density, ρ, and corner positions at x_i, y_j, and z_k, as derived by [15] is,

$$g = \gamma\rho \sum_{i=1}^{2}\sum_{j=1}^{2}\sum_{k=1}^{2}(-1)^i(-1)^j(-1)^k$$
$$\left[z_k \arctan\left(\frac{x_i y_j}{z_k R_{ijk}}\right) - x_i \log(R_{ijk} + y_j) - y_j \log(R_{ijk} + x_i)\right], \tag{2}$$

where γ is the gravitational constant, and $R_{ijk} = \sqrt{x_i^2 + y_j^2 + z_k^2}$. We note that the vertical co-ordinate of the bottom of the prism, z_2, is the unknown quantity to be recovered through our inversion.

As discussed in the preceding section, this inversion is ill-posed because we have only a finite number of inaccurate data on the surface, and we attempt to recover a basement relief that is more complex in structure than the smoothly varying gravity data. Consequently, there are a multitude of models that can fit the data to the same degree. In order to find a unique solution for interpretational purposes, we select one that is consistent with known information and is structurally simple. We choose to follow the Tikhonov regularization. This approach allows for the construction of different models by changing the form of the objective function according to prior information. We minimize a total objective function Φ, defined as a weighted sum of a model objective function Φ_m and a data misfit function Φ_d,

$$\Phi = \Phi_d + \mu\Phi_m, \tag{3}$$

where μ is the regularization parameter, and it determines the trade-off between the two terms. The data misfit function Φ_d is defined to be:

$$\Phi_d = \|W_d(g - g^o)\|^2, \tag{4}$$

where $W_d = diag\{1/\sigma_1, \ldots, 1/\sigma_N\}$, in which σ_i is the error standard deviation related to the ith observation, g^o represents the observed data, and g is the data predicted from the model. If the noise contaminating the data is uncorrelated and has a zero mean, the misfit Φ_d is a chi-squared variable with N degrees of freedom. The number of observations, N, therefore, becomes the target misfit (Φ_d^*) for the inversion since the expected value for a chi-squared distribution is N. In the case where the noise statistics are unknown, we must resort to different approaches to determine the optimal data misfit.

The model objective function Φ_m allows us to incorporate prior information about the model. The choice of prior information is problem-dependent, but in a general sense, the inverted model should be close to a reference model and be as smooth as the data allows in all directions. We, therefore, choose a model objective function having the following form:

$$\Phi_m = \alpha_s \int_S (h - h_0)^2 ds + \int_S \left[\frac{\partial(h - h_0)}{\partial x}\right]^2 ds + \int_S \left[\frac{\partial(h - h_0)}{\partial y}\right]^2 ds, \tag{5}$$

where h is the recovered model, h_0 is the reference model, and α_s is a coefficient that controls the relative importance of the first term to the others. In Equation (5), the first term provides a measure of the deviation from the reference model, whereas the remaining terms control the structural complexity of the model. Given the discretization used for the forward modeling, the recovered basement depth $h(x, y)$ becomes a piece-wise constant function, and it can be represented by a vector $h = (h_1, \ldots, h_M)^T$. When evaluating the integrals in Equation (5) according to the above-described discretization, we obtain a discrete form of the objective function:

$$\Phi_m = \|W_m(h - h_0)\|^2, \tag{6}$$

where W_m is the model weighting matrix.

The choice of the reference model is often left open in many publications since it is highly problem-dependent. In our inversion, however, the goal is to improve upon seismic interpretation by finding modifications using gravity data. We would like to find a model that deviates as little as possible from the seismic model while still fitting the gravity data. It is, therefore, optimal to use the seismic model as the reference model.

Since the unknown model to be recovered is the height of each prism, the relationship between the data and the model is nonlinear, as discussed earlier. Consequently, the misfit of the data in Equation (4) is not a quadratic function. As a result, we have a nonlinear inverse problem, and we choose to solve it iteratively through linearization. We assume that the thickness is $h^{(n)}$ at the n'th iteration, and a small perturbation δh can be added to improve the data misfit. By expanding the predicted gravity data using a Taylor series expansion in δh and ignoring higher order terms yield a linear relationship,

$$g_i\left(h^{n+1}\right) \approx g_i\left(h^{(n)}\right) + \sum_{j=1}^{M} \frac{\partial g_i\left(h^{(n)}\right)}{\partial h_j} \Delta h_j, i = 1,2,\ldots,N, \tag{7}$$

where $\Delta h_j = h_j^{(n+1)} - h_j^{(n)}$. Equation (7) can be compactly represented in a matrix form as:

$$g^{(n+1)} = g^{(n)} + J\Delta h, \tag{8}$$

where $g^{(n+1)}$ is the N—length vector of the predicted data, Δh is the M—length vector of model perturbations, and J is the $N \times M$ sensitivity matrix relating the predicted data to the changes in the model at each iteration according to Equation (2). Substituting Equation (8) into the discretized objective function yields the linearized form:

$$\Phi(\Delta h) = \|W_d\left(g^{(n)} + J\Delta h - g^{o}\right)\|^2 + \mu\|W_m\left(h^{(n)} + \Delta h - h_0\right)\|^2. \tag{9}$$

Minimizing Equation (9) with respect to the model perturbation yields the desired Δh, which allows us to update the model and proceed to the next iteration.

In addition to the smoothness and similarity to the seismic model, depth-to-basement information (from boreholes) is also available to constrain the solutions. The use of localized prior information as constraints is not new, and examples can be found in [8,16,17], among others. There are different means to introduce localized information, and the majority of methods rely on slightly different ways of minimizing the differences between the estimates and the known depths at well locations. In this paper, we have chosen to apply the logarithm barrier method (e.g., [11]), which has been used by [12,13] in the inversion of different geophysical datasets. One advantage is that this approach allows one to set different limits to every element of the model instead of only at those locations where depth-to-basement information is present. The log barrier method presents the additional advantage of allowing for the introduction of specific degrees of confidence (by narrowing or enlarging the barrier limits) to different information. In other words, it is possible to set very narrow limits at positions where reliable depth information is present and to relax the constraints in regions where the information is less accurate. A fundamental application of these advantages is the introduction of the well log information coming from those boreholes that have not reached the basement but that contain information about depths where the basement certainly is not present. In a very similar way, seismic information can be used in areas where no wells are available. In the log barrier method, such information can be easily incorporated into the inversion by setting the constraints to the minimum depth only. To the best of our knowledge, the use of such information as constraints in the inversion of potential field data is new. The logarithmic barrier method was implemented in our problem by adding a logarithmic term to the objective function of Equation (9) to form a new objective function:

$$\Phi = \|W_d\left(g^{(n)} + J\Delta h - g^o\right)\|^2 + \mu\|W_m\left(h^{(n)} + \Delta h - h_0\right)\|^2$$
$$-2\lambda\left[\sum_{j=1}^{M}\ln\left(\frac{h_j - a_j}{b_j - a_j}\right) + \sum_{j=1}^{M}\ln\left(\frac{b_j - h_j}{b_j - a_j}\right)\right], \quad (10)$$

where the last term is the barrier function, λ is the barrier parameter, a_j and b_j are, respectively, the minimum and maximum depth, and M is the total number of prisms in the model. The barrier term forms a barrier at the boundary of the feasible interval of the unknowns and prevents the minimization from producing unknowns outside their respective bounds. The value of λ is decreased during the minimization so that at the end, as λ approaches zero, the solution to Equation (10) approaches that of the original problem. Carrying out the complete minimization of Equation (10) for each value of λ is an expensive process, and it is also unnecessary. Instead, for each value of the barrier parameter λ, we take one Newton step towards minimizing Equation (10) to yield the model perturbation equation:

$$\left(J^TW_d^TW_dJ + \mu W_m^TW_m + \lambda X^{-2} + \lambda Y^{-2}\right)\Delta h =$$
$$J^TW_d^TW_d(g^o - g) + \mu W_m^TW_m(h_0 - h) + \lambda\left(X^{-1} - Y^{-1}\right)e, \quad (11)$$

where $X = diag\{h_1 - a_1, \ldots, h_m - a_M\}$, $Y = diag\{b_j - h_1, \ldots, b_m - h_M\}$, and $e = (1, \ldots, 1)^T$. The matrix system in Equation (11) is solved for Δh by using the conjugate gradient (CG) method. The model is then updated by a limited step-length:

$$h^{(n)} = h^{(n-1)} + \eta\beta\Delta h, \quad (12)$$

where β is the maximum permissible step length, and η is a parameter that limits the step length actually taken. The parameter β is given by:

$$\beta = \min\left(\min_{\Delta h_j > 0}\frac{b - h_j^{(n-1)}}{\Delta h_j}, \min_{\Delta h_j < 0}\frac{h_j^{(n-1)} - a}{|\Delta h_j|}\right). \quad (13)$$

The maximum step length is the value that will take the updated model to the bounds. Limiting it by the η prescribed within the interval (0, 1) ensures that the updated model remains within the bounds. After each iteration, the value of λ is reduced by:

$$\lambda^{n+1} = [1 - \min(\beta, \eta)]\lambda^n, \quad (14)$$

so that the barrier term becomes negligible as we move towards the final solution. The iterative process is terminated once the barrier term has become negligibly small and the original objective function has reached a plateau. This yields one solution for a given regularization parameter μ. The solutions for several values of μ are required to find the solution that produces the target misfit Φ_d^*.

3. Synthetic Example

We now apply our method to the synthetic dataset shown in Figure 2. The data simulates the gravity response of the model (Figure 2a) at 100 random locations (crosses). Gaussian noise with a zero mean and a standard deviation of 0.04 mGal was added to the entire set of synthetic measurements, resulting in the gravity response shown in Figure 2b. The synthetic gravity data were gridded using 500 m intervals for the purpose of display only.

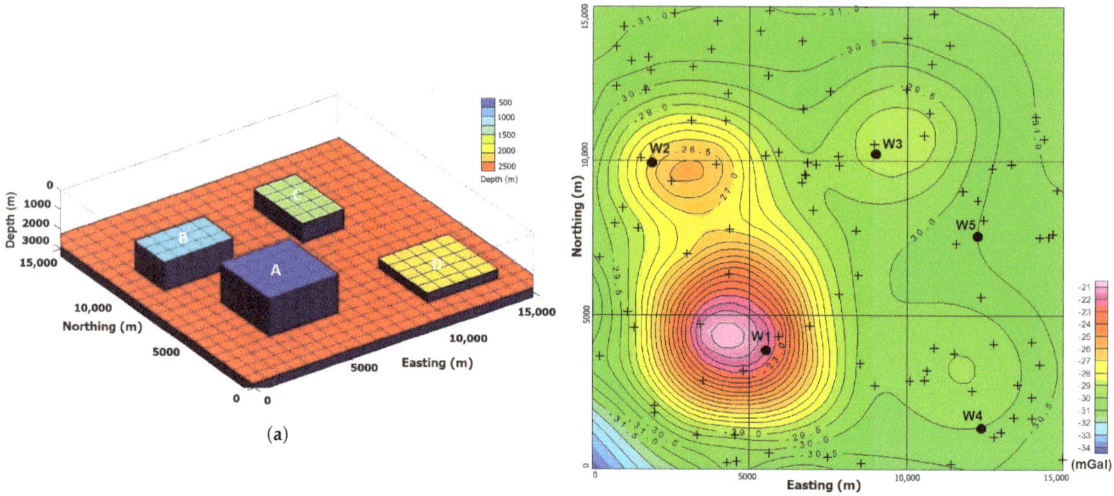

(b)

Figure 2. Synthetic model representing a restricted portion of a sedimentary basin and its gravity response. (a) The model is composed of rectangular features, marked A, B, C, and D, that intend to simulate four structural highs, for which the tops are positioned, respectively, at 500, 1000, 1500, and 2000 m. (b) The gravity response of the synthetic model is shown in (a), calculated for 100 randomly distributed stations (crosses) by using a density contrast of -0.30 g/cm^3. The black circles show the position of the five synthetic wells listed in Table 1. The data were gridded with 500 m intervals for the purpose of display only.

The synthetic model shown in Figure 2a simulates a small portion of a sedimentary basin covering an area of 15,000 m $\times$ 15,000 m. The basement structures are represented by four rectangular blocks (A, B, C, and D), for which the tops are at, respectively, 500, 1000, 1500, and 2000 m. The maximum depth in the model is 3000 m. The density contrast between the sediments and the basement is considered to be constant and equal to -0.30 g/cm^3. The model is discretized into 441 rectangular prisms, having a width of 750 m in x- and y-directions. Since the prisms represent the sedimentary section, the top of each prism is fixed at the surface, and its bottom will determine the depth to the basement at each location, as represented in Figure 1.

The well log constraints were imposed on the problem by assuming the depth-to-basement information at five locations (the black dots in Figure 2b), as listed in Table 1. The wells were incorporated into the model by setting the model's cells at the well locations to provide depth information and keep them fixed during the inversion. Except for the five positions where the depth to the basement is known, a model with a constant depth of 1500 m was chosen as the reference model.

Table 1. Location of the five synthetic wells used to constrain the inversion.

Wells	East Coord. (m)	North Coord. (m)	Depth (m)
1	5522	3849	500
2	1872	9943	1000
3	8987	10,232	1500
4	12,371	1345	2000
5	12,236	7562	2500

The final result of the inversion is shown in Figure 3a. It is clear that the inversion was not able to completely recover the model, but the results represent a satisfactory solution in terms of the location and average depth for all four structures. The histogram of the absolute data misfit in Figure 3c shows that 83% of misfits are smaller than 0.12 mGal, with 23% below 0.04 mGal, which is the standard deviation of the added noise. Such a result was expected mainly due to the noise and the limited number of observation points. As a comparison, Figure 3b shows the results of a new inversion that used 250 randomly spaced data points. In Figure 3d, the histogram of the absolute data misfit of the new inversion shows that all the misfits are below 0.12 mGal, with 83% of them below the standard deviation of the noise. The increase in the amount of observed data allows for a better definition of the gravity field by reducing ambiguity and helping to improve the final model.

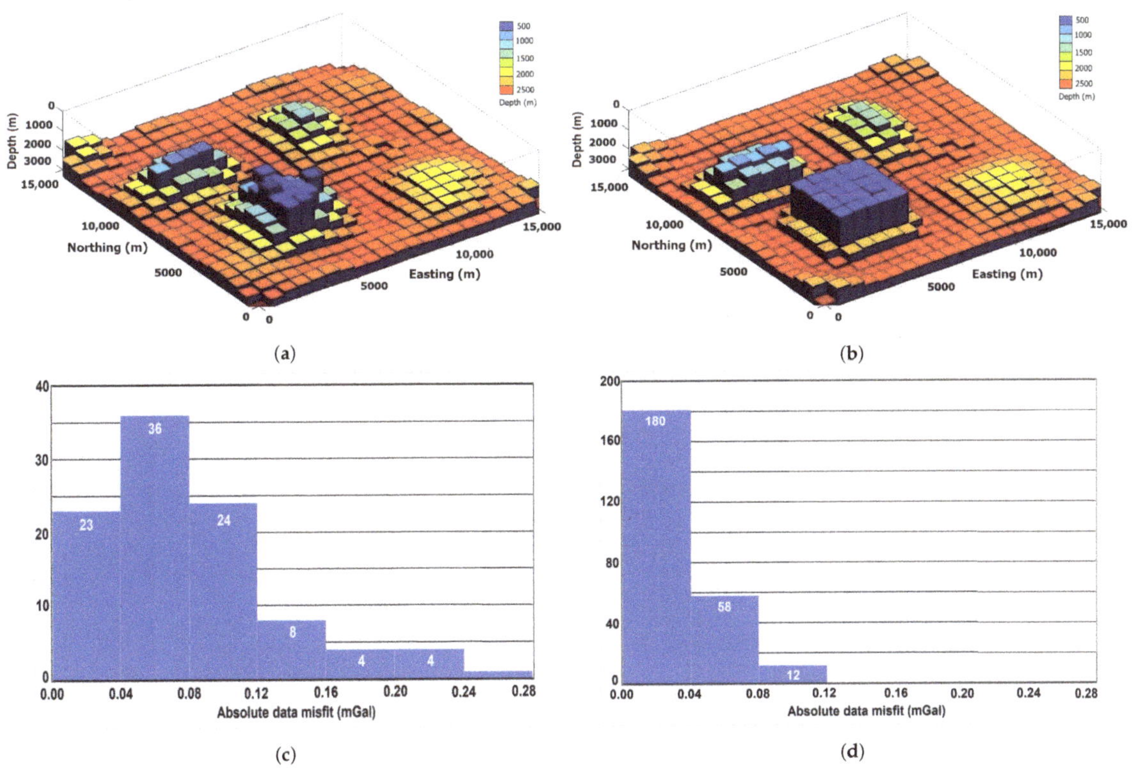

Figure 3. Inversion resulting models and respective data misfit histograms. (**a**) Inversion results for 100 noisy contaminated randomly distributed points showing reasonable estimates for locations and average depths for all structures. (**b**) Inversion results for 250 randomly distributed points, which allows better block definition. The histogram of the absolute data misfit for the inversion using 100 points is shown in (**c**), whereas (**d**) exhibits the histogram of the absolute data misfit for the 250 points inversion.

In inversion methods, the correct choice of parameters is usually problem-dependent, and there is no simple rule of thumb available. Therefore, in addition to showing the effectiveness of the proposed method, we also provide the reader with a short discussion on the effects of some of the parameters involved in this inversion process: the parameter η, the logarithm-barrier parameter (λ), and the regularization parameter (μ). Based on our experience, we hope that such a discussion can help the readers to develop a feeling for how to choose these parameters for their own problems.

The tests that used different values of the η-parameter showed that the influence of this parameter on the improvement of the solution is minor, and it is mainly restricted to the speed of convergence. Within the theoretically valid range, the number of iterations increases as η approaches zero since the actual step taken at each iteration is too small. As β approaches unity, the solution of Equation (11) becomes much more difficult. This is because of the disparity in the elements of matrices X and Y, which causes the matrix system to be poorly conditioned. Our tests indicate that values of η ranging from 0.9 to nearly 1.0 lead to similar convergence rates and computational costs. For the final solutions shown in Figure 3a,b, the η-parameter was chosen to be 0.99.

For the logarithm-barrier parameter, we usually start with a large value that must be reduced after each iteration (e.g., [11]). The tests that used different values for λ showed that, as expected, the initial choice of λ does not produce significant changes in the final solution, and this does not change the effectiveness of the depth constraints. We have chosen the approach in [12] to calculate the starting value of λ as:

$$\lambda = \frac{\Phi_d + \mu\Phi_m}{-2\sum_{j=1}^{M}\left[\ln\left(\frac{h_j - a_j}{b_j - a_j}\right) + \ln\left(\frac{b_j - h_j}{b_j - a_j}\right)\right]}. \tag{15}$$

The choice of the regularization parameter μ is the most important step towards a good inversion result. It should be noted that λ is an auxiliary parameter that does not directly change the final results, whereas μ is the parameter that determines the trade-off between model complexity and data misfit. Therefore, the parameter μ directly affects the final result, and its choice is crucial. The μ parameter is often chosen so that the misfit term reaches the target misfit at the final iteration. Such a criterion works well for cases where the noise is uncorrelated and zero-mean, and a good estimate of the standard deviation of this noise is available, as in the synthetic example presented here. Unfortunately, such cases are rare in practical applications.

When no information about data errors is available, other methods for estimating the regularization parameter must be used. [18] suggested the use of either GCV or L-curve criteria as an effective automatic estimator of the trade-off parameter in nonlinear inverse problems. We have found that the L-curve criterion produces good μ estimates for the synthetic examples in our problem. Therefore, we have incorporated this criterion in our inversion methodology by using the maximum curvature approach proposed by [19] to automatically locate the L-curve corner.

4. Recôncavo Basin Example

4.1. *Geologic Regional Settings*

As an example of application to a real problem, we have applied the proposed method to estimate the relief of the basement in a small portion of the Recôncavo Basin (RB), Brazil (Figure 4). Located in the country's northeast region, the mature RB is the oldest oil province in Brazil [20]. The RB is an aborted branch of a larger rift complex called the Recôncavo-Tucano-Jatobá rift, formed by a series of elongated half-grabens separated by oblique transfer faults extending over 620 km across the continent [20].

The origin and evolution of the rift are related to stresses that occurred in Gondwanaland during Mesozoic times, before continental drift [21]. The RB is characterized by a strong correlation between the distribution of oil fields and basement structures, which makes the correct understanding of the basement framework fundamental.

The stratigraphy of this basin can be simplified into two main sequences: the pre-rift and rift sequences [20]. The pre-rift sequence lays directly above the basement and is characterized by thick packages of sandstones, which are the principal reservoirs in the basin. Overlaying the pre-rift sequence is the rift sequence. This sequence is characterized mainly by shales, including the area's source rock. The thickness of the pre-rift sequence is almost constant along the entire work area, whereas the rift sequence is thicker in the structural lows of the basin. Two major fault systems, trending NE-SW and NW-SE, are responsible for the structural complexity of the area. Most faults directly connect reservoirs and source rocks, forming most oil fields over the internal structural highs. Figure 4 shows a simplified stratigraphic section in the work area.

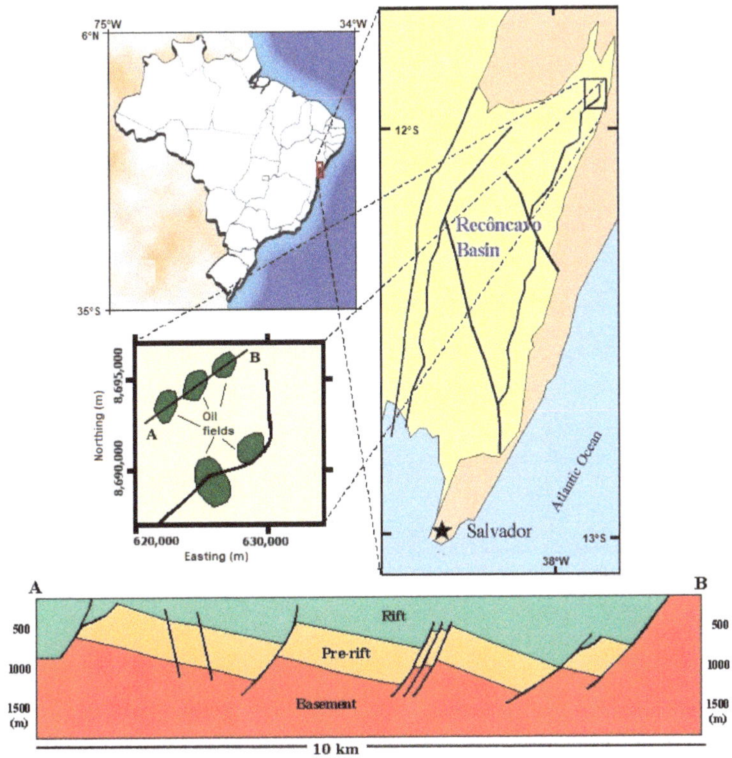

Figure 4. The geographic location of the Recôncavo Basin in Brazil. The detail shows the area chosen for the field example and the position of the schematic cross-section AB. The section shows the structural relationship between the three main stratigraphic features present in the work area.

4.2. Multiphysics Dataset

The studied area has been intensively explored, with several geophysical surveys acquired over several decades. The Brazilian National Petroleum Agency (ANP) provides free access to the public gravity and 2D seismic datasets used for the present interpretation.

The gravity data herein are interpreted as a subset of the onshore Debardenest regional gravity dataset made available by ANP. All land gravity stations were tied to the 1971 International Gravity Standardization Network (IGSN-1971) and were processed through a standard workflow to recover the Bouguer anomaly [22]. Our subset comprises 771 stations that are nearly uniformly distributed in a grid over the studied area, as shown by the crosses in the Bouguer map of Figure 5a. The data were interpolated to a regular grid using 250 m intervals.

The map in Figure 5b is the residual gravity anomaly that was used as the observed data for the inversion. These data were computed by removing a linear trend from the Bouguer anomaly map of Figure 5a. Although regional-residual procedures can change the amplitudes of anomalies in ways that can affect the depth estimates, this additional processing was required before the inversion for basement relief to remove the gravity effect of crustal thinning observed along the sedimentary basins of the northeast Brazilian continental margin (e.g., [17,23,24]).

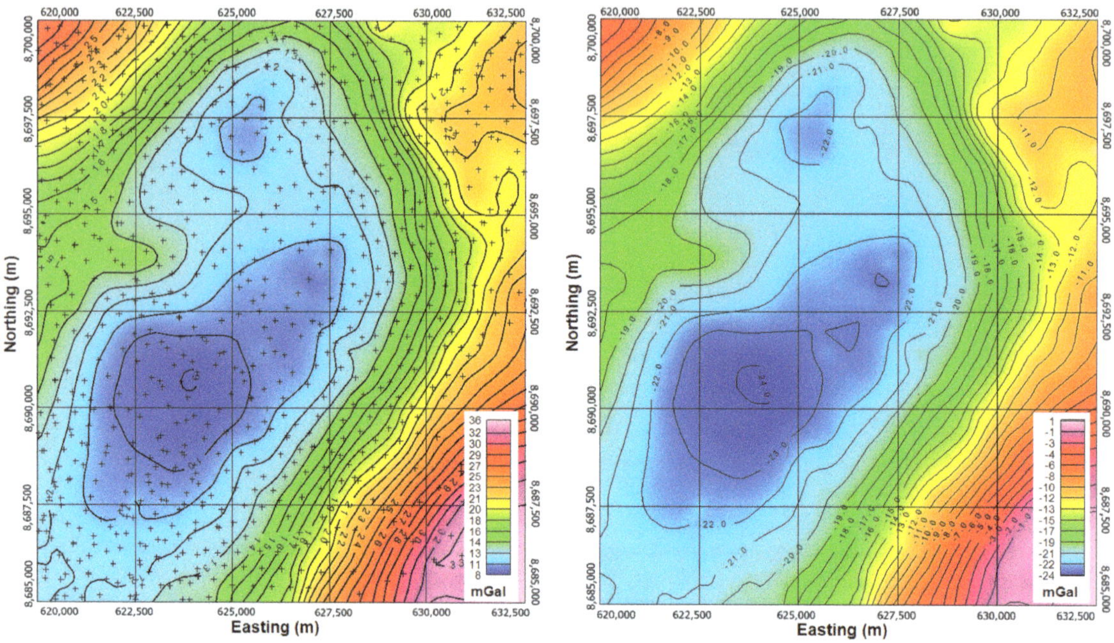

(a) (b)

Figure 5. (**a**) Bouguer anomaly map of the work area showing the position of the 771 gravity stations (crosses) where the measurements were made. (**b**) Residual anomaly map computed by removing a linear trend from the Bouguer anomaly. This residual is the observed data for the inversion. The data were gridded using 250 m intervals in both figures.

Despite the dense seismic coverage in this part of the basin, as shown in Figure 6a, no seismic basement map is available due to the poor quality of the seismic data. In a paper on the seismic characterization in Recôncavo Basin, [25] states that the large thickness of the recent sedimentary coverage associated with intense cultural activity in many areas makes it difficult to define the position of deep targets (Figure 6b). Besides, energy transmission difficulties and interfingering stratigraphy degrade the seismic signal in the studied area, affecting the signal-to-noise ratio and making the reconnaissance of basement reflections ambiguous. Because of that, reasonable basement estimates have been made from the top of the pre-rift sequence. The pre-rift sequence has two important characteristics in the study area: it is a well-defined seismic reflection that can be easily mapped and has a nearly constant thickness, averaging around 400 m. Due to these characteristics, it has been a typical and thriving practice in this portion of the basin to estimate the basement depths by adding 400 m to the top of the pre-rift sequence mapped from seismic data. We used this practice to get the basement estimate, as shown in Figure 7, which we used as a reference model.

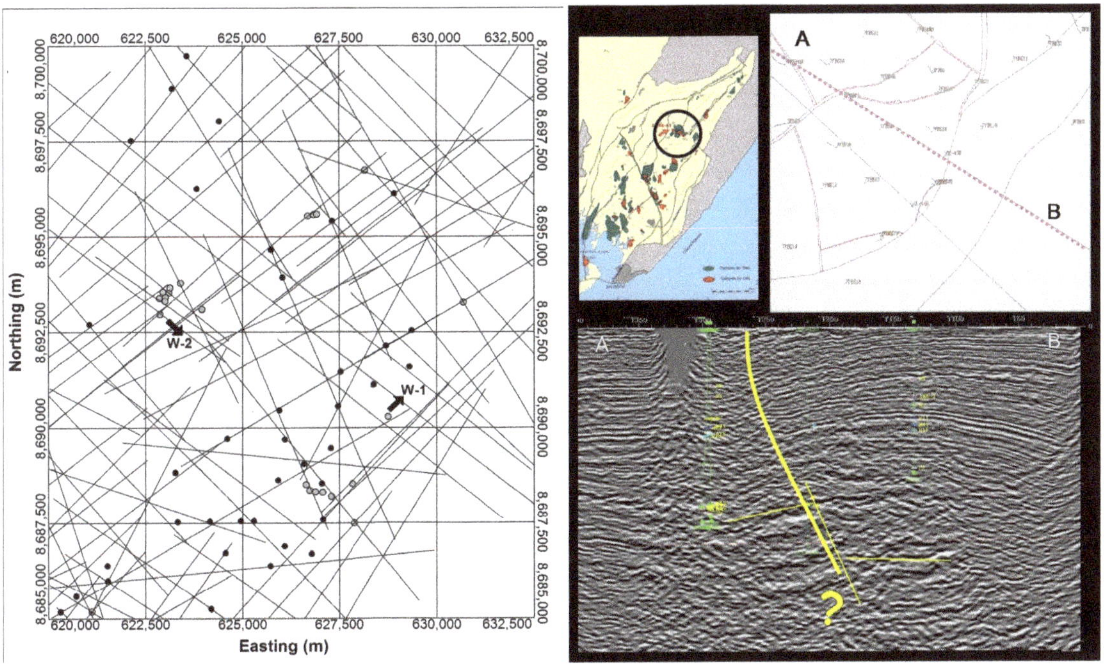

(**a**) (**b**)

Figure 6. (**a**) Position of seismic lines and wells used as constraints. The line segments are the 2D seismic lines used in the mapping of the top of the pre-rift sequence. The gray dots represent the location of the wells that provide reliable depth-to-basement information, and the black dots show the wells that provide constraints on the minimum basement depth only. (**b**) Example of the difficulties in defining the position of deep targets (reflections near the yellow question mark) caused by energy transmission problems, cultural interference, and complex stratigraphy, which degrade the seismic signal. The top left panel shows the location of this example in the Recôncavo Basin, whereas the top right panel shows the direction of the seismic section AB shown in the bottom panel (modified from [25]).

The depth constraints for this inversion come from 61 wells that are distributed throughout the area (Figure 6a). From this total, 23 wells provided direct depth-to-basement information (gray circles). The remaining 38 wells stopped inside the sedimentary section (black circles), providing only lower bounds to the basement depths at those locations. Although density logs were available for some of these wells, only the depth information was used to constrain the inversion.

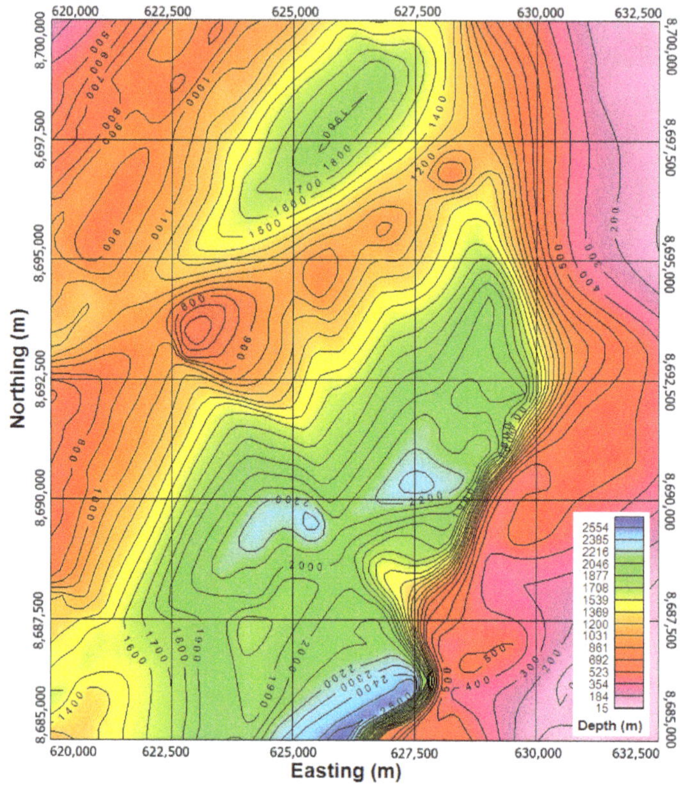

Figure 7. Structural map of the basement, as derived from the seismic mapping of the top of the pre-rift sequence (see text for details). These data represent the reference model for the inversion. The data were gridded using 250 m intervals.

4.3. Basement Relief Estimation

We assume an average contrast between the basement and sediments. The assumption of constant density contrast is a drawback of the technique since it is an approximation. However, we choose to adopt it because it is the most straightforward approach to be applied when the knowledge about the density distribution in the area needs to be improved, for example, in regions where the exploration is just beginning. In addition, in areas with reduced dimensions, like the study area, both the basement and sediments are not expected to change densities, and an average density contrast is a reasonable approach.

Despite the advanced geological knowledge in the study area, we decided to include only the minimum amount of information required by the inversion methodology. Such an approach allows us to better evaluate the technique's performance against the known geology. We emphasize that, in general, all available information should be used. For example, in cases where the density distribution is known to a certain depth, we suggest using techniques like gravity stripping [26] to remove the effect of known layers and then do the inversion for lower levels using a constant density contrast. Such a simple approach will reduce the complexity of the models, saving computer power and time during the inversion.

The average density contrast used during the inversion was estimated by considering the gravity response of the seismically derived model for different density contrasts at the exact position of those wells whose depth-to-basement is known. Since the seismically derived model honors the basement depths at the well locations, it is reasonable to assume that the most suitable average density contrast should be the one whose gravity response gives the smallest RMS error compared to the observed gravity at the well locations. According to this approach, the most appropriate density contrast is -0.39 g/cm^3. This average value is reasonable since it is within the range of density contrasts measured in several density logs distributed over the area.

The gravity response calculated for the seismically derived model using the estimated density contrast is shown in Figure 8a. The predicted data of Figure 8a should be similar to the observed data of Figure 5b if the seismically derived model were correct. There is a good resemblance between the two maps regarding shape, but the predicted data is a smooth version of the observed data. The discrepancies between predicted and observed data are better analyzed in the map of differences in Figure 8b. A quick statistical analysis shows that the amplitudes of the differences, ranging from -12 to 6 mGal, are very high. The same can be said about the standard deviation of 3.2 mGal. Analysis of the spatial distribution of the differences points to coherent features in the map, particularly an elongated positive feature, trending SW-NE at the lower center, which suggests an excessively deep basement. If the basement model was correct, the difference map should be dominated by incoherent features mainly related to noise. Therefore, although this model has been considered for a long time as a satisfactory estimate for the basement, it only partially justifies the gravity data and should be re-evaluated. We then applied gravity inversion to modify the basement depths and to reduce the discrepancies between predicted and observed gravity fields.

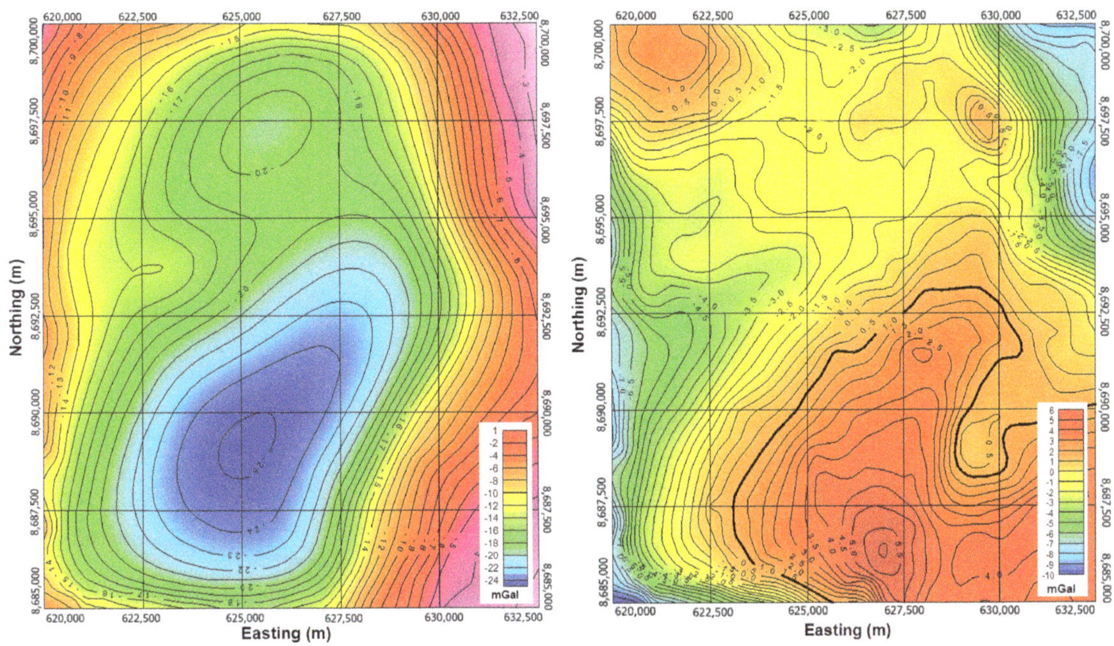

(**a**) (**b**)

Figure 8. (**a**) The residual gravity anomaly map, as calculated from the seismic-derived model of Figure 7 by using a density contrast of -0.39 g/cm^3. These are the predicted data for the seismic-derived model. (**b**) The differences calculated between the observed and predicted data. Notice that the presence of positive differences (greater than 1 mGal) in the SE corner (thick contour) indicates deficiencies in the seismic model that may be reduced by gravity inversion. The data were gridded using 250 m intervals in both figures.

Following the approaches developed in this paper, we select the inversion parameters as follows: γ-parameter is set to 0.99, initial λ-parameter is equal to 10^{-4}, and μ is equal to 10^{-10}. The result of inverting the observed data is shown in Figure 9a. The gravity-inverted basement is very similar in shape to the reference model of Figure 9a. This was already expected since the method requires the maximum possible similarity with the reference model. Figure 9b shows the differences between the reference and the model predicted by gravity inversion. Although the amplitudes of the differences range from −600 to 700 m, the average value is around −70 m. In the majority of the area, however, the amplitudes stay between −200 and 200 m, which indicates a reasonable agreement between the two models. The presence of large values at NE and SW corners can be related to some kind of border problems. Such effects may suggest that the first-order trend used for the regional-residual separation was not a reasonable approximation to the regional field at these locations. The most remarkable feature is the large positive difference situated at the central-south of the map. This feature is the most important contribution to the study area since the inversion has suggested the basement is over 300 m shallower than what was initially estimated from the seismically derived model.

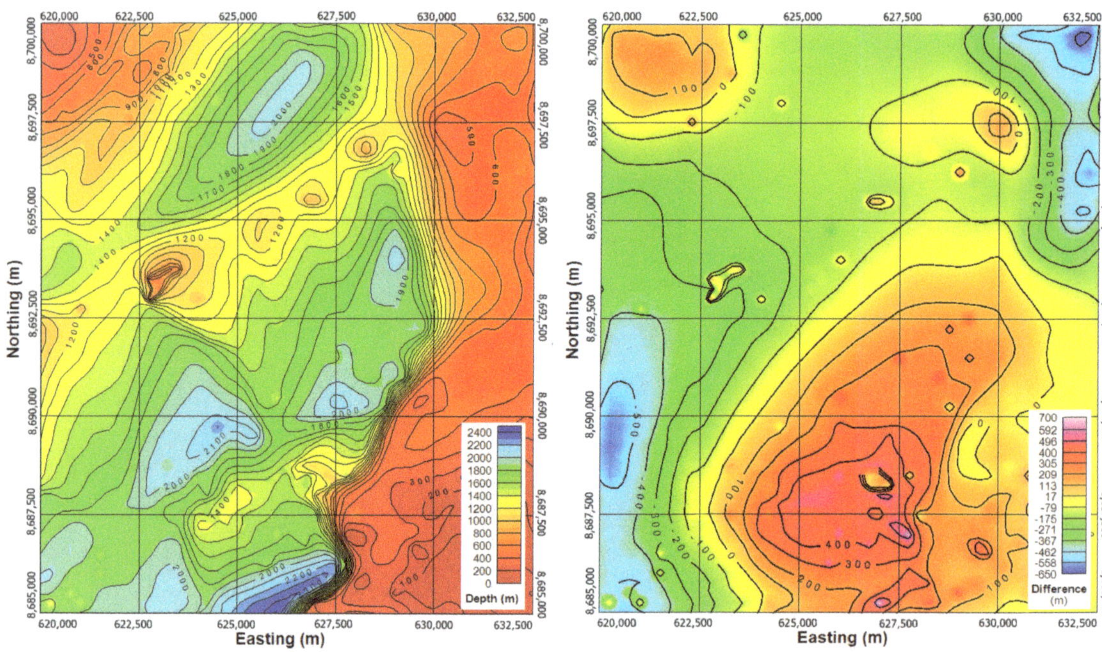

(**a**) (**b**)

Figure 9. (**a**) The new structural map of the basement after gravity inversion. Note the overall similarity to the reference model of Figure 7. (**b**) The differences between the seismic-derived and the gravity-inverted basement estimates. A larger difference at the SE corner (over 100 m) correlates with the feature found in the difference map of Figure 8b. The correlation between the two features seems to indicate that the inversion has considered the seismic basement as being too deep to fit the anomaly of Figure 8b and had to raise it in order to fit the anomaly. The data were gridded using 250 m intervals in both figures.

The predicted gravity field resulting from the inverted basement is shown in Figure 10a. In terms of shape, there is a great resemblance between this map and that of Figure 8a, the predicted field from seismic. Such behavior was already expected since only finer modifications were introduced in the seismic-derived model by inversion. The difference, however, is more significant in magnitude. As an example, the small gravity low toward the north is now deeper, whereas the large low at the south is shallower. The map of differences between observed and predicted gravity, in Figure 10b, also demonstrates the improvement in the fit. The amplitude of the differences now ranges from −10 to 6 mGal, with a standard deviation of 1.5 mGal, which is approximately half of that achieved with the seismically derived model.

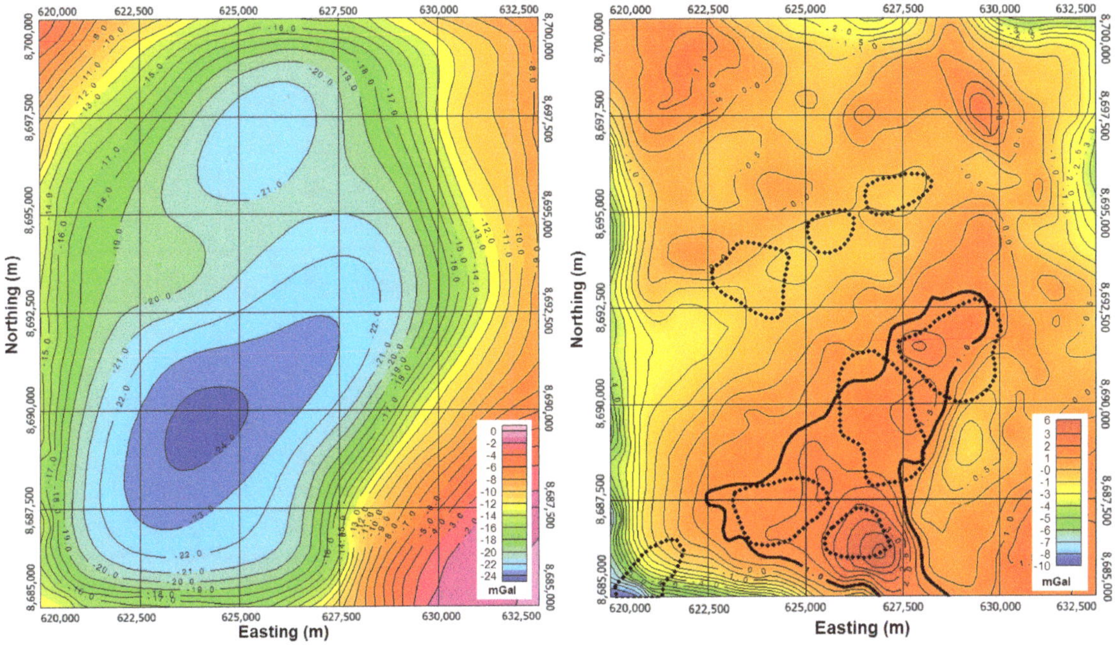

(a) (b)

Figure 10. (**a**) The predicted gravity field calculated by using inversion. Since only small changes were introduced into the model, the similarity to the field predicted by the seismic model (Figure 8a) is evident. (**b**) The differences between the observed and predicted data. Note the changes in the amplitude of the differences, especially in the two main lows, when compared to Figure 8b. The SW-NE structure (highlighted with a heavy line) in this map is justified by the presence of a trend of oil-bearing high-density sandstones (dotted lines) that are coincident with the structure. The discovery of such anomalous features gives important exploratory significance to this map. The data were gridded using 250 m intervals in both figures.

The improvement is more evident in the histograms of the absolute differences shown in Figure 11. Despite the general reduction in the misfit, the presence of the SW-NE structure in Figure 10b (heavy line), showing differences greater than 1 mGal, indicates that the raising of the basement imposed by inversion was not enough to completely fit the data. It is possible that the presence of constraints has prevented the inversion from raising the basement to justify the anomaly. If the basement is restricted to deeper parts and cannot be responsible for the gravity anomaly, then it is likely that density variations within the sedimentary section are the contributing factor. In fact, such a structure shows a strong correlation with a trend of oil-bearing sandstones (dotted lines in Figure 10b).

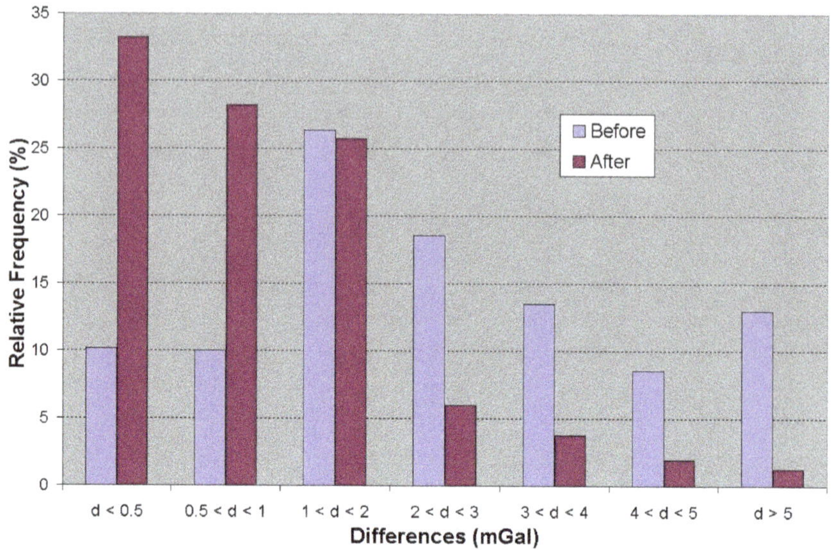

Figure 11. Comparison of the differences between the observed and predicted data before and after gravity inversion. The increase in the relative frequency for the absolute differences below 2 mGal after inversion, comparatively to before, proves the fitting improvement.

Detailed investigations about the characteristics of these reservoirs show they are formed by anomalous high-density sandstones located inside the rift section, as proved by the density log of Figure 12a. The region of the log that corresponds to the reservoirs shows densities similar to the basement. In contrast, in the case of a well located out of the trend (Figure 12b), the densities are lower, and there is significant contrast only with the basement at the bottom of the log. Since these intra-sedimentary density variations were not accounted for in the model, it was expected that the inversion would try to fit this anomaly, making the basement shallower in this area. However, the presence of the constraints, especially those limiting the minimum basement depth, avoided such compensation. In this case, therefore, the presence of residuals in the final result points to potential high-density regions correlated to oil-bearing stratigraphic features, which would have important exploratory significance.

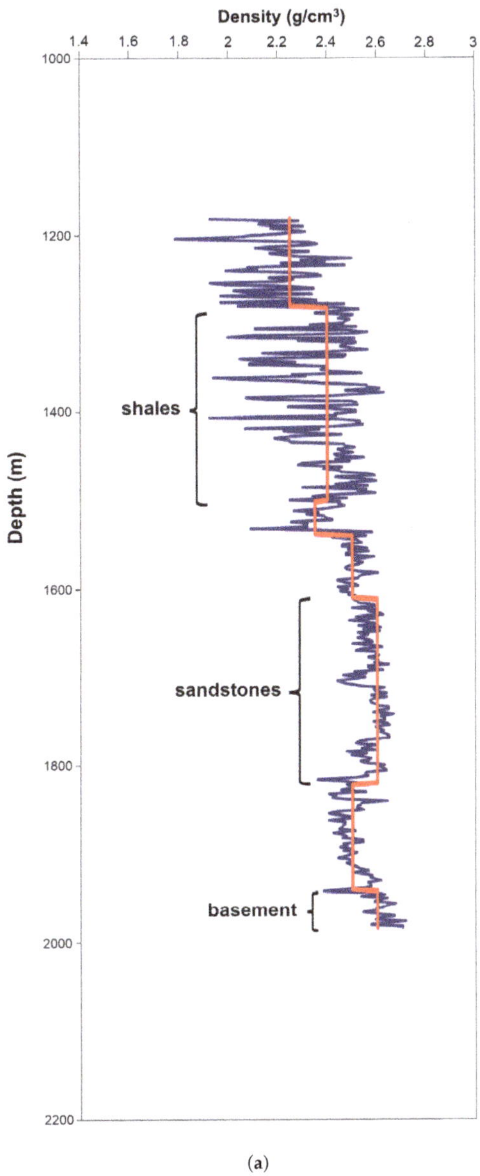

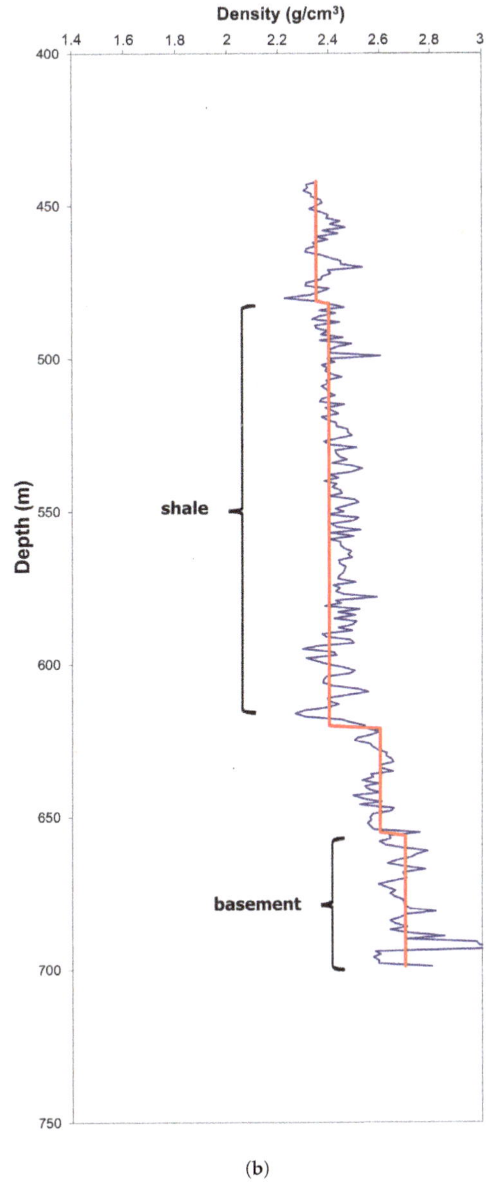

Figure 12. Density logs from the two wells at different positions: (**a**) inside the anomaly area (W-1), and (**b**) outside the anomaly area (W-2). Notice that the sandstones in (**a**) show a density similar to that of the basement. The position of the wells is given in Figure 6a. No density data were collected in the shallower portions of both wells.

5. Conclusions

We have proposed a new approach for estimating the relief of a surface separating two media of different densities. The method is based on inversion of gravity data that incorporate seismic interpretation and borehole logs. Similar to other inversion methods, the proposed method minimizes the objective function of the model that controls the data misfit and penalizes deviation from a reference model and the structural complexity of the model. The use of the logarithmic barrier method allows for the incorporation of well log information, even from wells that do not reach the basement. Since such wells are often more common than those penetrating the basement, the number of constraints increases. Furthermore, the logarithm barrier method enables the use of general bounds that vary with location within the basement model. The possibility of including such specific information in the inversion increases our confidence in the final model.

The proposed methodology was successfully tested on a synthetic dataset. The inversion has satisfactorily recovered both the position and the average depth-to-the-top of the structures present in the model. The errors in the recovered model seem to be caused by the fact that the inversion was applied to a limited number of noisy observations. Significantly better results can be achieved either by increasing the number of observations or by reducing the noise level.

The proposed approach was applied to a field gravity dataset acquired in the Recôncavo Basin, Brazil. This basin was chosen because it hosts a number of oil fields located at structural highs that correlate well with gravity anomalies. The dense gravity coverage in the area also makes it ideal for testing the new algorithm. Incorporating the seismic-derived model into the inversion and imposing well constraints on the depth to the basement produced a different basement model that significantly reduced the gravity data misfit. The method has been shown to be effective in the sense that it provides the necessary adjustment to the seismic model in order to produce a satisfactory fit to the observed gravity field.

In addition, the use of a constant density contrast (although for a very simplistic model for sedimentary basins) has proven to be appropriate in the case studied because it was constrained by a large amount of independent information, like seismic data and wells. By preventing the inversion to improve the fit, the presence of constraints led to inversion residuals that were related to density anomalies inside the sedimentary section. The mapping of these density anomalies is of significant importance for exploration in Recôncavo Basin because they are closely related to a specific kind of stratigraphic prospect.

Author Contributions: Conceptualization, J.C.S.O.L. and Y.L.; methodology, J.C.S.O.L. and Y.L.; software, J.C.S.O.L.; validation, J.C.S.O.L. and Y.L.; formal analysis, J.C.S.O.L.; investigation, J.C.S.O.L.; writing—original draft preparation, J.C.S.O.L.; writing—review and editing, Y.L.; visualization, J.C.S.O.L.; supervision, Y.L. All authors have read and agreed to the published version of the manuscript.

Funding: This research received no external funding.

Data Availability Statement: The data used in this research may be requested to the Brazilian National Petroleum Agency-ANP at https://www.gov.br/anp/pt-br/assuntos/exploracao-e-producao-de-oleo-e-gas/dados-tecnicos/legislacao-aplicavel/bndg-banco-nacional-de-dados-gravimetricos, accessed on 24 July 2023.

Acknowledgments: The authors thank the Special Issue Editor Paulo T. L. Menezes for the valuable suggestions. Julio Cesar S. O. Lyrio would like to thanks PETROBRAS for permission to publish this work.

Conflicts of Interest: The authors declare no conflict of interest.

References

1. Bain, J.; Scherrer, E.; Millegan, P.; Pearson, E.; Schulz, S.; Mulcahy, S. 3D visualization applied to integrated modeling of seismic, gravity and magnetics in the Gulf of Mexico. In Proceedings of the 72nd Annual International Meeting, Salt Lake City, UT, USA, 6–11 October 2022 ; Society of Exploration Geophysicists: Houston, TX, USA, 2002.
2. O'Brien, J.; Rodriguez, A.; Sixta, D.; Davies, M.A.; Houghton, P. Resolving the K-2 salt structure in the Gulf of Mexico: An integrated approach using prestack depth imaging and full tensor gravity gradiometry. *Lead. Edge* **2005**, *24*, 404–409. [CrossRef]
3. Colombo, D.; De Stefano, M. Geophysical modeling via simultaneous joint inversion of seismic, gravity, and electromagnetic data: Application to prestack depth imaging. *Lead. Edge* **2007**, *26*, 326–331. [CrossRef]
4. Cordell, L.; Henderson, R.G. Iterative three-dimensional solution of gravity anomaly data using a digital computer. *Geophysics* **1968**, *33*, 596–601. [CrossRef]
5. Oldenburg, D.W. The inversion and interpretation of gravity anomalies. *Geophysics* **1974**, *39*, 526–536. [CrossRef]

6. Last, B.J.J.; Kubik, K. Compact gravity inversion. *Geophysics* **1983**, *48*, 713–721. [CrossRef]
7. Guillen, A.; Menichetti, V. Gravity and magnetic inversion with minimization of a specific functional. *Geophysics* **1968**, *49*, 1354–1360. [CrossRef]
8. Barbosa, V.C.F.; Silva, J.B.C.; Medeiros, W.E. Gravity inversion of basement relief using approximate equality constraints on depths. *Geophysics* **1997**, *62*, 1745–1757. [CrossRef]
9. Moraes, R.A.V.; Hansen, R.O. Constrained inversion of gravity fields for complex 3-D structures. *Geophysics* **2001**, *66*, 501–510. [CrossRef]
10. Li, Y.; Oldenburg, D.W. 3-D inversion of gravity data. *Geophysics* **1998**, *63*, 109–119. [CrossRef]
11. Wright, S.J. *Primal-Dual Interior-Point Methods*; SIAM: Philadelphia, PA, USA, 1997.
12. Li, Y.; Oldenburg, D.W. 3-D inversion of induced polarization data. *Geophysics* **2000**, *65*, 1931–1945. [CrossRef]
13. Li, Y.; Oldenburg, D.W. Fast inversion of large-scale magnetic data using wavelet transforms and a logarithmic barrier method. *Geophys. J. Int.* **2003**, *152*, 251–265. [CrossRef]
14. Blakely, R.J. *Potential Theory in Gravity and Magnetic Applications*; Cambridge University Press: Cambridge, UK, 1995.
15. Plouff, D. Gravity and magnetic fields of polygonal prisms and application to magnetic terrain corrections. *Geophysics* **1976**, *41*, 727–741. [CrossRef]
16. Medeiros, W.E.; Silva, J.B.C. Geophysical inversion using approximate equality constraints. *Geophysics* **1996**, *61*, 1678–1688. [CrossRef]
17. Leão, J.W.D.; Menezes, P.T.L.; Beltrão, J.F.; Silva, J.B.C. Gravity inversion of basement relief constrained by the knowledge of depth at isolated points. *Geophysics* **1996**, *61*, 1702–1714. [CrossRef]
18. Farquharson, C.; Oldenburg, D. Automatic estimation of the trade-off parameter in nonlinear inverse problems using the GCV and L-curve criteria. In Proceedings of the 70th Annual International Meeting. Society of Exploration Geophysicists, Calgary, AB, Canada, 6–11 August 2000.
19. Li, Y.; Oldenburg, D.W. 3-D inversion of DC resistivity data using an L-curve criterion. In Proceedings of the 1999 SEG Annual Meeting, Houston, TX, USA, 31 October–5 November 1999; OnePetro: Charlottesville, VA, USA, 1999.
20. Magnavita, L.P.; Szatmari, P.; Cupertino, J.A.; Destro, N.D.G.R.; Roberts, D.G. The Recôncavo Basin. In *Regional Geology and Tectonics: Phanerozoic Rift Systems and Sedimentary Basins*; Elsevier: Amsterdam, The Netherlands, 2012; pp. 383–420.
21. dos Santos, C.F.; Cupertino, J.A.; Braga, J.A.E. Síntese sobre a geologia das bacias do Recôncavo, Tucano e Jatobá. In *Origem e Evolução de Bacias Sedimentares*; Raja Gabaglia, G.P., Milani, E.J., Eds.; Petrobras: Rio de Janeiro, Brazil, 1990; pp. 235–266.
22. Coelho, A.C.Q.M.; Menezes, P.T.L.; Mane, M.A. Gravity data as a faulting assessment tool for unconventional reservoirs regional exploration: The Sergipe–Alagoas Basin example. *J. Nat. Gas Sci. Eng.* **2021**, *94*, 104077. [CrossRef]
23. Barbosa, V.C.F.; Menezes, P.T.L.; Silva, J.B.C. Case History Gravity data as a tool for detecting faults: In-depth enhancement of subtle Almada's basement faults, Brazil. *Geophysics* **2007**, *72*, B59–B68. [CrossRef]
24. Adriano, L.B.; Menezes, P.T.L.; Adriano, M.S.; Cunha, A.S.; Cabrera, M.H.; Silva, D.S.; Moura, L.P. Jequitinhonha Basin: Structural aspects, relationship with igneous activity, and hydrocarbon exudations. *Interpretation* **2018**, *6*, T51–T60. [CrossRef]
25. Barreto, J.; Schinelli, M. Novos desafios para caracterização sísmica em bacia madura. In Proceedings of the 10th International Congress. Brazilian Geophysical Society, Rio de Janeiro, Brazil, 19–23 November 2007.
26. Hammer, S. Deep gravity interpretation by stripping. *Geophysics* **1963**, *28*, 369–378. [CrossRef]

Disclaimer/Publisher's Note: The statements, opinions and data contained in all publications are solely those of the individual author(s) and contributor(s) and not of MDPI and/or the editor(s). MDPI and/or the editor(s) disclaim responsibility for any injury to people or property resulting from any ideas, methods, instructions or products referred to in the content.

minerals

MDPI

Article

Using Large-Size Three-Dimensional Marine Electromagnetic Data for the Efficient Combined Investigation of Natural Hydrogen and Hydrocarbon Gas Reservoirs: A Geologically Consistent and Process-Oriented Approach with Implications for Carbon Footprint Reduction

Max A. Meju [1,2,*] and Ahmad Shahir Saleh [1]

[1] Multiphysics Department, Petronas Upstream, Petronas Twin Towers, Kuala Lumpur 50088, Malaysia; ahmadshahir_saleh@petronas.com.my

[2] Crossgradients Geo-Decisioneering, The Orchard, Stanmore Drive, Lancaster LA1 5BL, UK

* Correspondence: maxmeju@gmail.com; Tel.: +44-7845530644

Abstract: The recycling or burial of carbon dioxide in depleted petroleum reservoirs and re-imagining exploration strategies that focus on hydrogen reservoirs (with any associated hydrocarbon gas as the upside potential) are a necessity in today's environmental and geopolitical climate. Given that geologic hydrogen and hydrocarbon gases may occur in the same or different reservoirs, there will be gains in efficiency when searching for both resources together since they share some commonalities, but there is no geophysical workflow available yet for this purpose. Three-dimensional (3D) marine controlled-source electromagnetic (CSEM) and magnetotelluric (MT) methods provide valuable information on rock-and-fluid variations in the subsurface and can be used to investigate hydrogen and hydrocarbon reservoirs, source rocks, and the migration pathways of contrasting resistivity relative to the host rock. In this paper, a process-oriented CSEM-MT workflow is proposed for the efficient combined investigation of reservoir hydrocarbon and hydrogen within a play-based exploration and production framework that emphasizes carbon footprint reduction. It has the following challenging elements: finding the right basin (and block), selecting the right prospect, drilling the right well, and exploiting the opportunities for sustainability and CO_2 recycling or burial in the appropriate reservoirs. Recent methodological developments that integrate 3D CSEM-MT imaging into the appropriate structural constraints to derive the geologically robust models necessary for resolving these challenges and their extension to reservoir monitoring are described. Instructive case studies are revisited, showing how 3D CSEM-MT models facilitate the interpretation of resistivity information in terms of the key elements of geological prospect evaluation (presence of source rocks, migration and charge, reservoir rock, and trap and seal) and understanding how deep geological processes control the distribution and charging of potential hydrocarbon, geothermal, and hydrogen reservoirs. In particular, evidence is provided that deep crustal resistivity imaging can map serpentinized ultramafic rocks (possible source rocks for hydrogen) in offshore northwest Borneo and can be combined with seismic reflection data to map vertical fluid migration pathways and their barrier (or seal), as exemplified by the subhorizontal detachment zones in Eocene shale in the Mexican Ridges fold belt of the southwest of the Gulf of Mexico, raising the possibility of using integrated geophysical methods to map hydrogen kitchens in different terrains. The methodological advancements and new combined investigative workflow provide a way for improved resource mapping and monitoring and, hence, a technology that could play a critical role in helping the world reach net-zero emissions by 2050.

Keywords: marine electromagnetic methods; electrical resistivity; joint geophysical inversion; hydrocarbons; native hydrogen; exploration guide; reservoir mapping and monitoring; net-zero emissions and energy transition

Citation: Meju, M.A.; Saleh, A.S. Using Large-Size Three-Dimensional Marine Electromagnetic Data for the Efficient Combined Investigation of Natural Hydrogen and Hydrocarbon Gas Reservoirs: A Geologically Consistent and Process-Oriented Approach with Implications for Carbon Footprint Reduction. *Minerals* **2023**, *13*, 745. https://doi.org/10.3390/min13060745

Academic Editor: Paulo T. L. Menezes

Received: 4 April 2023
Revised: 27 May 2023
Accepted: 28 May 2023
Published: 30 May 2023

Copyright: © 2023 by the authors. Licensee MDPI, Basel, Switzerland. This article is an open access article distributed under the terms and conditions of the Creative Commons Attribution (CC BY) license (https://creativecommons.org/licenses/by/4.0/).

1. Introduction: Challenges and Conceptual Models

1.1. Net-Zero Emissions and Geological Complexity Challenges

Easy-to-find oil and gas accumulations no longer exist, and the deepening climatic emergency is driving a global transition to low-carbon energy sources, such as native hydrogen and geothermal reservoirs. It is now widely believed that hydrogen could significantly abate CO_2 emissions, and, hence, finding cheap clean sources of molecular hydrogen will play a central role in helping humanity reach net-zero emissions by 2050 and limit global warming to 1.5 degrees Celsius. Native hydrogen is cheaper to produce than current methods (i.e., green hydrogen from renewable or nuclear-powered electrolysis, grey/blue hydrogen from natural gas-powered steam-methane reforming, and brown/blue hydrogen from coal-powered gasification), with an estimated cost of less than USD 1 per kg. It is also attractive because of its sustainability since it is generated by geological processes, can be directly used in industrial applications, and does not require fossil fuels for production. Understanding the deep geological controls on the genesis and distribution of native hydrogen is, thus, of the utmost importance in our quest to exploit this vital natural resource and help reduce our carbon footprint. Moreover, geologic hydrogen and hydrocarbons may occur in the same or different reservoirs, and there will, therefore, be efficiency gains in searching for both resources together, unlike the conventional practice. Indeed, re-imagining exploration strategies to focus on hydrogen reservoirs with the associated hydrocarbons as the upside potential is called for in today's environmental and geopolitical climate.

The focus of geophysical investigations into natural hydrogen and hydrocarbon gas reservoirs has now moved into frontier areas where structural complexity, heterogeneous overburden, and resource system fundamentals pose significant challenges [1]. The conventional hierarchical approach in hydrocarbon reservoir exploration (Figure 1a) and development (Figure 1b) have to be tailored to meet the challenges in this new environment. The main industry challenges in the exploration of commercial marine hydrocarbon and native hydrogen accumulations include the following: finding the right basin and play (requiring basin-scale investigations), selecting the right prospects (requiring prospect-scale investigations), and drilling the right well (requiring reservoir-scale investigations). These technical challenges derive mostly from geological complexity and, hence, require the robust integration of geophysical and geological data to reduce the exploration risk [1].

The seismic, controlled-source electromagnetic (CSEM), magnetotelluric (MT), gravity, and magnetic methods of geophysics play major roles in marine exploration for hydrocarbon and/or native hydrogen reservoirs. On their own, individual geophysical methods provide non-unique models of subsurface properties and the fluid type present, but integrating them [2,3] together with geological models maximizes accuracy, minimizes uncertainty in a "Shared Earth", and leads to a consistent and possible indication of a working resource or surveillance system [1,4,5]. Optimizing this process has huge implications for time and cost savings, and there are opportunities for carbon footprint reductions at all spatial scales, as advocated for in this paper. It is also proposed here that any new hydrocarbon and hydrogen exploration plan must consider, as a top priority, whether such reservoirs can sequester CO_2 for 100 years for net-zero emissions compliance (Figure 1). The traditional measurements in the search for hydrocarbons do not include hydrogen gas surveys, although hydrogen has been found to be associated with hydrocarbons in some wells. Perhaps now is the time to redress this imbalance in data collection. Along these lines, the appropriate conceptual models for electromagnetic investigations of natural hydrocarbon and hydrogen gas reservoirs are described in Section 1.2, and a new practical workflow is developed for the efficient combined investigation of these reservoirs in Section 1.3.

1.2. Conceptual Models for Electromagnetic Investigation of Hydrocarbon and Hydrogen Reservoirs

Important and, in some respects, unique information about a hydrocarbon gas-charged or hydrogen-filled reservoir can be obtained by measuring its electrical resistivity. Resistivity is a function of porosity, fluid saturation, temperature, and chemical composition, and hence the accurate 3D mapping and monitoring of this property in a reservoir is of practical importance and is remotely possible if the subsurface structural configuration is correctly specified using seismic, borehole, and other types of a priori information [2,6,7]. Figure 2 shows the relationship between bulk resistivity and resistive gas saturation ([8] Archie 1942) for different porosities (0.15%–0.3%), representing typical unconventional reservoirs (shales and tight sands) and porous clastic reservoirs (permeable sands and carbonates). It is obvious that the measurable bulk resistivity increases only marginally with low gas saturation but quite significantly with high gas saturation. This provides a justification for using electromagnetic methods for mapping and monitoring offshore hydrocarbon reservoirs [9,10].

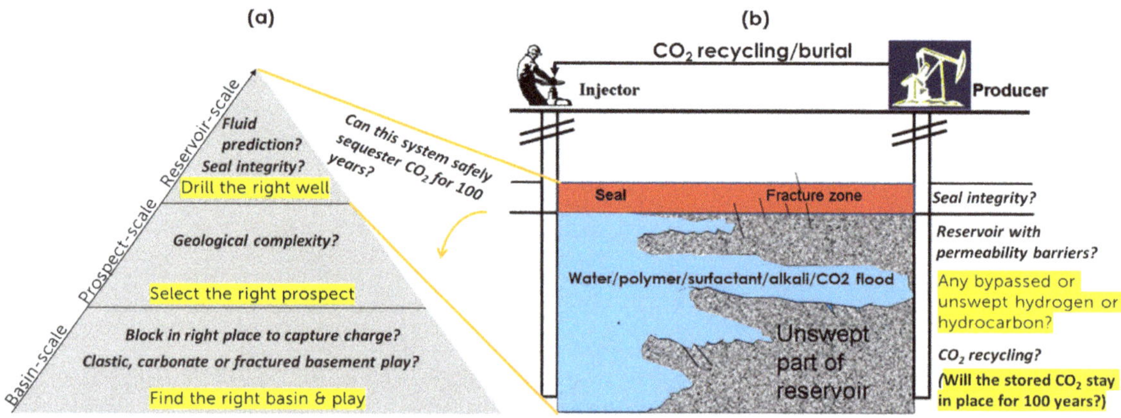

Figure 1. Multi-scale challenges in reservoir exploratory and monitoring investigations with opportunities for carbon footprint reduction. The main objectives are highlighted in yellow, and the key technical challenges are in italics. (**a**) Conventional multi-scale hierarchical exploration approach. (**b**) Monitoring of hydrocarbon (or hydrogen) production and enhanced recovery aided by carbon dioxide recycling and/or chemical flooding of reservoirs. For completeness, the reservoirs should be investigated for permanent CO_2 storage potential.

The CSEM and MT methods are the principal resistivity-measuring tools for the remote investigation of hydrocarbon or native hydrogen reservoirs offshore, and three-dimensional (3D) surveying is the method of choice in frontier regions [1]. These methods probe from the near-seabed surface to several km in depth, making them ideal for investigating prospective sedimentary sequences above and below thick salt or igneous layers and any associated deep structural controls. Marine CSEM-MT 3D surveying requires setting up arrays of electric and magnetic sensors (or receivers) on the seafloor (Figure 3). A horizontal electric dipole (HED) source is towed about 20–50 m above the seafloor or 10–15 m below the sea surface when only CSEM measurements are required. The receivers (Rx) are laid out on a grid (Figure 3), typically staggered with the line and Rx spacing being ~1–3 km, depending on the target size and depth. The HED source is towed along each survey line starting in one direction (in-tow) from ~20–30 km away from one edge of the area of exploration interest to ~20–30 km beyond the other edge (out-tow) and transmitting signals of multiple frequencies (typically 0.1 to 10 Hz) whilst the boat is traveling at ~2 knots (3.704 km/h) [1]. The source may also be towed in between the lines of receivers (Figure 3). For studies of deep structural control or magmatic impact on the distribution of reservoirs, the receivers are typically left on the seafloor for 7 to 14 days in order to record sufficient MT periodicities and good quality data, depending on the water depth [11].

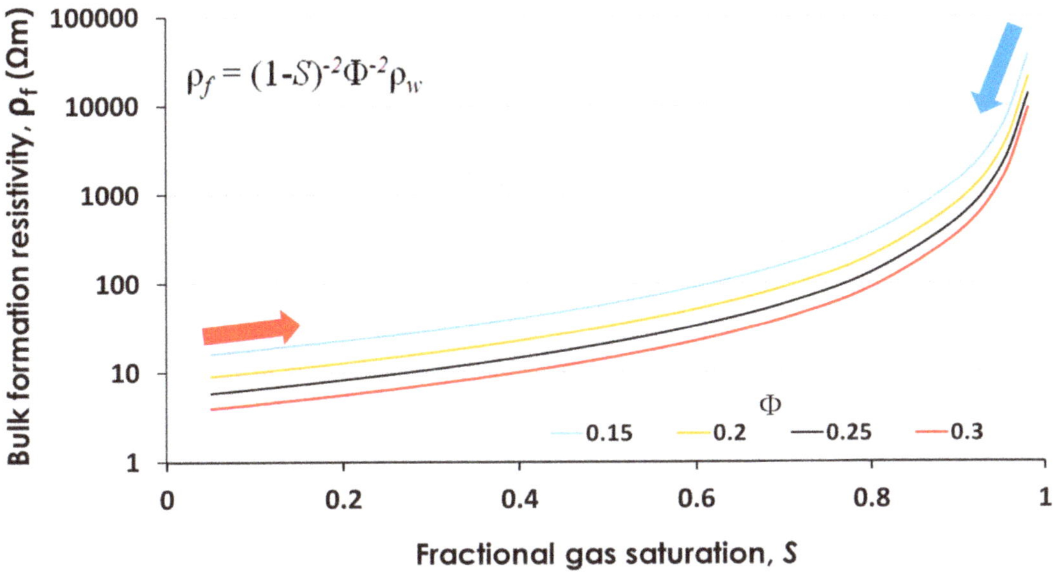

Figure 2. A simple physical basis for mapping and monitoring bulk resistivity changes with fluid saturation in tight or porous reservoirs [8]. The relationship between the bulk resistivity of reservoir materials (ρ_f) and gas saturation (*S*) for four different porosities (Φ = 0.15, 0.2, 0.25 & 0.3) is shown, assuming a value of 0.33 for the fluid resistivity, ρ_w. The inset shows the equation used for calculating ρ_f [8]. For a given reservoir, the effect of injecting resistive fluids (e.g., CO_2, natural hydrogen, and hydrocarbon gas) is to move up along the curve towards higher resistivity (as indicated by the red arrow), providing that the porosity does not change; if porosity is reduced due to new mineralization over time, we would expect one to move across the curves. In the case of brine or surfactant injection, it would move one down the curve rather than across the curves (depicted by the light blue arrow) unless the porosity changed in the process, for which one then moves across the curves.

Electrical anisotropy is a major challenge in marine CSEM-MT exploration [1,12]. A conceptual model of a hydrocarbon-filled structural trap in the anisotropic ground is shown in (Figure 3). Localized 3D zones of chemosynthetic communities and patches of small-size bodies may abound in the near-surface, imparting strong heterogeneity, while at the reservoir level, anisotropy may be present in the form of laminations, microcracks, or other structural alignments. Inverting the complete, sufficient, and consistent marine CSEM-MT data to reveal the resistivity structure, taking into account 3D heterogeneity and anisotropy, is thus important for successful offshore reservoir mapping and monitoring. Industry benchmarking in 2017 revealed inconsistencies in horizontal and vertical resistivity distributions from anisotropic 3D CSEM inversion models produced by the major service providers for a given field dataset used for the test; invited academic practitioners abstained from the test [13]. This provided the motivation for the development of a crossgradient-based anisotropic-resistivity-imaging algorithm that directly enforces structural similarity between the horizontal and vertical resistivity models without control from seismic or well data [4,14,15]. The seismic, image-guided anisotropic 3D inversion of CSEM and MT data is an adaption of this crossgradient philosophy, in which the structural controls from seismic or other images are infused in the anisotropic resistivity inversion process [7,16–18].

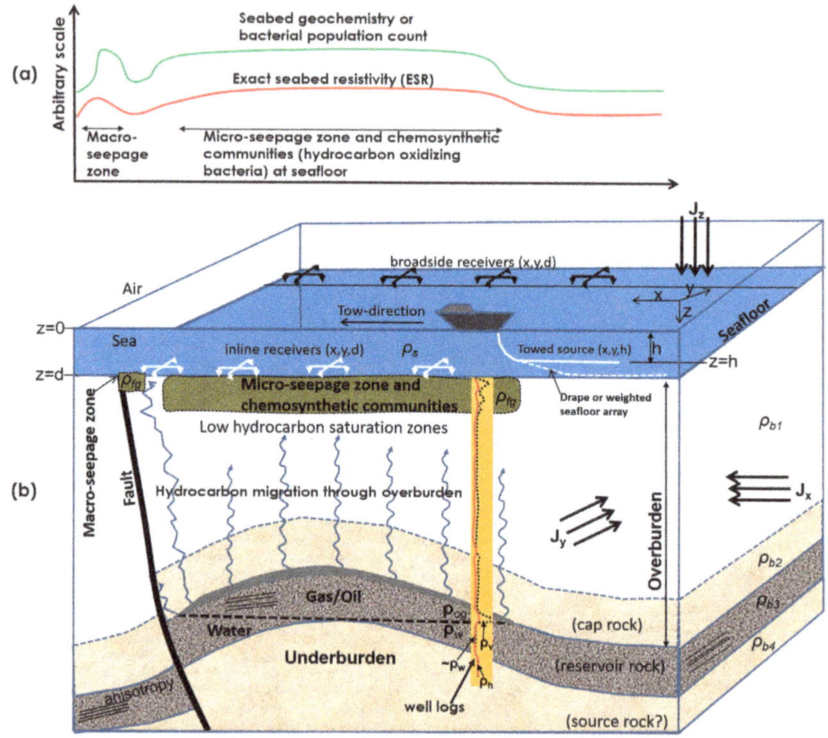

Figure 3. A conceptual model and the marine CSEM-MT field set-up for mapping or monitoring a hydrocarbon-charged reservoir in heterogenous and anisotropic host sediments. (**a**) Cartoon showing the expected seabed profiles of electrical resistivity, geochemical anomalies, and bacterial population counts over a leaky hydrocarbon structural trap. (**b**) Cartoon showing a section through the leaky structural trap with seepage-related features near the seafloor and the typical 3D CSEM-MT survey layout [1]. Depth variations of background resistivity (ρ_{b1}, ρ_{b2}, ρ_{b3}, and ρ_{b4}) and anomalous 3D resistivities (ρ_{og} and ρ_{fg}) are indicated by the inset hypothetical well logs of the horizontal and vertical resistivities (ρ_h and ρ_v). Resistivity changes in the direction of the applied current for anisotropic materials, and J_x, J_y, and J_z are the associated current densities.

The petroleum system model (source, reservoir, trap, and seal) is generally well-known. The presence and effectiveness of hydrocarbon source rocks, migration and charge, reservoir rock, and trap and seal (see Figure 3) are the critical geological attributes that must be satisfied for successful exploration and hydrocarbon discovery [19]. While 3D surveying is now well-established in marine CSEM-MT exploration in frontier regions, how to relate the information to the key elements of geological prospect evaluation (presence of source rocks, migration and charge, reservoir rock, and trap and seal) still remains a non-trivial technical challenge [1].

Surface seeps of native hydrogen are common in volcanoes, geothermal springs, and deep-rooted faults, but the native hydrogen system (source, reservoir, trap, and seal) is less well-known than the petroleum system model. The known favorable subsurface environments for the occurrence of native hydrogen include serpentinized ultramafic complexes in mid-ocean ridges, land-based ophiolite-peridotite intrusives (remnants of oceanic crust) (Figure 4a), and fossil arc-continent collision orogen (Figure 4b) [20–22]. The source rocks include ultramafic igneous rocks and iron-rich craton rocks [23,24]. The serpentinized mafic rocks at the base of the fossil subduction, obduction, and major thrust zones are known to produce fluids rich in native hydrogen [21,25]. Reservoir rocks include permeable, fractured basements and carbonate and marl or clayey rocks. They are often associated with ultrabasic or hyperalkaline waters (pH > 9) [26] and deep faults that tap basement or mantellic sources of hydrogen. The generation rate and preservation are unknown, and there is uncertainty about the size of the resource, but accidental discoveries suggest that subsurface accumulations are viable, such as those found on land in Mali, and the geological association with Precambrian crystalline shields was established [27–29].

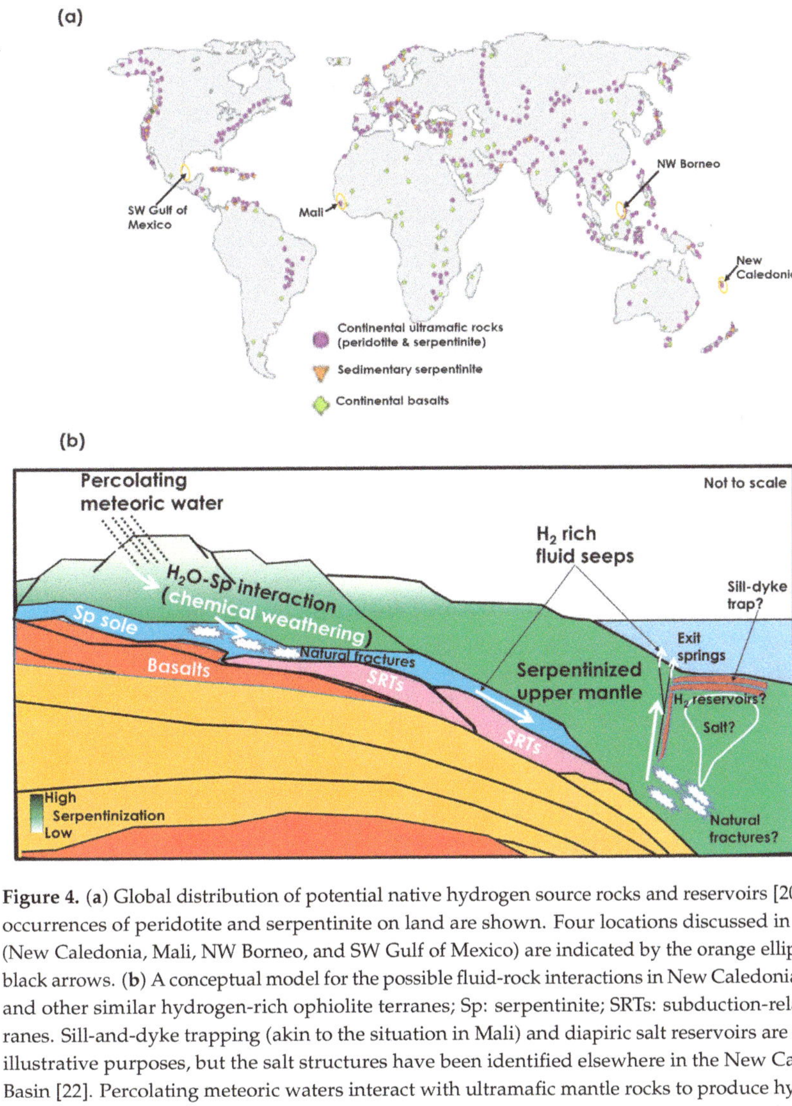

Figure 4. (**a**) Global distribution of potential native hydrogen source rocks and reservoirs [20]. Main occurrences of peridotite and serpentinite on land are shown. Four locations discussed in the text (New Caledonia, Mali, NW Borneo, and SW Gulf of Mexico) are indicated by the orange ellipses and black arrows. (**b**) A conceptual model for the possible fluid-rock interactions in New Caledonia [21,22] and other similar hydrogen-rich ophiolite terranes; Sp: serpentinite; SRTs: subduction-related terranes. Sill-and-dyke trapping (akin to the situation in Mali) and diapiric salt reservoirs are only for illustrative purposes, but the salt structures have been identified elsewhere in the New Caledonia Basin [22]. Percolating meteoric waters interact with ultramafic mantle rocks to produce hydrogen-rich fluids. The natural fractures and faults provide pathways for the transport of the fluids and their eventual vertical migration into suitable reservoirs or exit springs on the surface. It is possible that there will be a meteoric fluid-serpentinite-evaporite hydrogen kitchen in this basin and similar geological settings around the world [22].

So far, it is known that methane and native hydrogen are produced in the competing processes of serpentinization and magmatism in slow- and ultraslow-spreading mid-ocean ridges, continental margins, and forearc settings of the subduction zones [21,30–32]. Serpentinization occurs when ultramafic rocks react chemically with invading seawater (or meteoric water) without biotic mediation, generating hydrogen, which can reduce carbon dioxide to methane. This is akin to the chemical weathering of the seafloor [24], and a similar process or an electrochemical mechanism may be operating on land to explain the discovery of native hydrogen in Mali [27]. This abiogenic chemical reaction leads to volume increases by up to 30% and the destruction of primary magmatic minerals [33] and, hence, new physical properties measurable by geophysical methods. Serpentinized mantle rocks are generally more conductive than mafic crustal rocks, such that electrical and electromagnetic methods can be used for their mapping, depending on their depth of occurrence. The hydrogen released during the serpentinization of ultrabasic rocks is one of the main fuels for chemosynthetic life [24], and the expected seafloor resistivity profile will be as in Figure 3.

The distribution of hydrogen and methane in the known reservoirs at the Bourakebougou field in Mali is shown in Figure 5. Hydrogen is lighter than methane, and its maximum concentration in these wells is found above that of methane. This fluid stratification based on density differences is similar to the arrangement in petroleum reservoirs. The Bourakebougou field consists of multiple layers of dolerite (diabase or micro-gabbro) sills serving as cap rock for multiple reservoirs, and the system is segmented by near-vertical fault-bounded dolerite dykes [27]. This mafic rock is chemically and mineralogically similar to basalt but is coarser-grained and contains silica; it may be no coincidence that basalt and gabbro form important layers of the crystalline crust in the known hydrogen-rich oceanic terranes [23]. The dominant reservoir at Bourakebougou is marble, but marl is an important reservoir (see Bougou-13 well in Figure 5). This reservoir specificity, the chemical weathering, evidenced by the thick lateritic cover in the area [27], and the fact that the total gas-rich column heights are identical for the wells, which are more than 4 km apart (red double-headed arrows in Figure 5), suggest the operation of a specific large-scale fluid-rock interaction process (perhaps involving meteoric alkaline waters and mafic rocks, with a possible contribution from along the steep faults tapping deeper sources). Electrical anisotropy will be expected across the prospective target depth here. Resolving the deep 3D electrical structure in such environments is important for understanding the geologic evolution of the hydrogen system [11].

1.3. Adaptive Play-Based Exploration Workflow for Combined Investigations of Hydrocarbon and Hydrogen Reservoirs

Although surface seeps and subsurface occurrences of native hydrogen are known, there is no framework for the geophysical exploration of reservoir hydrogen in potential hydrogen-rich geological settings. Geologic hydrogen and hydrocarbons may accumulate in the same or different reservoirs, and there will be efficiency gains in searching for both resources together, especially in offshore frontier regions. A template is provided in Figure 6 as a framework for the systematic investigation of hydrocarbon and/or native hydrogen reservoirs and their related structures in favorable geological settings. Here, the multi-scale three-step play-based exploration concept used for traditional petroleum exploration (Figure 1a) is appropriately adapted to the electromagnetic mapping and monitoring of hydrocarbon and native hydrogen reservoirs, with an emphasis on reducing carbon footprint. In this paper, it is stressed that the CSEM and MT methods can effectively contribute to optimizing all the phases of play-based exploration for hydrocarbon and/or native hydrogen reservoirs and their eventual monitoring (Figure 6), facilitated by new developments in data attribute analyses and three-dimensional (3D) anisotropic resistivity inversions of large-size survey data on high-performance computers. Since these methods provide valuable information on rock-and-fluid variations in the subsurface (Figure 2), they can also be used to monitor reservoirs, and a practical algorithm for 4D time-lapse resistivity imaging is proposed. It is also shown here that EM methods can contribute significantly to the understanding of deep geological controls on the distribution of hydrocarbon and native hydrogen reservoirs, leading to improved resource mapping and monitoring, hence a technology that could play a critical role in helping the world reach net-zero emissions by 2050.

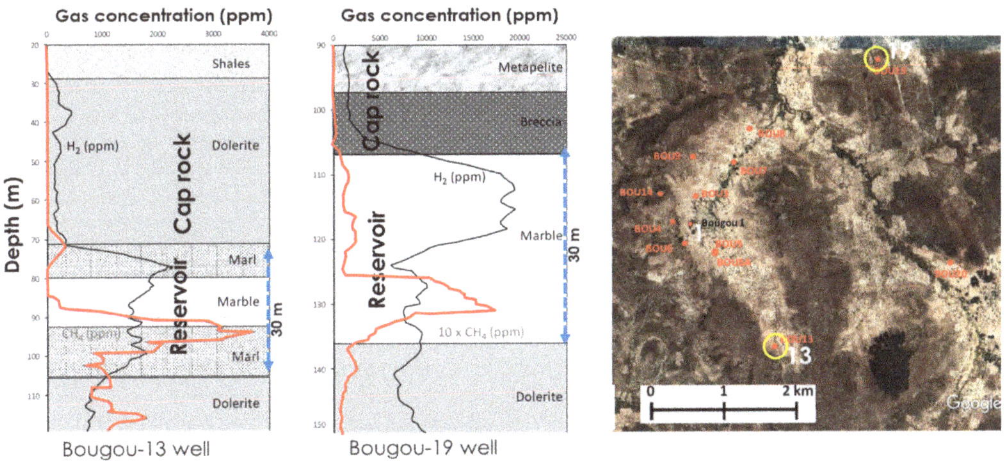

Figure 5. Distribution of native hydrogen (H_2) and methane (CH4) in two logged wells at the Bourakebougou field in Mali, modified from [28]. The variation of gas concentrations with depth in the host rocks is shown for hydrogen (black line) and methane (red line) for well Bougou-13 (**left** panel) and well Bougou-19 (**middle** panel). Although both wells are more than 4 km apart, the total gas-rich column heights (blue double-headed arrows) are identical. The **right** panel is a map showing the locations of the two wells, amongst others including Bougou-1, modified from [28].

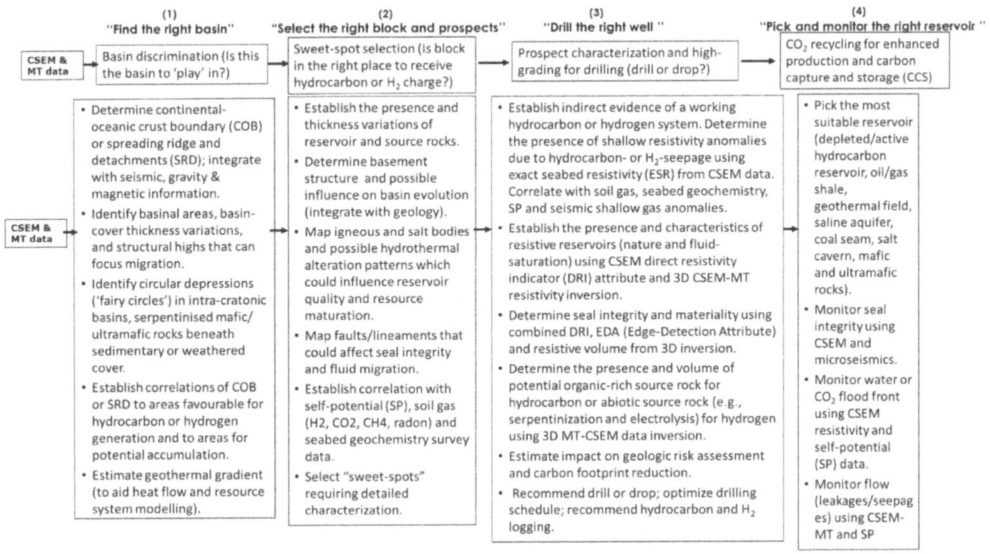

Figure 6. A multi-scale process-oriented framework for combined investigations of hydrocarbon and hydrogen reservoirs using CSEM-MT methods. It extends the conventional three-phase play-based exploration approach (Figure 1a) to allow for reservoir development and monitoring that are beneficial to carbon footprint reductions.

2. Quantitative Methodological Developments

2.1. Geologically Consistent CSEM Multi-Attribute Analysis: Effective Prior Settings for Optimized 3D Inversion

A recent practical development in marine CSEM data processing is the multi-attribute analyses of field data that enable a simple, rapid assessment of the key elements of geological prospect evaluation: the presence of source rocks, migration and charge, reservoir rock, and trap and seal [1] and is particularly relevant for the optimized mapping of stratigraphic traps. A simple geometrical normalization of CSEM amplitude data yields what is termed a direct resistivity indicator (DRI) attribute and can facilitate the retrieval of the exact seafloor resistivity (ESR) and layered background resistivities in ideal conditions [1]. The DRI attribute is also used to obtain edge-detection attributes (EDA) for finding the horizontal position and shape of anomalous 3D bodies [1]. While CSEM is a proxy method for reservoir detection, unlike the more direct seabed geochemistry and seismic methods, combining DRI, ESR, and EDA can help improve objectivity or modify decision-making based on the conventional subjective geological risk evaluation of petroleum ventures [19].

The main benefits for geological venture risk evaluation can be summarized as

- DRI attribute shows the existence of a regional resistive play or carrier bed (characterized by symmetric in-tow and out-tow response profiles) and the desirable localized 3D resistors (characterized by asymmetric in-tow and out-tow response profiles). This meets the geological requirement for identifying the presence of reservoirs (regional plays or localized traps of exploration interest) in frontier regions. This attribute facilitates the rapid screening of CSEM data to select areas warranting follow-up, expensive, and rigorous 3D inversion, leading to significant efficiency gains and computational cost savings [1]. It is best used for the rapid polarization (ranking) of a portfolio of leads and prospects in offshore regions where there are competing targets previously identified using seismic data [1,34,35];
- The effective area or size of a hydrocarbon-charged 3D reservoir is a key requirement in reserve estimation [19]. EDA permits the accurate determination of the exact lateral boundaries of 3D resistive targets and, hence, the effective area necessary for reserve estimation;
- The ESR attribute provides a link to geochemical seabed data, and together these permit a judgment regarding the presence of a working petroleum system (i.e., presence of charge). The ESR profiles sample the near-surface area (top 10–50 m of the seabed based on skin-depth considerations) and should ideally correlate with seabed geochemistry profiles or seismic shallow gas clouds and, thus, provide a basis for comparing or integrating these disparate datasets to confirm the presence of a working petroleum system [1];
- The results from the analyses of these attributes make up the initial model $\mathbf{m}_0$ (i.e., robust model priors) for the subsequent tailored 3D-depth imaging of targeted reservoirs [1].

2.2. Geologically Consistent 3D Anisotropic Resistivity Inversion: Robust Depth Conversion

It is emerging that integrating 3D anisotropic CSEM-MT data imaging into seismic or structural constraints leads to geologically robust models [4,7,15]. Two methods of geologically consistent 3D anisotropic CSEM and MT resistivity inversion (using crossgradient constraints) have recently been developed [7,15,18] and are briefly described below.

In the first method, the MT and CSEM data are inverted for the vertical and horizontal resistivities directly while imposing a crossgradient constraint to ensure the structural similarity between the models [14,15]. The objective function to minimize can be stated as

$$\Phi(\mathbf{m}) = \|\mathbf{W}(\mathbf{d} - \mathbf{f}(\mathbf{m}_{\mathrm{v,h}}))\|^2 + \alpha \int_V \|\mathbf{R}\mathbf{m}_{\mathrm{v,h}}\|^2 dV + \beta \int_V \|\nabla \mathbf{m}_{\mathrm{v}} \times \nabla \mathbf{m}_{\mathrm{h}}\|^2 dV + \int_V \|\mathbf{C}_{\mathrm{v,h}}^{-1}\left(\mathbf{m}_{\mathrm{v,h}} - \mathbf{m}_{\mathrm{v,h(ref)}}\right)\|^2 dV \quad (1)$$

where $\mathbf{d}$ represents the observed MT and/or CSEM amplitude and phase data, $\mathbf{f}(\mathbf{m}_{\mathrm{v,h}})$ are the model responses predicted by 3D anisotropic forward theory, $\mathbf{W}$ is a weighting matrix whose entries are the inverse variance or covariance of $\mathbf{d}$, and $\mathbf{m}_\mathrm{v}$ and $\mathbf{m}_\mathrm{h}$ are the sought vertical and horizontal resistivity models, respectively. $\mathbf{R}$ is the Tikhonov regularization operator, and the pre-multipliers α and β are weights selected by initial trial-and-error testing [17,36]. The third term on the right-hand side is the non-linear crossgradients operator, or structure-promoting regularization [37], where $\nabla\mathbf{m}_\mathrm{v}$ and $\nabla\mathbf{m}_\mathrm{h}$ are the gradients of the vertical and horizontal resistivity models, respectively. The fourth term on the right-hand side contains the statistically independent a priori information $\mathbf{m}_{\mathrm{v,h(ref)}}$, such as from seismic or well log data that are wished to be retained in the solution, and $\mathbf{C}_{\mathrm{v},h}^{-1}$ contains the variances of the sought anisotropic model parameters. More details of this inversion approach are given elsewhere [15].

In the second method, the CSEM and MT data are jointly inverted using the nonlinear crossgradients operator to relate the horizontal and vertical resistivities as before, but they additionally imprint structural fabric from 3D seismic reflection images onto the resulting models using a linear crossgradients operator [7]. This is equivalent to the use of statistically independent a priori information to constrain the resistivity inversion process. The resistivity models are guided by structure tensors computed from good quality 3D pre-stack depth migrated (PSDM) seismic data instead of using interpreted seismic horizons that might be subject to the interpreter's error. The composite objective function to minimize can be stated as

$$\Phi(\mathbf{m}) = \|W(\mathbf{d} - \mathbf{f}(\mathbf{m}_{\mathrm{v,h}}))\|^2 + \alpha\int_V \|\mathbf{R}\mathbf{m}_{\mathrm{v,h}}\|^2 dV + \beta\int_V \|\nabla\mathbf{m}_\mathrm{v} \times \nabla\mathbf{m}_\mathrm{h}\|^2 dV + \tau\int_V \|\nabla\mathbf{m}_{\mathrm{v,h}} \times \nabla\mathbf{m}_{\mathrm{SEIS}}\|^2 dV \quad (2)$$

where the gradient field derived from seismic structure tensors $\nabla\mathbf{m}_{\mathrm{SEIS}}$ is fed into the last term on the right-hand side and represents a linear operation since these seismic gradients are fixed as statistically independent from the a priori information. The other terms have the same meaning as in Equation (1). The pre-multipliers α, β, and τ are also selected by initial trial-and-error testing. More details of this 3D seismic image-guided inversion method can be found in [7]. Equations (1) and (2) are solved in the adopted inversion program RLM3D using a nonlinear conjugate gradients algorithm [38] that is well-suited for this type of large-scale optimization problem.

2.3. *Electrostratigraphic Imaging: Accurate Post-Inversion Reservoir Boundary Mapping*

It is also emerging that the gradient information in 3D resistivity images can be used to relate the structure of the objects in the $\mathbf{m}_\mathrm{v}$ and $\mathbf{m}_\mathrm{h}$ models that are derived from the above anisotropic inversion methods or simply for identifying features of interest (e.g., edge detection) for direct boundary recognition [11]. For simplicity, one can exploit the magnitude of the directional derivative of either $\mathbf{m}_\mathrm{h}$ or $\mathbf{m}_\mathrm{v}$ implemented by differing between the neighboring pixels in the 3D model. The sought maximum change in pixel values for boundary detection in smoothness-constrained 3D resistivity models is defined as [11]

$$\left|\overrightarrow{\Delta\rho}\right| = \sqrt{I_x^2 + I_y^2} \quad (3)$$

where the partial derivatives $I_\mathrm{x} = \frac{\partial\rho_x}{\partial z}$ and $I_\mathrm{y} = \frac{\partial\rho_y}{\partial z}$ are the pixel-based estimates of the components of the gradient vector computed using the forward or central differences between neighboring pixels in the given 3D subsurface resistivity model. The output of Equation (3) is termed the "total gradient" in order to distinguish it from the first vertical derivative ("1VD") at a point, which was also shown to be equally useful for boundary detection [11].

2.4. *Structurally Consistent 4D Time-Lapse Electromagnetic Imaging: Robust Fluid Tracking*

Equations (1) and (2) can be adapted to any multi-set electromagnetic data and, hence, can be used to estimate the changes $\Delta\mathbf{m}_1$ due to dynamic processes in the subsurface monitored using the baseline data ($\mathbf{d}_0$) and monitor data ($\mathbf{d}_1$) acquired at different times: t_0 and t_1 [39]. Here, the reference anisotropic resistivity models are the models ($\mathbf{m}_{\mathrm{base\ (v,h)}}$)

from the crossgradient inversions of the baseline 3D survey data ($\mathbf{d}_0$). In the case analogous to Equation (1), for any given base and monitor surveys, the basic problem statement is posed as minimize:

$$\Phi(\mathbf{m}) = \|W(\mathbf{d}_1 - \mathbf{f}(\mathbf{m}_{v,h}))\|^2 + \alpha \int_V \|\mathbf{R}\mathbf{m}_{v,h}\|^2 dV + \beta \int_V \|\nabla\mathbf{m}_v \times \nabla\mathbf{m}_h\|^2 dV + \int_V \|\mathbf{C}_{v,h}^{-1}\left(\mathbf{m}_{v,h} - \mathbf{m}_{base(v,h)}\right)\|^2 dV \quad (4)$$

where the various parameters are as previously defined. Note that the last term on the right-hand side of Equation (4) can be replaced by the measure $\tau \int_V \|\nabla\mathbf{m}_{v,h} \times \nabla\mathbf{m}_{base(v,h)}\|^2 dV$ as the equivalent implementation of image-guided inversion (see Equation (2)). Equation (4) is solved using a nonlinear conjugate gradients algorithm [38] for this large-scale optimization problem.

In the geologically complex areas where the search for new energy-fluid reservoirs is taking place, it is critical to integrate various geophysical, geological, and environmental modeling tools in three dimensions so as to increase accuracy and, hence, reduce uncertainty in the conventional 3D subsurface predictions [2,40]. The integrated tools can also be used to monitor reservoirs during CO_2 recycling and/or sequestration (Figure 1b), which are important for meeting the global net-zero emissions target. Improving the methodologies for geologically consistent multi-physics time-lapse imaging and uncertainty quantification [2,39,41,42] should, therefore, be seen as the way forward. For the n-type (multi-modality) geophysical measurements, we can apply crossgradient joint inversion to assure structural consistency between the respective model parameter changes $\Delta\mathbf{m}$ derived from the multi-modality baseline and monitor datasets. For notational simplicity, let us denote our anisotropic models for the multi-modality geophysical systems at the base survey time t_0 as $\widehat{m}_b$ and those at the monitor survey time t_1 as $\widehat{m}$. The joint multi-physics inversion problem is then posed as minimize [2,39]:

$$\Phi\left(\widehat{m}^{(1)}, \ldots, \widehat{m}^{(n)}\right) = \sum_{j=1}^{n}\left\{\|\mathbf{d}_1^{(j)} - \mathbf{f}^{(j)}\left(\widehat{m}^{(j)}\right)\|^2_{\mathbf{C}^{-1}_{d_j d_j}{}^{(j)}} + \alpha^{(j)}\|\mathbf{R}\widehat{m}^{(j)}\|^2 + \tau\|\widehat{m}^{(j)} - \widehat{m}_b^{(j)}\|^2_{\mathbf{C}^{-1}_{m_b m_b}{}^{(j)}}\right\} \quad (5)$$

subject to the condition

$$\nabla\widehat{m}^{(g(x,y,z,t1))} \times \nabla\widehat{m}^{(j)} = 0 \quad \forall j \neq g(x,y,z,t1),\ j \leq n \quad (6)$$

where $\widehat{\mathbf{m}} = \widehat{\mathbf{m}}_b + \Delta m_j$, and the desired model changes are $\Delta m_j = \widehat{m} - \widehat{m}_b$. Equation (6) holds for all the possible crossgradient combinations equal to zero, as long as the selected g-pivot image is set to $\nabla\widehat{\mathbf{m}}^{(g(x,y,z,t1))} \neq 0$. This constraint will impose a resemblance between the multiple geophysical images as long as the g-pivot image has no null gradient [2]. The geophysical method with the highest subsurface resolution capability may be selected as the g-pivot set (e.g., seismic reflection, if this is available for this quantitative integration), but the pivot image can be varied for each position in the 3D model if the use of the morphology of a single-image property is not intended. Equations (5) and (6) can be solved using a nonlinear conjugate gradients algorithm [38]. Overall, the benefits of such a combined approach can be summarized as

1. Risk mitigation in frontier ventures with no wells or good quality seismic data;
2. Reduces the need to drill unnecessary wells, leading to significant cost savings;
3. Potential to reduce prospect evaluation/maturation period;
4. Maximize accuracy and reduce uncertainty, especially with regard to reservoir properties;
5. Applying such a quantitatively integrated methodology can contribute significantly to a better understanding of the geological controls of the transport and distribution of fluids in clean-energy (especially geothermal and hydrogen) reservoirs, leading to improved resource mapping and monitoring, and hence, a technology that could play a critical role in helping the world meet the net-zero emissions target.

3. Instructive, Process-Oriented Case Studies

The recent results of 3D inversions of CSEM and MT survey data from large-scale industrial surveys in deep waters in NW Borneo [43] and the southwestern Gulf of Mexico [11] are revisited and interpreted here in the context of the proposed extended play-based framework for finding and monitoring hydrocarbon and hydrogen reservoirs (Figure 6), and the new lessons learned are discussed.

3.1. Play-Based Exploration in Deep Water in NW Borneo

3.1.1. Find the Right Basin and Play

A good place to select a new exploration block will be in a basin with proven plays, or that is near an existing field. The area of study is located in the fold-and-thrust belt of offshore northwest Borneo in water depths ranging from about 500 to 1600 m in the South China Sea (Figure 7a). It is known to be structurally complex, and the sedimentary rocks are mostly stacks of thin, bedded sand in shale, characterized by low resistivity and low contrasts [15]. The Miocene-Pliocene stratigraphy and the main turbidite fan intervals in this region [44] are shown in Figure 7b. At least seven turbidite fans have been identified as the reservoir targets. These were deposited in basin floor systems [44]. The structural style here is characterized by northeast-southwest trending ridges that are formed by elongated thrust anticlines, with intervening mini basins and toe thrust zones [15,17,44]. The hanging wall anticline with four-way dip closure is the main sought-after trap style in this area, while the stratigraphic plays located in the syncline remain under-explored.

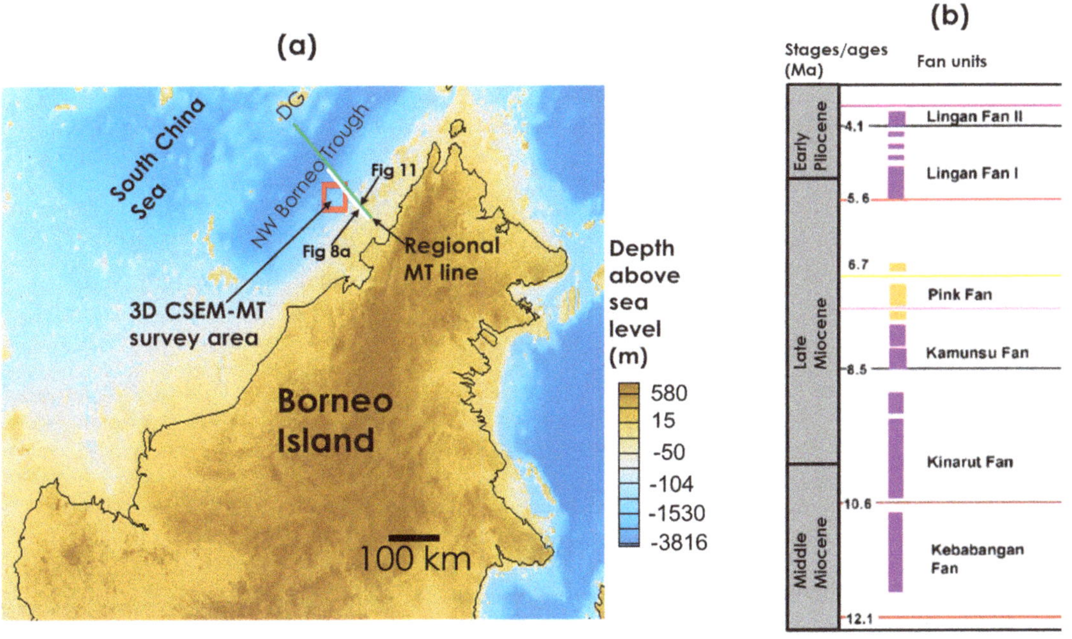

Figure 7. (**a**) Map showing the location of the 3D CSEM-MT survey area in offshore NW Borneo. The thick white line is a 120 km-long regional MT survey line, for which the result is summarized in Figure 8a. The thick green line is the location of the model shown later. DG: Dangerous Grounds. (**b**) Mio-Pliocene stratigraphy and the main turbidite fan intervals in the study area [44].

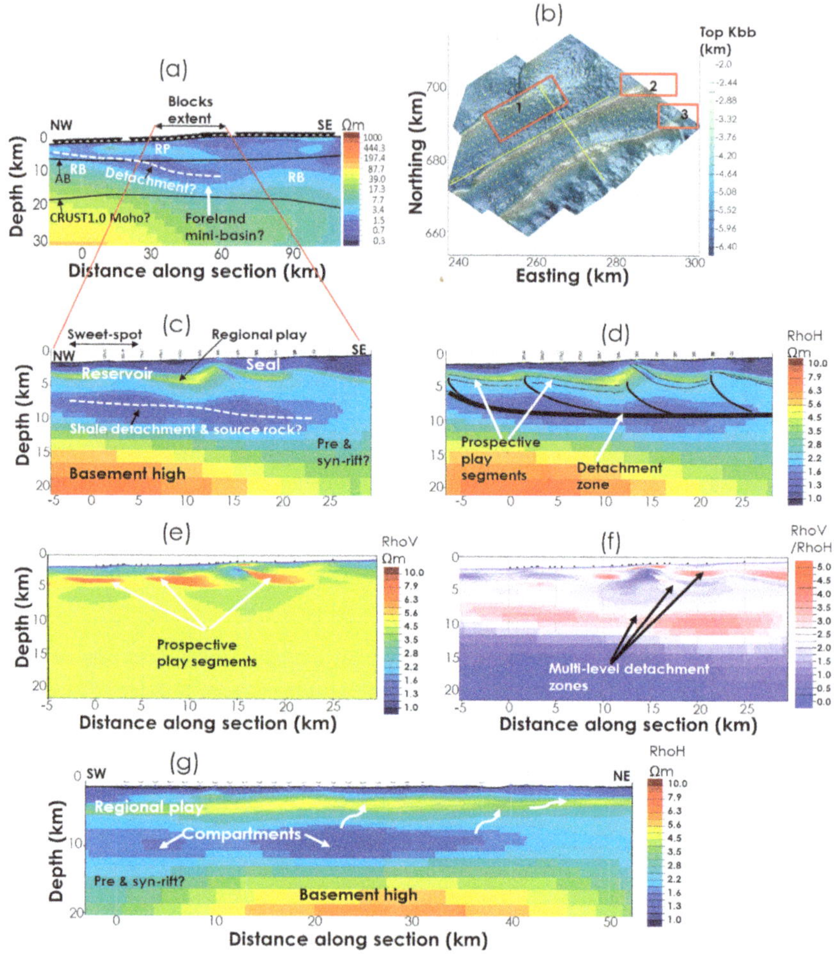

Figure 8. Concept of 'finding the right play' and 'selecting the right blocks and prospects' is demonstrated using anisotropic resistivity models from 3D MT and CSEM inversion [43]. (**a**) Horizontal resistivity cross-section extracted along the white line in Figure 7a from a 3D model generated by an anisotropic inversion of regional MT data. The actual MT field data and the model verification using resistivity logs from several exploration wells are presented in [7,17]. RP is a resistive play (clastics). The independently interpreted acoustic basement (AB) and the crust-mantle interface (Moho) from the CRUST1.0 seismic compilation [45] are shown only for comparison. (**b**) Seismic depth map of the top of the Kebabangan reservoir, showing the structural pattern. Yellow dots are CSEM-MT stations [17]. Yellow lines are the dip line and strike line, along which were extracted the sections from joint CSEM-MT inversion, as shown in (**c**–**g**). The red boxes (1, 2, and 3) are potential sweet spots selected for detailed investigation. (**c**) Dip-section extracted from the 3D horizontal resistivity volume resulting from joint CSEM-MT inversion. (**d**) Our structural interpretation of the image shown in (**c**), together with independently interpreted seismic horizons (thin black lines) clearly demonstrates the geological consistency between CSEM-MT and the seismic results. The thick black lines show the interpreted subhorizontal detachment zone and steep-thrust faults demarcating sedimentary wedges or segments. (**e**) Vertical resistivity section and (**f**) anisotropy section extracted along the dip line from the anisotropic joint CSEM-MT inversion models. (**g**) Strike section extracted from the 3D horizontal resistivity volume, resulting from joint CSEM-MT inversion.

The results from the recent use of regional MT inversion for basin-scale play mapping in this petroliferous deep-water fold-and-thrust belt are shown in Figure 8a. The objectives were to understand the basement structure and the distribution of potential hydrocarbon plays in the basin [43]. The full tensor data for the MT line were inverted in 3D, taking into account seafloor topography and electrical anisotropy. The inversion was unsupervised (i.e., initiated with half-space featureless starting anisotropic models) to also test the MT data resolution capability. The interpreted acoustic basement and the crust-mantle interface (Moho) from the CRUST1.0 seismic compilation [45] are plotted on the MT image of Figure 8a for comparison. The MT imaging detects a resistive basement (RB in Figure 8a), for which the top varies in depth from 10 to 15 km, and defines a mini foreland basin beneath the exploration blocks of interest. The basement cover rocks contain a resistive play (RP in Figure 8a) that is overlain by a conductive layer (possible cap rocks) and underlain by a thick, conductive layer (possible source rocks and detachment zone). Note the interpreted subhorizontal detachment zone into which root the steep-thrust faults demarcating sedimentary wedges or segments. Notice the apparent discrepancies between the regional MT result and the predicted acoustic basement and Moho in Figure 8a, especially the foreland mini-basin (flexural moat?) and major subhorizontal detachment suggested by MT imaging in the region, where seismic data are of poor quality.

3.1.2. Select the Right Block and Prospects

An example result of large-scale 3D CSEM-MT surveying for multi-block mapping and evaluation in this basin is presented in Figure 8b–e. An important objective was to better understand the effect of basement control on the surface deformation and hydrocarbon distribution in the blocks. Two vintage MT-CSEM surveys were acquired for the client and were processed by EMGS in 2015 and 2016 over the area of the study [7,17]. The combined number of receivers from both surveys was 647, with a receiver spacing of 1.5 to 2 km in the 2015 prospect-specific localized acquisitions and 3 km for the large-scale regional acquisition in 2016 [7,17]. The 2016 3D EM survey has the largest 3D grid coverage of EM data ever recorded in a single survey in Malaysia and was used to generate the results presented here (Figure 8b). There are four wells (A, B, C, and D in Figure 9a) in the area, and these provided resistivity logging data for verification of the CSEM-MT result. The electric field CSEM data for five frequencies (0.125, 0.375, 0.625, 2, and 2.125 Hz) at source-receiver offsets ranging from 1000 to 10,000 m, and the full bandwidth of the recorded MT frequencies (0.000667 to 1.443 Hz) were utilized in the 3D inversion aimed at rapid regional prospectivity scanning. The model grid size was $312 \times 199 \times 145$ cells in x, y, and z directions (excluding air layers) with cell sizes of $250 \times 250 \times 25$ m with incorporated bathymetry. The cell sizes in the z direction increased by some factors as the deepest bathymetry was reached. The seawater resistivity values were derived from a sound velocity profile measured by EMGS during data acquisition and were kept fixed during the inversion. This dataset was used in testing the seismic image-guided joint CSEM-MT inversion [7], but the focus was not on petroleum system evaluation or understanding the influence of the deep basement on the gravity-driven deformation of the play, as attempted here. A structurally guided 3D inversion of all the MT data is reported elsewhere, but the focus here was on determining the best structural control to use via a comparison with the well log data [17]. These previous studies show the quality of the data and how well they are matched by the 3D model responses, and it will serve no further purpose to reproduce them here. What is new here is the play-based geological analyses of the joint inversion results in terms of petroleum and hydrogen systems and net-zero emission considerations. The results presented here are important for optimized resource system evaluations because

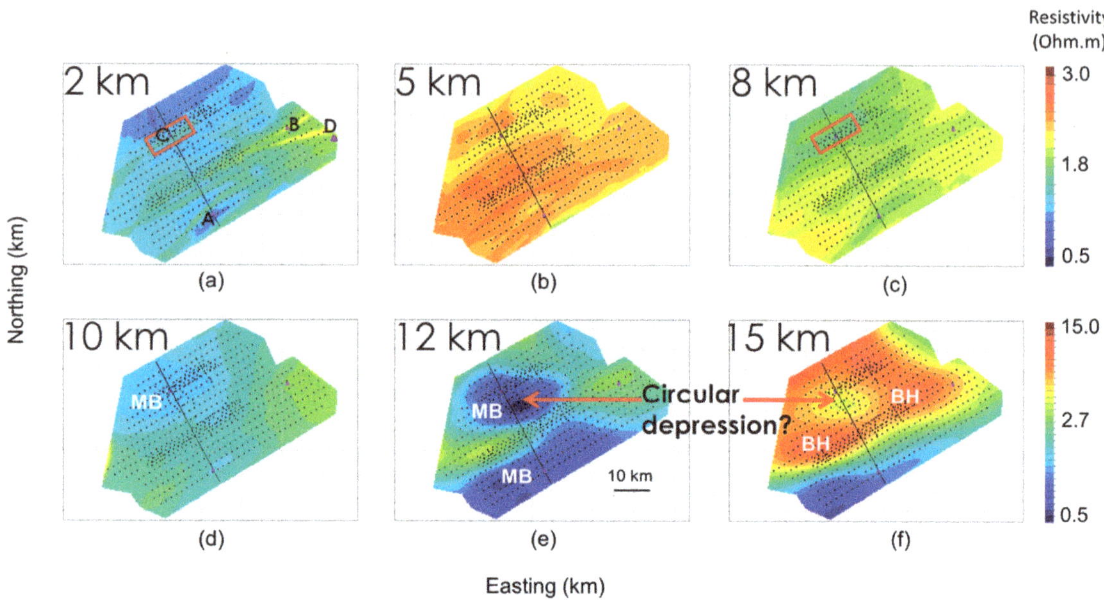

Figure 9. Concept of selecting the right prospect to drill in the right mini basin or structural compartment. Shown are the horizontal slices at six different depths (2, 5, 8, 10, 12, and 15 km) in the 3D resistivity model derived from the seismic structure-constrained inversion of MT data [17] to show basin evolution in space and time and facilitate the selection of the best sweet spot for drilling. MB: mini basin; BH: basement high. A, B, C and D are the locations of drilled hydrocarbon exploration wells. The black dots are marine MT stations. The red rectangle was picked as the sweet spot with the best potential for finding a hydrocarbon-charged reservoir.

- The main plays in the blocks were clearly mapped based on their resistivity and anisotropy characteristics, as illustrated in Figure 8c,e–g. The 1–3 km thick conductive overburden (potential cap rock), resistive units (potential reservoirs in the turbidite fans), thick conductive underburden (possible source rock and shale detachment zone, as shown in Figure 8f) forming the post-rift play and rifted resistive basement with possible syn-rift play are evident;
- Potential migration pathways could be predicted along the steep faults and up-dip along dipping carrier beds (Figure 8g), but there are significant multi-level detachment zones (Figure 8f) that could prevent vertical migration and, hence, form effective seals;
- Structural compartmentalization is present in both the dip and strike directions (Figure 8c,e,g), which suggests the presence of different thrust wedges that are possibly separated by transfer faults, allowing relative block motions, with each compartment associated with a different prospectivity;
- The relatively rugged seabed topography suggests a recent deformational event and is structurally related to the geometry of the detachment zones in the upper 5 km in the resistivity anisotropy model (Figure 8f). The consistency between seafloor deformation and subsurface resistivity anisotropy suggests the active deformation of the sedimentary pile above the basement [11];
- The structure with the highest seabed expression is underlain by a steep southerly dipping basement edge, forming a buttress against which the transported sediments appear to have deformed considerably; this suggests a significant risk to any hydrocarbon accumulation; hence, the preferred sweet spot is the less disturbed fold structure properly centered on the basement high, which will focus hydrocarbon migration;

- The resistivity model is geologically consistent with seismic structure (Figure 8d), assuring confidence in the above interpretations and allowing the selection of the prospective play segments, dubbed "sweet spots" (Figure 8c,e), warranting further detailed integrated geological and geophysical analysis.

3.1.3. Drill the Right Well

Detailed examination of the time-space variation in resistivity provides further insights on basin evolution to aid the selection of the right block segment or 'sweet spot' to drill, as demonstrated in Figure 9. The key messages from this figure are the following:

- The horizontal or seismic horizon-controlled resistivity slices, extracted at different depths from the 3D horizontal resistivity model, revealed how the basement structure influenced basin evolution (the vertical resistivity model could also be used for this);
- The presence of a circular depression beneath the sedimentary cover. A localized circular mini basin can be seen in the 15, 12, and 10 km depth slices (possibly a mini foreland basin due to flexural response to loading and/or igneous intrusion). This is overlain by a clastic play comprised of resistive (sandy or carbonate-bearing reservoir rocks) and conductive (shaley source and/or cap rocks) materials, as revealed by the horizontal resistivity slices at 8, 5, and 2 km depths. Pockmarks were observed on the seabed in this locality. Well C (Figure 9a) was drilled in 2016 and encountered hydrocarbons at the predicted reservoir. Unfortunately, the measurement of hydrogen gas was not considered at the time;
- Could this circular basin be a hydrogen prospect? The observed pockmarks suggest the presence of chemosynthetic communities, analogous to what is observed across 'fairy circles' on land, characterized by seepages of native hydrogen. It will be interesting to survey this area for native reservoir hydrogen in a future effort to help reduce the carbon footprint. A seafloor 3D self-potential survey and geochemical soil gas sampling for H_2, CO_2, CH_4, and radon, followed by the appropriate gas sampling in the wells, will be the recommended way forward here.

3.1.4. Pick and Monitor the Right Reservoir

An important consideration in any future investigations in this area will be the possibility of re-using the hydrocarbon reservoirs or saline aquifers for carbon sequestration. Figure 10 shows the result of the crossgradient 3D CSEM inversion for structurally consistent vertical and horizontal resistivity distributions for one of the three selected sweet spots (sweet-spot 3 in Figure 8b) [4]. The key messages from this figure are the following:

- Two wells sampled hydrocarbon at a depth correctly predicted by the 3D CSEM inversion (Wells D and E in Figure 10). So, this technology is useful for detecting and mapping potential hydrocarbon-charged reservoirs in this geologic setting;
- Note the strong presence of shallow resistive gas (evidence of a working petroleum system). A future 3D SP survey will be useful for assessing the gas flow pattern here;
- There is no major resistive cover at the crest of the anticlinal structure below this resistive shallow gas zone. This suggests that it is likely a blown trap allowing hydrocarbons to migrate vertically upward to form the observed shallow gas body. Parts of this reservoir may, therefore, be sub-optimal for future CO_2 storage.

3.2. Understanding Deep Geologic Controls on the Genesis and Distribution of Hydrogen in the Investigated Area

It is widely believed that there is a fossil subduction system in the region of the MT surveys in offshore Borneo (Figure 7a). It was proposed that during the Paleogene, the Proto-South China Sea was subducted beneath northern Borneo and that subduction ended with the collision of the Dangerous Grounds block (Figure 7a) and the Sabah–Cagayan Arc in the Early Miocene [46]. Figure 11 shows the result of the 3D MT inversion of a regional survey line crossing the NW Borneo Trough (see the green line of the section in Figure 7a). Notice the conductive zone above the resistive mantle in the northern half of the profile.

This is possibly serpentinized mantle. There are also possible deep-rooted faults extending from below the seismically inferred Moho. These could allow for the tapping of mantellic sources of hydrogen. At the top basement level, there is a major detachment zone. These are the features expected in a favorable terrane for the genesis and migration of natural hydrogen see [21–25]. So, the same electromagnetic dataset acquired for hydrocarbon exploration can be used for an extended search for potential natural hydrogen targets. The implications from Figure 11 are the following:

- MT imaging can detect anomalous, electrically conductive basement rocks. There are two interesting conductive bands (see the dotted white lines 1 and 2 in Figure 11) in the western half and a shallow conductive detachment in the eastern half of the transect. The deeper conductive band may be thrust-related and associated with serpentinized mantle rocks, which are important sources of native hydrogen. Note the possible structural similarity with the serpentinite sole in Figure 4b;
- MT imaging can robustly map deep-rooted steep faults (thick solid white lines in Figure 11) that may be tapping deep mantellic sources and act as a migration pathway to potential reservoirs at higher levels;
- There appear to be suitable cap rocks for hydrogen accumulation, such as electrically resistive igneous rocks (sills?) and conductive claystone or detachment zones, as suggested in Figure 11. Clays and igneous rocks are known to form cap rocks for significant hydrogen reservoirs elsewhere (e.g., Figure 5);
- The zones of relatively high anisotropy in the crust and mantle shown in Figure 11c may be multi-level detachment zones, suggesting active deformation involving a combination of tectonic- and gravity-driven processes.

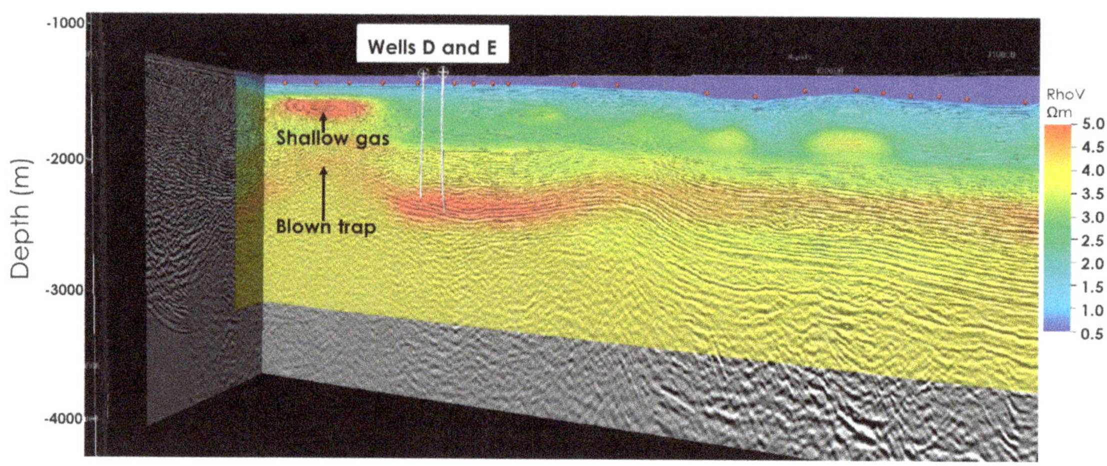

Figure 10. Step 3: Drilling the right well, as demonstrated for sweet-spot 3 [4]. Shown here is the result of prospect-scale 3D CSEM resistivity inversion integrated with the seismic PSDM data for improved decision-making [4]. Note the blown trap beneath the shallow gas accumulation. Parts of this reservoir may, therefore, not be selected for future CO_2 storage.

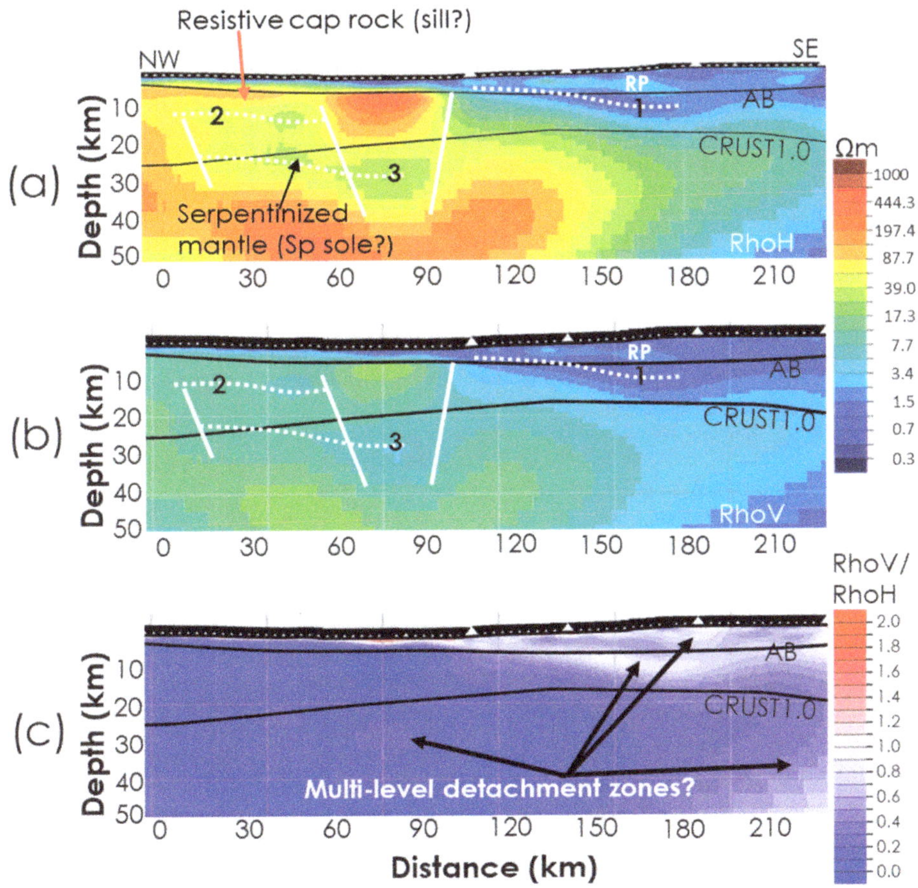

Figure 11. MT resistivity and anisotropy sections across a proposed fossil subduction zone in offshore NW Borneo. The transect is shown in Figure 7a. (**a**) Horizontal resistivity. (**b**) Vertical resistivity. (**c**) Resistivity anisotropy. The thick white lines in Figure 8a,b are potential deep-rooted faults that can aid fluid migration; RP is the same regional clastic play shown in Figure 8, and the dashed white lines, 1, 2, and 3, are conductors and are possibly thrust-related since they have relatively high anisotropy (Figure 8c). The black interfaces are the interpreted regional acoustic basement from pre-stack depth-migrated seismic data (AB) and the CRUST1.0 Moho [45]. Sp: serpentinite.

4. Discussion

4.1. Extended Play-Based Workflow for Combined Hydrocarbon and Hydrogen Investigations

An important thrust of this paper is the demonstration that a given large-size marine electromagnetic survey dataset can efficiently contribute to the understanding of deep geological controls on the genesis and distribution of both hydrocarbon and native hydrogen targets, leading to improved efficiencies in resource mapping and monitoring, and represents a technology that could play a critical role in meeting the net-zero emissions target. A framework is provided in this paper for the systematic combined investigation of hydrocarbon and hydrogen reservoirs and how their genesis and distribution are influenced by deep subsurface processes in favorable geological settings (Figure 6). The well-known concept of a 'petroleum kitchen' and the emerging hypotheses on the 'hydrogen kitchen' are integrated with a play-based exploration concept (Figure 1a) to yield a practical framework for the electromagnetic mapping and monitoring of hydrocarbon and native hydrogen reservoirs with implications for carbon footprint reduction (Figure 6). It is stressed that electromagnetic methods can effectively contribute to optimizing all the phases of play-based exploration for these reservoirs and their eventual monitoring facilitated by new developments in CSEM multi-attributes analyses and the 3D anisotropic resistivity inversion of large-size CSEM-MT survey data on high-performance computers. The case studies presented in this paper specifically show how marine CSEM and MT data imaging facilitates decision-making in the industry from the basin scale through to the prospect scale and the reservoir scale. The past works in this basin [7,17,43] focused on deriving resistivity structures that were structurally consistent in a hydrocarbon exploration context and considered neither the full geological implications for resource mapping and monitoring nor the possibility for natural hydrogen exploration and carbon footprint reduction. Unlike the common practice, this paper also demonstrates the interpretation of 3D resistivity information in terms of the key elements of geological prospect evaluation (presence of source rocks, migration and charge, reservoir rock, and trap and seal) and the use of electromagnetic models to understand the influence of geological processes on the genesis and distribution of potential hydrocarbon and native hydrogen reservoirs. This is a somewhat radical CSEM and MT approach but has been tested in several geological terrains using expensive industry-type data and is particularly relevant as the deepening climatic emergency now drives a global transition to low-carbon energy sources.

4.2. Wider Implications for Mapping Low-Carbon Reservoirs, Source Rocks, and Migration Paths

Understanding the deep crustal structure of geologically favorable environments for the genesis and accumulation of natural hydrogen should be considered a necessity in any practical exploration campaign since hydrogen is considered to be a key commodity in the low-carbon energy transition. Abundant geologic evidence shows that mafic and ultramafic rocks at the base of fossil subduction-obduction and major thrust zones, serpentinized ultramafic complexes in mid-ocean ridges, land-based remnants of oceanic crust (Figure 4a), and fossil arc-continent collision orogen (Figure 4b) produce geological fluids rich in native hydrogen [21,24,25]. It is known that surface seeps of native hydrogen are common in volcanoes and geothermal springs. It is also accepted that the source rocks for native hydrogen include ultramafic igneous rocks and iron-rich craton rocks [23,24]. However, the lithospheric structure across such terrains is often not well understood due to a lack of 3D pre-stack depth-migrated (PSDM) seismic data that have been the mainstay of the oil and gas industry [11]. It has been demonstrated (in Figure 11) that the deep marine MT imaging of fossil subduction or thrust zones is feasible using legacy MT data acquired for hydrocarbon exploration. Evidence of deep-rooted (extensional?) faults, possible serpentinized mantle, and resistive caps below the sedimentary cover are apparent in Figure 11. It will be instructive to examine whether the same can be achieved for fossil mid-ocean ridges, where serpentinization and magmatism are competing processes for the generation of hydrogen-rich fluids.

The results from a recent 3D marine MT investigation on the Mexican Ridges fold belt in the southwest of the Gulf of Mexico are re-analyzed here in Figures 12–14. The area of study lies at the apex of the extinct Jurassic seafloor spreading center in the Gulf of Mexico [47,48] (Figure 12a). A banded crystalline basement structure was found across the fossil-spreading center comprising WSW-ENE trending, 6–10 km wide, electrically resistive subvertical sheets with conductive and anisotropic borders (Figure 12b), which merge into a basal resistive stock-like body at a depth of 15–20 km. NNW trending major faults were also found to cut the steep banded system. The resulting MT-evinced WSW and NNW structural trends, if rotated clockwise by 25–30 degrees, correlate with the previously interpreted transform and normal faults that formed during the Late Jurassic opening of the Gulf of Mexico [11]. The conductive bands spatially coincide with possibly fluid-filled vertical fracture sets in the overlying sediments, as seen in the seismic data (Figures 13 and 14), and were, therefore, interpreted as hydrothermal fluid pathways [11]. In terms of trapping mechanisms, it is interesting that an electrically conductive subhorizontal detachment zone in Eocene shale deposits at a depth of about 5–6 km was found to act as an effective barrier to vertical fluid migration (Figures 13 and 14). There are also microseismic activities in the area, and Meju et al. [11] inferred that a magmatic body recently intruded the area, its ascent controlled by pre-existing basement structures, and this influenced the deformation of the basement cover rocks. This is effectively an offshore geothermal prospect, but the message here is that providing suitable source rocks exists, the migration pathways and seal for potential hydrogen-rich or hydrothermal fluids can be mapped by combining MT and seismic imaging. We suggest that geothermal environments on land could host significant accumulations of native hydrogen, and conventional datasets should be re-appraised for hydrogen upside potential. Future efforts should be directed at incorporating micro-earthquake data in the 3D marine MT inversion to better constrain the likely reservoir targets as performed for land investigations of geothermal energy resources [49], which are widely acknowledged as critical components of the low-carbon energy transition challenge.

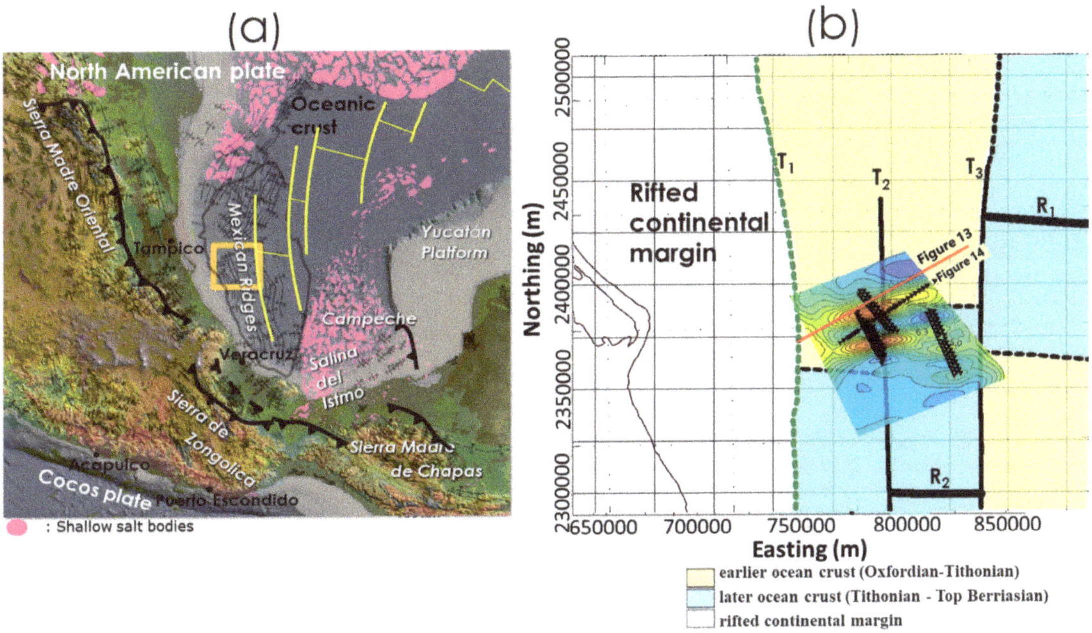

Figure 12. Examining the deep resistivity structure of a fossil mid-ocean ridge and a possible geothermal system [11]. (**a**) Map showing the Mexican Ridges fold belt and the location of the 3D MT study area in the southwest of the Gulf of Mexico. The orange box is the MT study area. The yellow lines represent ridge-transform segments of the extinct Jurassic spreading center [47]. (**b**) Map showing the proposed ridge-transform system in the MT study area, after [48]. T_1, T_2, and T_3 are the transform faults. R_1 and R_2 are the spreading ridge segments; earlier ocean crust (Oxfordian–Tithonian) and later ocean crust (Tithonian: Top Berriasian) are the proposed crustal types [48]. The embedded image is a resistivity depth slice at the top basement level from the 3D MT horizontal resistivity model [11]. The locations of the SW-NE sections in Figure 13 (red line) and Figure 14 (diamond symbols) are shown.

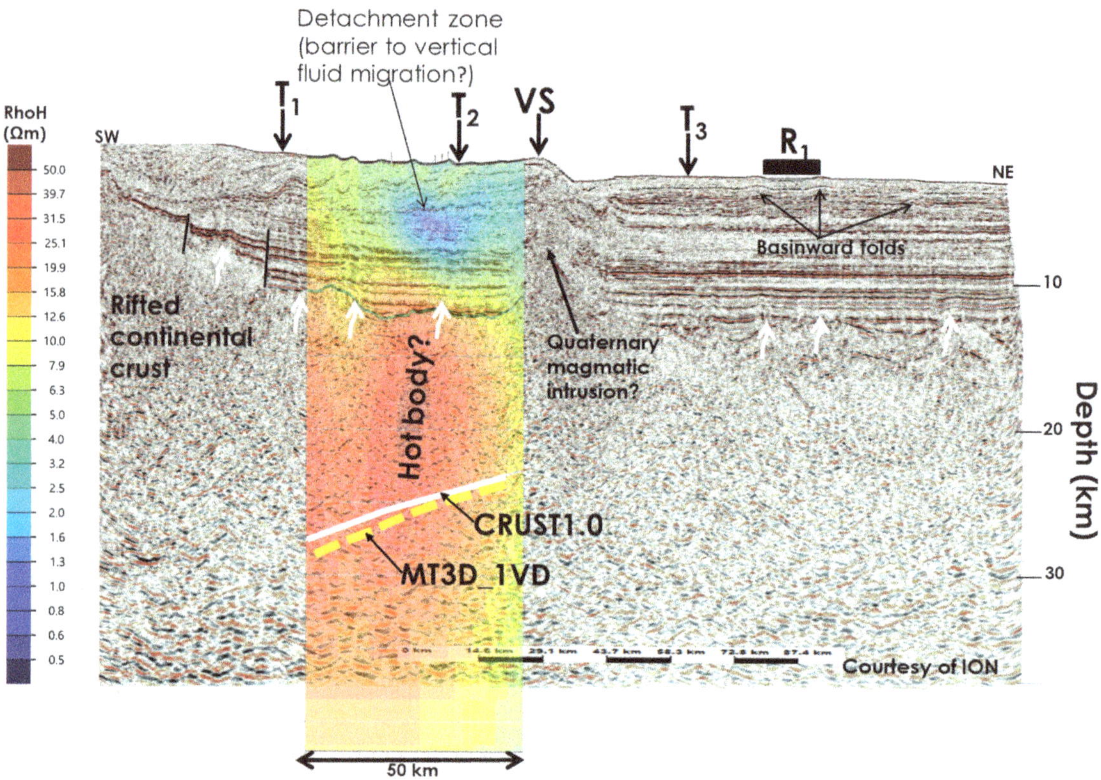

Figure 13. Integrating seismic and MT information to identify vertical fluid migration paths and barriers [11]. Ion BasinSPAN 2D seismic PSDM data for a line crossing the survey area (green line in Figure 12b), draped by MT horizontal resistivity extracted along this line from the 3D crossgradient anisotropic MT inversion model [11]. The thick dashed yellow line is the interpreted crust-mantle interface from post-inversion electrostratigraphic imaging using the first vertical derivative (1VD) of the 3D resistivity model (see Section 2.3). The thick white line is the CRUST1.0 Moho [45]. The thick white wiggly arrows show possible seismic evidence of upward fluid migration along steep faults extending from the basement. The locations where this Ion 2D seismic line intersected features T_1, T_2, T_3, VS (a possible volcanic seamount), and R1 (Figure 12b) are indicated at the top. MT stations are indicated by grey bars on the seafloor.

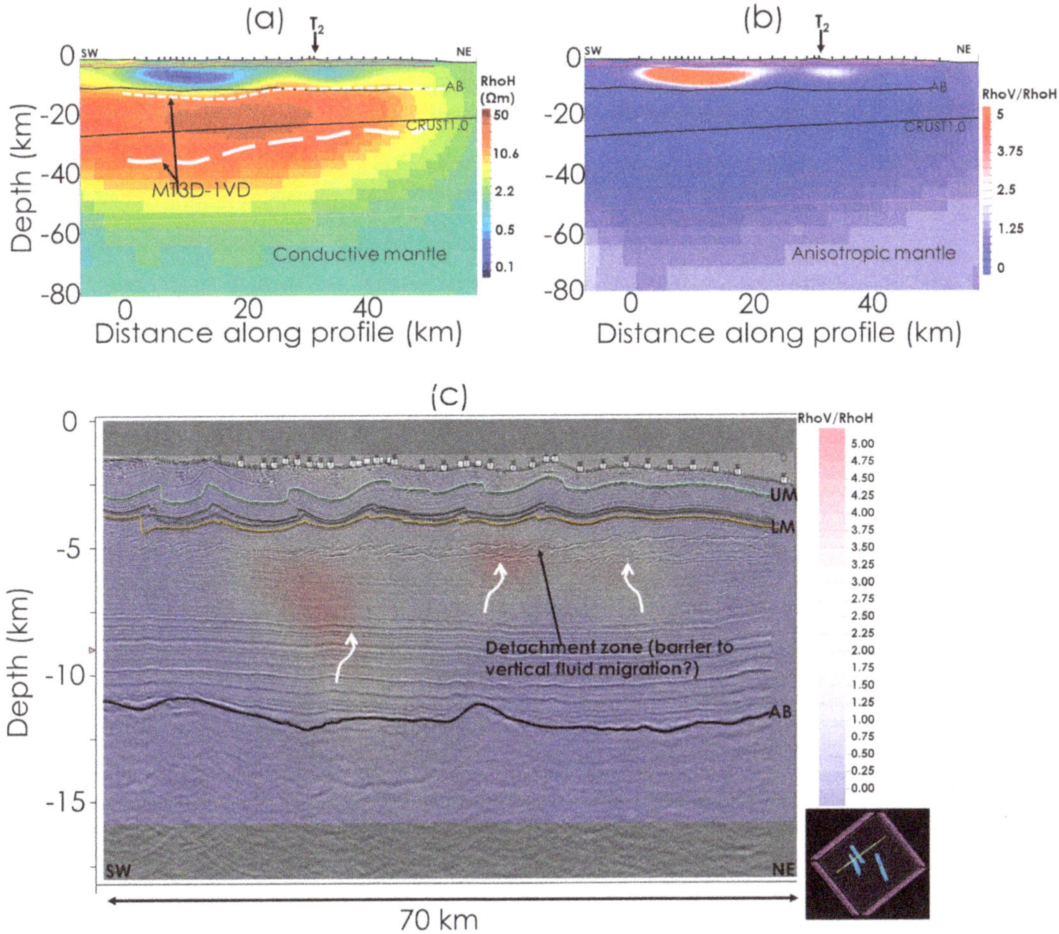

Figure 14. Integrating information from 3D anisotropic MT models and 3D seismic data to verify vertical fluid migration paths and barriers. (**a**) Lithospheric horizontal resistivity section extracted along the SW-NE regional MT survey line [11]. (**b**) Lithospheric electrical anisotropy results were derived as the ratio of the vertical-to-horizontal resistivities along the regional line [11]. (**c**) Upper crustal seismic image along the regional MT line extracted from the 3D seismic cube with the resistivity anisotropy overlaid for comparison. Seismic-derived boundaries (solid black lines) and those from post-inversion electrostratigraphic imaging using the 1VD of the 3D horizontal resistivity model (white dashed lines) are shown for comparison. AB, acoustic basement. CRUST1.0 is seismic Moho [45]. The UM (Upper Miocene) and LM (Lower Miocene) are interpreted seismic horizons. Thick white wiggly arrows show possible anisotropy and seismic evidence of upward fluid migration along the steep faults extending from the basement. The other symbols are as in Figure 13. MT stations are located on the seabed (white squares and inverted black diamonds).

5. Conclusions

The search for hydrocarbon and native hydrogen reservoirs has moved to complex frontier regions where resource system fundamentals pose significant challenges, and it is critical to integrate various geophysical, geological, and environmental modeling tools in three dimensions to increase accuracy and, hence, reduce uncertainty in subsurface predictions. It is demonstrated in this paper that marine electromagnetic geophysical

41. Meju, M.A. Regularized extremal bounds analysis (REBA): An approach to quantifying uncertainty in nonlinear geophysical inverse problems. *Geophys. Res. Lett.* **2009**, *36*, L03304. [CrossRef]
42. Mackie, R.L.; Miorelli, F.; Meju, M.A. Practical methods for model uncertainty quantification in electromagnetic inverse problems. In Proceedings of the Expanded Abstract, SEG International Exposition and 88th Annual Meeting 2018, Anaheim, CA, USA, 14–19 October 2018. [CrossRef]
43. Saleh, A.S.; Meju, M.; Mackie, R.; Andersen, E.; Ismail, N.A.; Nawawi, M. Seismic-Electromagnetic Projection attribute: Application in integrating seismic quantitative interpretation and 3D CSEM-MT broadband data inversion for robust ranking and sweet-spotting of hydrocarbon prospects in offshore NW Borneo. *Geophysics* **2023**, *in review*.
44. Ingram, G.M.; Chisholm, T.J.; Grant, C.J.; Hedlund, C.A.; Stuart-Smith, P.; Teasdale, J. Deepwater North West Borneo: Hydrocarbon accumulation in an active fold and thrust belt. *Mar. Pet. Geol.* **2004**, *21*, 879–887. [CrossRef]
45. Laske, G.; Masters, G.; Ma, Z.; Pasyanos, M.E. CRUST1.0: An updated global model of Earth's crust. *Geophys. Res. Abstr.* **2012**, *14*, 743.
46. Hall, R. Contraction and extension in northern Borneo driven by subduction. *J. Asian Sci.* **2013**, *76*, 399–411. [CrossRef]
47. Nguyen, L.A.; Mann, P. Gravity and magnetic constraints on the Jurassic opening of the oceanic Gulf of Mexico and the location and tectonic history of the Western Main transform fault along the eastern continental margin of Mexico. *Interpretation* **2016**, *4*, SC23–SC33. [CrossRef]
48. Pindell, J.; Miranda, C.; Ceron, A.; Hernandez, L. Aeromagnetic map constrains Jurassic-Early Cretaceous synrift, break up, and rotational seafloor spreading history in the Gulf of Mexico. Mesozoic of the Gulf Rim and Beyond: New Progress in Science and Exploration of the Gulf of Mexico Basin. In Proceedings of the 35th Annual Gulf Coast Section SEPM Foundation Perkins-Rosen Research Conference. GCS SEPM Foundation, Houston, TX, USA, 8–9 December; 2016; pp. 123–153.
49. Soyer, W.; Mackie, R.; Hallinan, S.; Pavesi, A.; Nordquist, G.; Suminar, A.; Intani, R.; Nelson, C. Geologically consistent multiphysics imaging of the Darajat geothermal steam field. *First Break* **2018**, *36*, 77–83. [CrossRef]

Disclaimer/Publisher's Note: The statements, opinions and data contained in all publications are solely those of the individual author(s) and contributor(s) and not of MDPI and/or the editor(s). MDPI and/or the editor(s) disclaim responsibility for any injury to people or property resulting from any ideas, methods, instructions or products referred to in the content.

minerals

MDPI

Article

New Concept Ocean-Bottom Multiphysics (OBMP) Nodes for Reservoir Monitoring

Paulo T. L. Menezes [1,2,*], Andrea Zerilli [3], Jorlivan L. Correa [1], Everton N. Menor [1], Sergio M. Ferreira [1] and Tiziano Labruzzo [3]

1 PETROBRAS-EXP/PEN/AB, Av. Henrique Valadares 28, 4 andar, Rio de Janeiro 20231-030, Brazil
2 Departamento de Geologia Aplicada, Faculdade de Geologia/UERJ, Rua São Francisco Xavier, 524, sala 2009 Bloco A, Rio de Janeiro 20550-900, Brazil
3 Zlemlink Ltda., Rua Taylor 39 sala 805, Rio de Janeiro 24241-060, Brazil
* Correspondence: ptarsomenezes@petrobras.com.br or paulo.menezes@uerj.br

Abstract: Marine-controlled source electromagnetics (CSEM) have been extensively applied to various exploration scenarios worldwide. However, its perceived value and cost relative to seismic and the scarcity of realistic case studies have limited the industry's interest in time-lapse reservoir-monitoring (4D) applications. A feasible way to make demand for CSEM for 4D-monitoring programs would be to increase the value of information and reduce survey costs by performing joint operations where seismic and CSEM data are acquired during the same survey and at equivalent spatial densities. To this end, we propose a new multiphysics ocean-bottom nodes (OBN) concept and show the industry that CSEM can be a cost efficient and effective integrators to 4D seismic projects. To this end, we conducted a feasibility study demonstrating that horizontal magnetic field components have the required sensitivities and can be used instead of horizontal electric field components in mapping the 3D resistivity distribution and 4D fluid change responses in a given reservoir. This makes engineering a new OBN class simpler and cheaper, as various miniaturized magnetic field sensors are available off-the-shelf or readily working along with packaging and coupling solutions.

Keywords: reservoir monitoring; CSEM; ocean-bottom nodes; multiphysics acquisition

Citation: Menezes, P.T.L.; Zerilli, A.; Correa, J.L.; Menor, E.N.; Ferreira, S.M.; Labruzzo, T. New Concept Ocean-Bottom Multiphysics (OBMP) Nodes for Reservoir Monitoring. *Minerals* **2023**, *13*, 602. https://doi.org/10.3390/min13050602

Academic Editor: Georgy Cherkashov

Received: 10 January 2023
Revised: 26 February 2023
Accepted: 8 March 2023
Published: 27 April 2023

Copyright: © 2023 by the authors. Licensee MDPI, Basel, Switzerland. This article is an open access article distributed under the terms and conditions of the Creative Commons Attribution (CC BY) license (https://creativecommons.org/licenses/by/4.0/).

1. Introduction

During the last 20± years, marine CSEM has been applied for de-risking deep-water high-cost drilling decisions in many basins worldwide. From the early days, CSEM has been expanded to a broader range of geographic areas, geological environments, and application scenarios. Ongoing development efforts have demonstrated the potential to enhance cost-effective deep-water reservoir-monitoring applications. Indeed, several pioneering feasibility studies (e.g., [1–3]) have revealed that time-lapse (4D) CSEM can play a role in improving our knowledge of the reservoir structure, fluid flow, and fluid saturation changes when appropriate acquisition and advanced integrated interpretation workflows are applied.

Nonetheless, a single 4D CSEM survey has been reported in the literature [4]. We attribute the perceived value and cost of CSEM, relative to seismic aquisition, to the lack of realistic case studies, and, most of all, the lack of new developments and investments in effective 4D acquisition systems as the main reasons the industry has not yet considered CSEM in 4D reservoir-monitoring programs.

CSEM uses a horizontal electric dipole source towed along a line by a surface vessel, usually as close to the seafloor as possible (commonly 30–50 m). The source transmits an EM signal through the seafloor at user-defined frequencies, usually in the 0.1–10 Hz range. The ocean-bottom electromagnetic (OBEM) receivers, which measure the horizontal electric and/or magnetic fields, are placed on the seafloor either in a free-fall operation along a line (for 2D) or grid (for 3D) pattern, with typical OBEM spacing in the 1–3 km range for hydrocarbon exploration purposes (Figure 1).

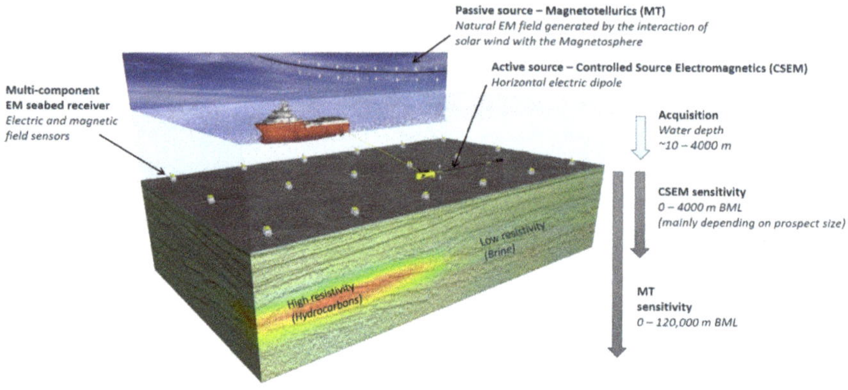

Figure 1. Acquisition scheme and sensitivity below mud line (BLM) of MT and CSEM methods. Courtesy of EMGS.

The current industry's state-of-the-art CSEM nodes are equipped with electric field dipoles of 8 to 10 m and two to three induction coils for the magnetic field measurements [5,6]. The nodes are attached to 200 kg concrete anchors, used to sink the instruments from the surface to the ocean bottom. Present-day CSEM ships can store hundreds of receivers [7]. Surveying large areas and/or denser receiver spacings often affects the efficiency and, consequently, the costs of a given project.

Dedicated 4D ocean-bottom node (OBN) surveys comprise most of the current time-lapse seismic projects. Acquisitions are designed to maximize position repeatability, and timing is planned to impact reservoir management. The OBN surveys have seen exponential progress in cost efficiency, with the cost per trace reducing by approximately 50% every two years [8]. Ultra-compact sensors (1 m) drastically increase node count onboard (1000–10,000 nodes), enabling far denser sampling and significantly improving ROV (remotely operated vehicles) operational efficiency (Figure 2).

A possible way to step up CSEM into a cost-efficient integrator to a 4D seismic program might be to conduct multiphysics surveys, where seismic and CSEM data are jointly acquired during the same survey, comparable to marine gravity and magnetic–seismic joint acquisitions. Going back to basics, reassessing every aspect of CSEM, from node redesign and deployment to operational efficiency, is required to show the industry that CSEM could be a cost-efficient and effective integrator.

According to [7], one of the keys to success is the development of seismic-EM bundled-integrated ocean-bottom multiphysics (OBMP) nodes inspired by and following the current seismic OBN industry searching for better data, efficiency, cost, and simplicity. Ultra-compact nodes facilitate far denser sampling and greatly enhanced operational efficiency, which demands optimization, higher sensitivities, and miniaturization of the EM sensors to integrate them within the current state-of-the-art seismic nodal technology. This may be the case to acquire the magnetic field only, neglecting the electric field, as the magnetic sensors could be easier and faster to incorporate into an ultra-compact node.

A potential issue with our proposition is that CSEM quantitative interpretations are, presently, performed solely with the horizontal electric field components. The magnetic field components recorded in the OBEM nodes are used mainly to compute magnetotelluric responses. Nevertheless, few published studies show that the magnetic field has comparable sensitivity to the electric field in the CSEM interpretation.

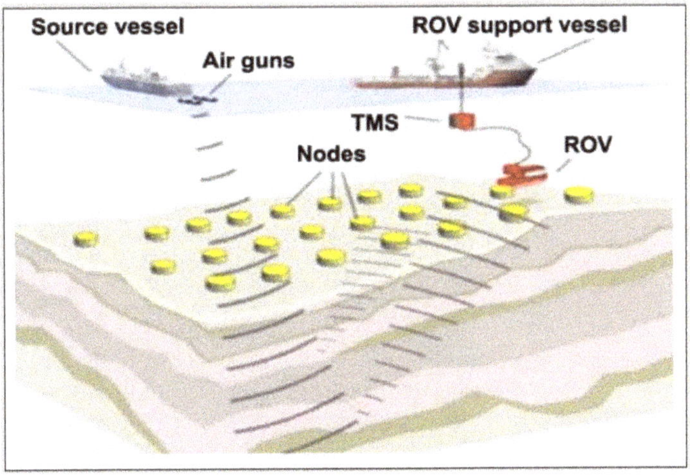

Figure 2. Seismic OBN concept. The tether management system (TMS) separates the ROV (remotely operated vehicles) from the vessel heave and the surface umbilical mass, which allows the ROV more freedom of movement. Modified from [9].

Using a synthetic canonical 1D reservoir model, Key (2009) [10] showed that separate 1D inversions of the electric and magnetic fields perform equally well at recovering the reservoir. Key (2009) [10] also concluded that recording only a single field type would be adequate for specific exploration applications.

Constable et al. (2019) [11] interpreted a CSEM dataset acquired at the Scarborough gas field, Australia. They carried out 2D inversions of various data combinations to define how well they recovered the expected reservoir's properties and shape. As a result, they showed that individual inversions of the electric and magnetic components generate almost identical models, suggesting that these datasets do not carry independent information.

3D inversions of CSEM data were conducted by Tharimela et al. (2019) [12] to investigate gas hydrate accumulations in the Pelotas Basin, Brazil. They showed that the magnetic field inversions provided smoother models that were more compatible with seismic events when compared with electric field inversions. These results were further confirmed by synthetic modelling and inversion studies using simple theoretical models. They attributed the spurious anomalies of the electric inversions to the static shift phenomena that affect the electric field only.

Herein we develop the findings of [13] and demonstrate that the horizontal magnetic field components (Hx, Hy) can be effectively used for 3D appraisal studies and 4D-monitoring applications. We conducted a time-lapse feasibility study based on the MR3D realistic turbiditic model [14] employing the horizontal magnetic components of the 4D CSEM dataset calculated by [3]. We show that the horizontal magnetic field inversion outcomes are comparable to the horizontal electric field inversions and are sensitive to map time-lapse resistivity changes associated with oil–water substitution in the studied reservoir.

2. MR3D Project

Marlim R3D (MR3D) is an open-source project aiming to supply a realistic geoelectric model to be employed as a benchmark for CSEM studies of the Brazilian continental margin deep-water turbidites reservoir systems. These turbidites are considered analogues to several others worldwide, such as those found in the African continental margin. Recently, refs. [14,15] made the MR3D geoelectric model and the synthetic-associated CSEM dataset public, respectively.

The MR3D VTI model was constructed based on the geological knowledge of the mature Marlim oilfield [16], located in the northeastern portion of the offshore Campos Basin, Brazil (Figure 3a). Marlim is a heavy oil field discovered in 1985 at water depths ranging from 600 to 1200 m (Figure 3a). Oil production began in 1991 and seawater injection also started by that year [17].

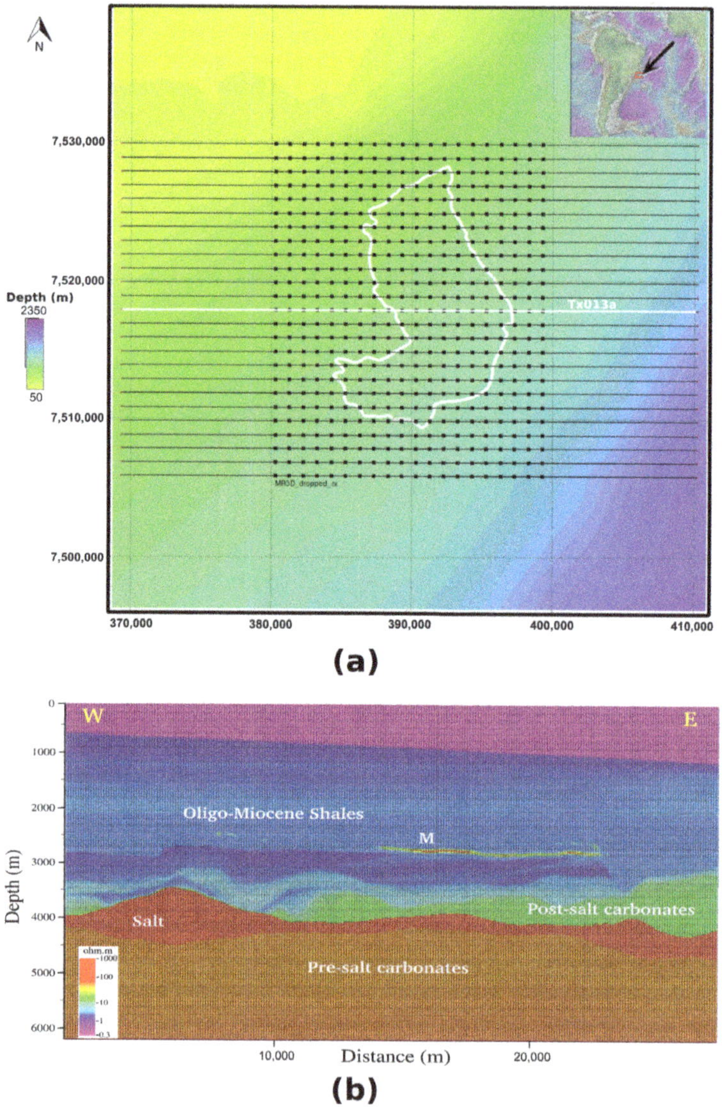

Figure 3. (**a**) Marlim field with the CSEM acquisition geometry superimposed on bathymetry. The reservoir outline is displayed as a white line. Black dots show the receiver locations distributed in a 1000 m spacing regular grid. Black lines are source west–east towlines evenly spaced at 1000 m. Towline Tx013a is highlighted as the white line. (**b**) Cross-sections of the MR3D model along the east–west towline Tx013a. Marlim oil-prone turbidites (M) appear as the thin resistive body embedded in the Oligocene–Miocene shales sequence. Modified from [13].

The MR3D model mixes fine-scale stratigraphy and complex oil-filled reservoirs in a complex geological environment consisting of Oligocene–Miocene shales, post-salt carbonates, a thick Aptian salt layer, and pre-salt carbonates (Figure 3b). The reservoir facies are formed by high porosity, in the 26–32% range, and clean sandstones [16]. The reservoir has a variable thickness of up to 125 m, with an average of 80 m [17].

The heavy oil reservoirs of the Marlim field are composed of an Oligocene–Miocene deepwater turbidite system of the Carapebus Formation (Figure 4). They include a set of amalgamated sandstone bodies recognized as Marlim sandstone [18]. The turbiditic sedimentation that has occurred was influenced by a regional northwest–southeast trending transfer faults system [19]. This system built the pathway for the deposition, reworking, and displacement of many deepwater turbiditic reservoirs in the Campos Basin [20].

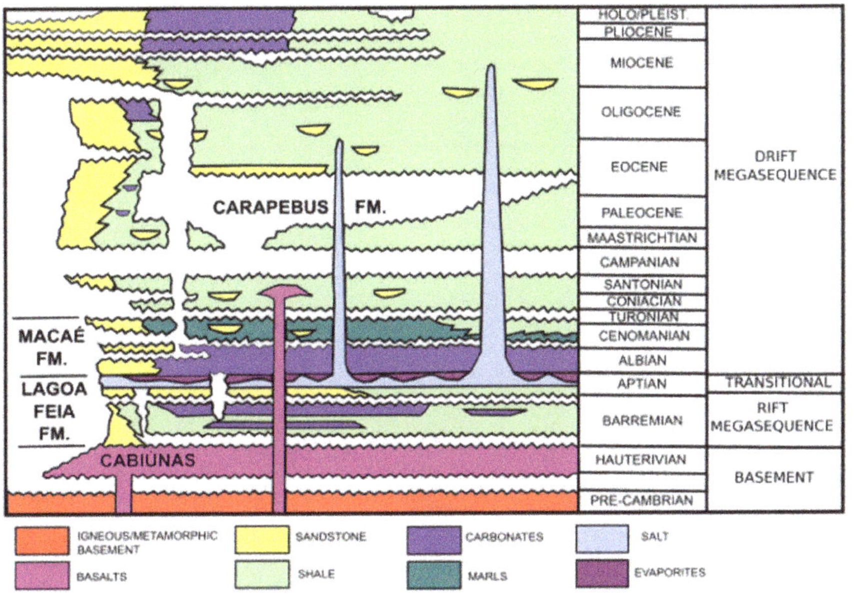

Figure 4. Campos Basin simplified stratigraphic chart, modified from [21].

Figure 5 exhibits the isopach map of the MR3D reservoir with four major thick northwest–southeast trending turbidite lobes (A–D in Figure 5) that comprise the main oil-producing zones of the Marlim field [22].

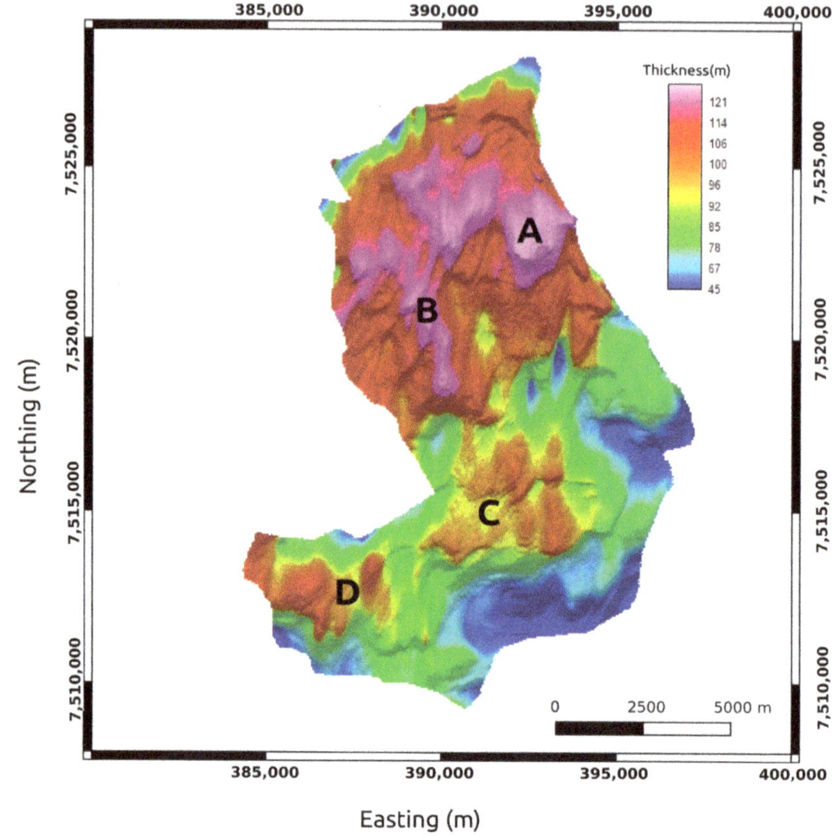

Figure 5. Isopach map of the MR3D reservoir. A–D are thick turbidite lobes and the main oil-producing zones.

3. 4D Feasibility

3.1. Forward Modelling—MR4D Data

Menezes et al. (2021) [3] conducted a 4D feasibility study to assess the CSEM effectiveness in detecting resistivity changes associated with water saturation changes within the Marlim reservoir. The starting point of that study was the pre-production scenario (1991) of the MR3D model (Figure 3b). Then, a fluid flow simulator was used to simulate the oil–water substitution for the 2021 and 2031 production years representing 30 and 40 years of production, and seawater injection to build subsurface cellular resistivity models and perform a forward-modelling study.

The flow simulator yielded saturation variations that were then converted into resistivity. The reservoir connection, the relative scarcity of gas, the relative permeability curves, and the field location led to a seawater injection of 600,000 barrels/day as the most viable method for pressure maintenance and oil recovery [17].

In the present study, we extend the findings of [3] by investigating the ability of the magnetic fields to image a 3D oil-filled reservoir and detect time-lapse changes due to fluid replacement within that reservoir. We assume that the resistivity changes are only associated with the fluid substitution within the reservoir's boundaries (Figure 3a), while the background resistivity remains unchanged. Other superficial effects are not considered as they have been thoroughly investigated by [2].

Figure 3a shows the design of the synthetic CSEM survey in MR3D. The survey consists of 25 E–W 1 km spaced towlines and a dense 1 × 1 km grid of 500 OBEM receivers. A 3D forward modelling parallelized commercial fast finite-difference time-domain (FDTD) CSEM modelling algorithm [23] was used to compute all six electromagnetic field components Ex, Ey, Ez, Hx, Hy, and Hz for six source frequencies of 0.125, 0.25, 0.5, 0.75, 1, and 1.25 Hz.

The noise-free data's electromagnetic fields were computed with 0.1 s time steps, at a maximum number of 200,000 time steps, and a convergence accuracy of 10^{-4}. The electric fields were normalized by the dipole moment. Multiplicative noise was added with a 1% standard deviation following the procedure of [24]. A noise floor of 10^{-15} V/Am^2 for the electric fields, and 10^{-12} m^{-2} for the magnetic fields was applied.

We modelled a horizontal electric dipole source towed 30 m above the seabed. The transmitter current was directed along the towing direction with inline and broadside data registered up to 12 km offsets [14].

The synthetic CSEM data were generated for the 2021 and 2031 monitoring years described above. The modelling exercise delivered high-quality magnetic data to a source-receiver offset up to 12 km for the lower frequencies and maximum ranges of 5 to 6 km for the higher frequencies (Figure 6).

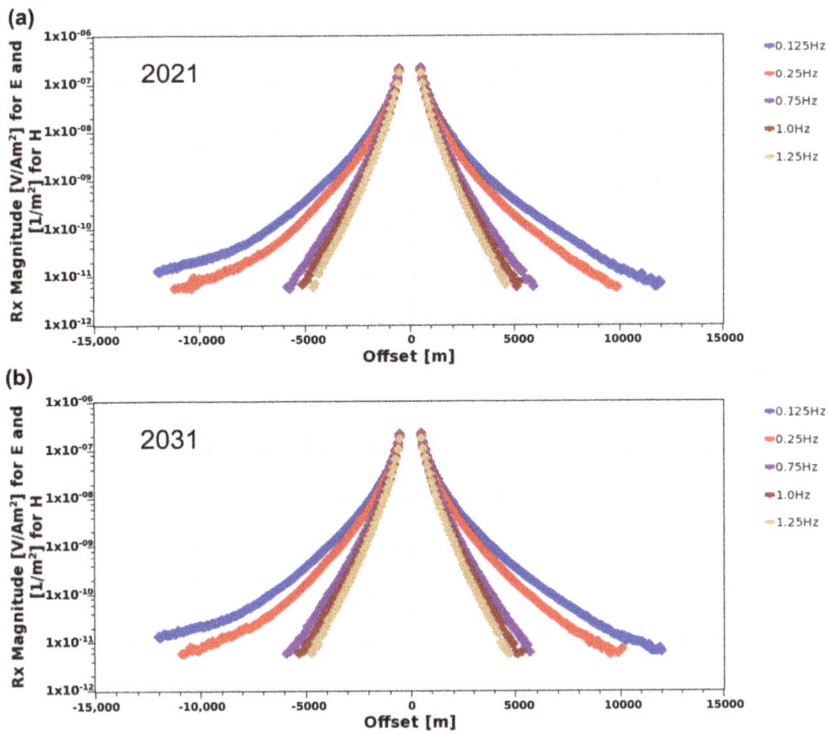

Figure 6. Typical data for the MR3D model showing the inline magnetic field. Site 04Rx251a. (**a**) 2021 base survey; (**b**) 2031 monitoring survey.

3.2. Inversions for 3D Appraisal and 4D Responses

We then conducted unconstrained 3D quasi-Newton BFGS anisotropic inversions [25] of the horizontal magnetic data (Hx-Hy) for the 2021 and 2031 datasets. To compare with the electric data (Ex-Ey) inversions results [3], we duplicated the inversion parametrization for both cases: the L1 smoothness norm utilized for the logarithm of the resistivity values, a mesh of approximately 9 million cells (see Table 1), and the identical starting model, based on previous seismic-EM integrated interpretations [3]. The inversions converged in less than 50 iterations to a satisfactory RMS target misfit of 1.

Figure 7 displays the comparison between the recovered Rv final model of the base year's (2021) unconstrained anisotropic inversion for the horizontal electric fields (Figure 7a) and the horizontal magnetic fields (Figure 7b). The Ex-Ey and Hx-Hy models were similar, demonstrating that these datasets do not carry independent information [11].

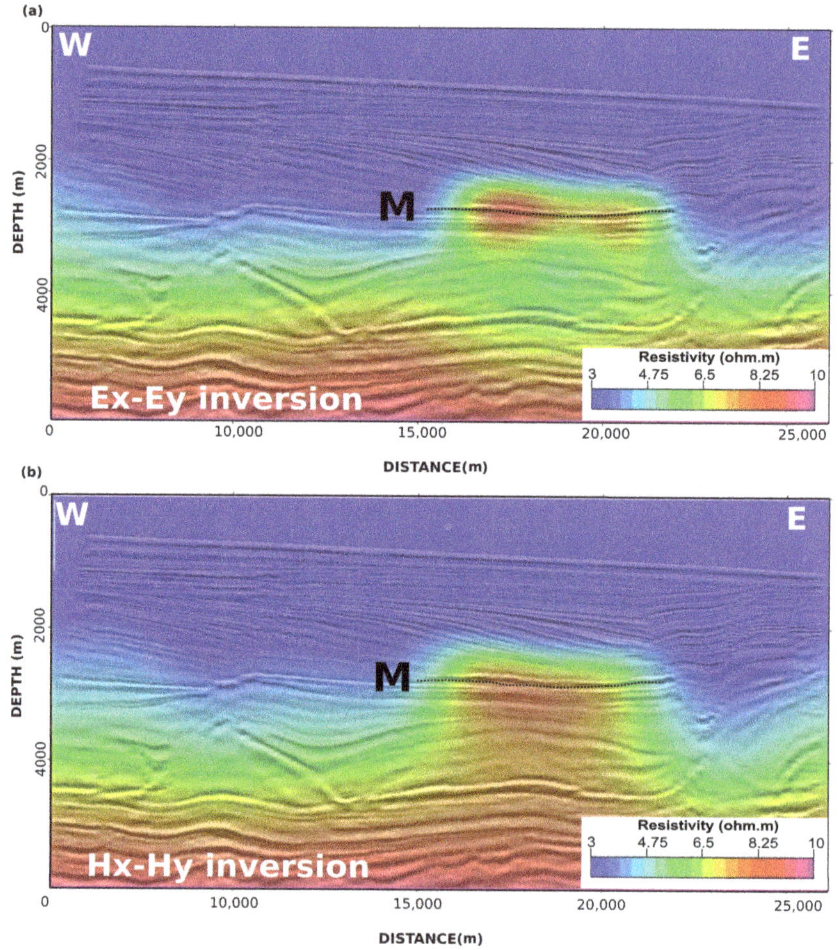

Figure 7. Cross-section along line Tx013a (location in Figure 3a—vertical resistivity Rv for 2021 base year inversions). (**a**) Horizontal electric fields. (**b**) Horizontal magnetic fields. Black dotted line defines the lateral extension of the MR3D reservoir (M).

Both extracted cross-sections along the Tx013a line display smooth Rv anomalies with low vertical resolution around the thin MR3D turbidite horizon at 2650 m depth (M in Figure 7). The resistivity anomalies are reasonably isolated from the resistive background below 3800 m depth. This deep resistor is associated with the deep carbonates and autochthonous salt layers (Figure 3b).

Table 1. Mesh used in the MR3D time-lapse BFG inversions.

Mesh Parameters	Dimensions
Cell size - X	200 m
Cell size - Y	200 m
Cell size - Z	50 m
Number of cells - X	220
Number of cells - Y	195
Number of cells - Z	110
Total number of cells	8,910,000.00

Conversely, the horizontal resolution provided by the CSEM method is much higher than the vertical resolution. Consequently, the correlation of anomalies with the associated resistive oil reservoirs is much better imaged when performed along constant depth slices or stratigraphic horizons [26].

Herein we followed the [3] guideline to interpreting the EM attributes extracted within the Marlim reservoir depth interval. We estimated the RMS attribute (Rv_{RMS}) of the retrieved Rv in the [−50, +250 m] window around the top of the 2650 m depth slice (Figure 8). As expected, the computed Rv_{RMS} for the 2021 base year shows an increased horizontal resolution compared to the vertical resolution (Figure 7). The resistive anomalies are fully enclosed within the reservoir's boundaries, and the anomaly borders match the MR3D reservoir outline. Furthermore, one can observe that the magnetic field inversions respond similarly to the electric field inversions, and both respond nicely to the four main producing zones of the MR3D reservoir (Figure 5).

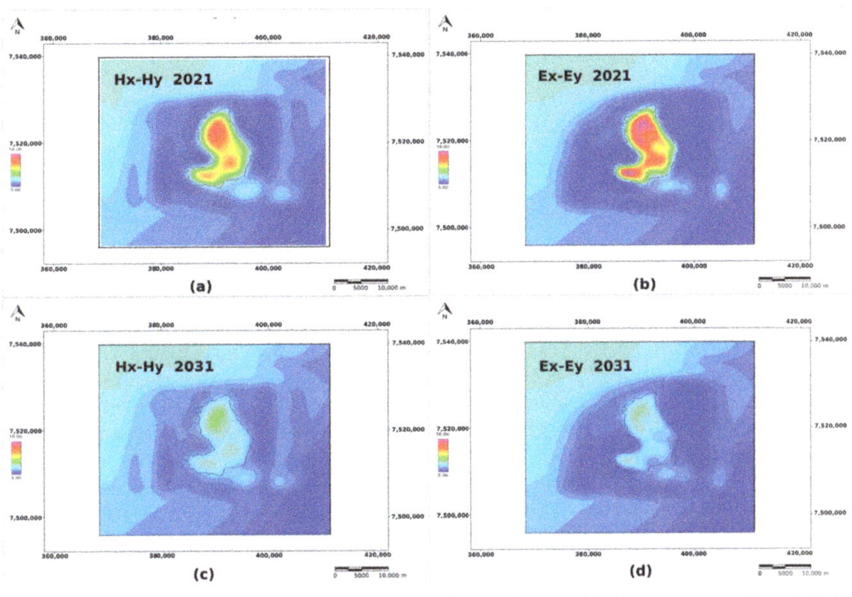

Figure 8. Rv_{RMS} attribute computed at 2650 m depth slice. (**a**) 2021 Hx-Hy inversion. (**b**) 2021 Ex-Ey inversion. (**c**) 2031 Hx-Hy. (**d**) 2031 Ex-Ey inversion.

We extracted the Rv_{RMS} attributes for the 2031 inversions at the top of the 2650 m depth slice to demonstrate the time-lapse effect on the CSEM data. Figure 8c shows the Rv_{RMS} attribute for the Hx-Hy inversion, and Figure 8d the Rv_{RMS} outcome for the Ex-Ey inversion. Both 2031 inversions also yielded similar models showing a resistive anomaly confined within the MR3D reservoir's boundaries.

By comparing the 2021 base year with the 2031 monitoring year results, we observe a reduction in resistivity anomalies that is in agreement with the flow simulator results of [3], which showed drops in the reservoir's resistivity associated with the fluid substitution for up to 40 years of seawater injection.

These results demonstrate that horizontal magnetic field components can also be effectively used for long-term 4D studies. For the 2031 example (Figure 8c), the stronger anomalies are correlated with the main thickness oil zones within the MR3D reservoir (Figure 5). Indeed, the recovered magnetic anomalies (Figure 8c) are up to 15% stronger than their electric correspondent (Figure 8d).

4. Discussion

Dedicated seismic OBN node surveys make up most of the oil industry expenditure for ongoing and future reservoir-monitoring programs. Although this study claims that the case for CSEM as a promising integrator to 4D seismic is strong, the perceived value and cost needed for additional CSEM measurements inhibit these additional acquisitions in monitoring projects.

Going back to basics, reassessing every aspect of CSEM, from instrument layout and deployment to operational efficiency, is required to demonstrate to the industry that CSEM can be the cost-efficient integrator to 4D seismic as the monitoring requirements become more challenging owing to the high drilling and intervention costs of deep-water reservoir exploration.

This change should begin from the operational side, where most of the cost and efficiency issues reside, by bringing a re-engineered integrated OBMP seabed node (seismic-EM bundled), enabling a single deployment/positioning/recovery cycle and comparable data density inspired and following the current ocean-bottom seismic industry strive for better data, efficiency, cost, and simplicity. The CSEM data would significantly reduce total project costs in such joint seismic-EM investments since the highest expenditures for 4D-monitoring programs are associated with operations and logistics.

5. Conclusions

The Marlim case study reveals that magnetic fields have equivalent sensitivity to that of electric fields to map the 3D resistivity distribution and 4D responses efficiently. Resistivity models in agreement with those estimated from fluid simulators applied to the years 2021 and 2031 were retrieved, thus portraying time intervals of 30 and 40 years, respectively, referring to the pre-production/pre-injection year of 1991.

The use of magnetic sensors could only make re-engineering nodes simpler and cheaper. Various miniaturized magnetic field sensors, along with more straightforward packaging and coupling solutions, are available off-the-shelf or readily operatational.

When the global deep-water oil industry is set to emerge stronger and prepared for the subsequent phase of improvements, the time could be suited for an industry-funded OBMP reservoir-monitoring system that will add substantial value to reservoir surveillance decisions guiding more efficient recovery, diminished costs, and raised valuable information. Our feasibility study is encouraging as we envision the needs, potential, and challenges when we come together with the seismic industry and might further persuade us to feel securer about our return on investments.

Author Contributions: Conceptualization, P.T.L.M. and A.Z.; methodology, P.T.L.M., J.L.C. and A.Z.; validation, P.T.L.M., J.L.C. and E.N.M.; writing—original draft preparation, P.T.L.M.; writing—review and editing, P.T.L.M., A.Z., S.M.F., T.L. and E.N.M.; supervision, S.M.F.; project administration, S.M.F. All authors have read and agreed to the published version of the manuscript.

Funding: This research was funded by Petrobras.

Data Availability Statement: Data associated with this research are confidential and cannot be released.

Acknowledgments: We are thankful to the management of Petrobras for the support and permission to publish this work.

Conflicts of Interest: The authors declare no conflict of interest. The funders had no role in the design of the study.

Abbreviations

The following abbreviations are used in this manuscript:

CSEM	Controlled-source electromagnetic
BML	Below mud line
EM	Electromagnetic
OBEM	Ocean-bottom electromagnetic
OBMP	Ocean-bottom multiphysics
VTI	Vertical transverse isotropy
RAR	Resistivity anisotropy ratio
Rv	Vertical resistivity
Rh	Horizontal resistivity
ATR	Anomalous transverse resistance
OWC	Oil–water contact
BFGS	Broyden–Fletcher–Goldfarb–Shannon
RMS	Root-mean-square

References

1. Black, N.; Wilson, G.A.; Gribenko, A.V.; Zhdanov, M.S.; Morris, E. 3D inversion of time-lapse CSEM data based on dynamic reservoir simulations of the Harding field, North Sea. In *SEG Technical Program Expanded Abstracts 2011*; Society of Exploration Geophysicists: San Antonio, TX, USA, 2011; pp. 666–670.
2. Shantsev, D.V.; Nerland, E.A.; Gelius, L.J. Time-lapse CSEM: How important is survey repeatability? *Geophys. J. Int.* **2020**, *223*, 2133–2147. [CrossRef]
3. Menezes, P.T.; Correa, J.L.; Alvim, L.M.; Viana, A.R.; Sansonowski, R.C. Time-Lapse CSEM Monitoring: Correlating the Anomalous Transverse Resistance with SoPhiH Maps. *Energies* **2021**, *14*, 7159. [CrossRef]
4. Ziolkowski, A.; Parr, R.; Wright, D.; Nockles, V.; Limond, C.; Morris, E.; Linfoot, J. Multi-transient electromagnetic repeatability experiment over the North Sea Harding field. *Geophys. Prospect.* **2010**, *58*, 1159–1176. [CrossRef]
5. Dong, Z.; Zhang, J.; Yang, G.; Cai, Z.; Lu, Y.; Chen, K. Micro-ocean-bottom electromagnetic receiver for controlled-source electromagnetic and magnetotelluric data acquisition. *Rev. Sci. Instrum.* **2021**, *92*, 044705. [CrossRef] [PubMed]
6. Constable, S. Instrumentation for marine magnetotelluric and controlled source electromagnetic sounding. *Geophys. Prospect.* **2013**, *61*, 505–532. [CrossRef]
7. Zerilli, A.; Menezes, P.T.L.; Crepaldi, J.L.; Correa, J.L.; Viana, A.R. Ocean-bottom multiphysics—Path to superior reservoir monitoring. In *Proceedings of the SEG/AAPG/SEPM First International Meeting for Applied Geoscience & Energy*; OnePetro: Houston, TX, USA, 2021. [CrossRef]
8. Bunting, T. The Evolution of the OBN Acquisition Solution. In *Proceedings of the EAGE Seabed Seismic Today: From Acquisition to Application*; European Association of Geoscientists & Engineers: Kosterijland, The Netherlands, 2020; Volume 2020, pp. 1–4.
9. Walker*, C.; McIntosh, S. Autonomous Nodes–The Future of Marine Seismic Data Acquisition? In Proceedings of the 12th International Congress of the Brazilian Geophysical Society & EXPOGEF, Rio de Janeiro, Brazil, 15–18 August 2011; Society of Exploration Geophysicists and Brazilian Geophysical Society: Riode de Janeiro, RJ, Brazil, 2011; pp. 1–6.
10. Key, K. 1D inversion of multicomponent, multifrequency marine CSEM data: Methodology and synthetic studies for resolving thin resistive layers. *Geophysics* **2009**, *74*, F9–F20. [CrossRef]
11. Constable, S.; Orange, A.; Myer, D. Marine controlled-source electromagnetic of the Scarborough gas field—Part 3: Multicomponent 2D magnetotelluric/controlled-source electromagnetic inversions. *Geophysics* **2019**, *84*, B387–B401. [CrossRef]
12. Tharimela, R.; Augustin, A.; Ketzer, M.; Cupertino, J.; Miller, D.; Viana, A.; Senger, K. 3D controlled-source electromagnetic imaging of gas hydrates: Insights from the Pelotas Basin offshore Brazil. *Interpretation* **2019**, *7*, SH111–SH131. [CrossRef]
13. Menezes, P.T.L.; Correa, J.L.; Menor, E.N.; Ferreira, S.M.; Zerilli, A.; Labruzzo, T. Ocean-bottom multiphysics for reservoir monitoring—Increasing the value of information. In Proceedings of the Second International Meeting for Applied Geoscience & Energy, Houston, TX, USA, 28 August–1 September 2022; pp. 692–696. [CrossRef]
14. Correa, J.L.; Menezes, P.T. Marlim R3D: A realistic model for controlled-source electromagnetic simulations—Phase 2: The controlled-source electromagnetic data set. *Geophysics* **2019**, *84*, E293–E299. [CrossRef]
15. Carvalho, B.R.; Menezes, P.T.L. Marlim R3D: A realistic model for CSEM simulations-phase I: Model building. *Braz. J. Geol.* **2017**, *47*, 633–644. [CrossRef]

16. Nascimento, T.M.; Menezes, P.T.; Braga, I.L. High-resolution acoustic impedance inversion to characterize turbidites at Marlim Field, Campos Basin, Brazil. *Interpretation* **2014**, *2*, T143–T153. [CrossRef]
17. Shecaira, F.S.; Daher, J.S.; Gomes, A.C.A.; Capeleiro Pinto, A.C.; Filoco, P.R.; Gomes, J.A.T.; Freitas, L.C.S.; Balthar, S. Injeção de Água no Campo de Marlim—Uma Abordagem Integrada para o Gerenciamento de Reservatórios. In *Proceedings of the Rio Oil & Gas Expo and Conference*; Brazilian Petroleum Institute—IBP: Rio de Janeiro, RJ, Brazil, 2000; pp. 1–9.
18. Peres, W.E. Shelf-fed turbidite system model and its application to the Oligocene deposits of the Campos Basin, Brazil. *AAPG Bull.* **1993**, *77*, 81–101.
19. Lourenço, J.; Menezes, P.T.L.; Barbosa, V.C.F. Connecting onshore-offshore Campos Basin structures: Interpretation of high-resolution airborne magnetic data. *Interpretation* **2014**, *2*, SJ35–SJ45. [CrossRef]
20. Cobbold, P.R.; Meisling, K.E.; Mount, V.S. Reactivation of an obliquely rifted margin, Campos and Santos basins, southeastern Brazil. *AAPG Bull.* **2001**, *85*, 1925–1944. [CrossRef]
21. Guardado, L.R.; Spadini, A.R.; Brandão, J.S.L.; Mello, M.R. Petroleum System of the Campos Basin, Brazil. In *Petroleum systems of South Atlantic margins: AAPG Memoir 73*; Mello, M.R., Katz, B.J., Eds.; AAPG: Tulsa, OK, USA, 2000; Chapter 22, pp. 317–324.
22. Mutti, E.; Cunha, R.S.; Bulhoes, E.; Arienti, L.M.; Viana, A.R. Contourites and turbidites of the Brazilian marginal basins. In Proceedings of the AAPG Annual Convention & Exhibition, Houston, TX, USA, 6–9 April 2014.
23. Maaø, F.A. Fast finite-difference time-domain modeling for marine-subsurface electromagnetic problems. *Geophysics* **2007**, *72*, A19–A23. [CrossRef]
24. Mittet, R.; Morten, J.P. Detection and imaging sensitivity of the marine CSEM method. *Geophysics* **2012**, *77*, E411–E425. [CrossRef]
25. Zach, J.; Bjørke, A.; Støren, T.; Maaø, F. 3D inversion of marine CSEM data using a fast finite-difference time-domain forward code and approximate Hessian-based optimization. In *SEG Technical Program Expanded Abstracts 2008*; Society of Exploration Geophysicists: Las Vegas, NV, USA, 2008; pp. 614–618.
26. Lyrio, J.C.S.O.; Menezes, P.T.L.; Correa, J.L.; Viana, A.R. Multiphysics anomaly map: A new data fusion workflow for geophysical interpretation. *Interpretation* **2020**, *8*, B35–B43. [CrossRef]

Disclaimer/Publisher's Note: The statements, opinions and data contained in all publications are solely those of the individual author(s) and contributor(s) and not of MDPI and/or the editor(s). MDPI and/or the editor(s) disclaim responsibility for any injury to people or property resulting from any ideas, methods, instructions or products referred to in the content.

minerals

MDPI

Article

Source Rock Evaluation from Rock to Seismic Data: An Integrated Machine-Learning-Based Work Flow and Application in the Brazilian Presalt (Santos Basin)

Maria Anna Abreu de Almeida dos Reis *, Andrea Carvalho Damasceno , Carlos Eduardo Dias Roriz, André Leonardo Korenchendler , Atilas Meneses da Silva, Eric da Silva Praxedes and Vitor Gorni Silva

Petróleo Brasileiro S.A.—PETROBRAS, Av. República do Chile, 65, Rio de Janeiro 20031-170, Brazil; adamasceno@petrobras.com.br (A.C.D.); roriz@petrobras.com.br (C.E.D.R.); andre.korenchendler@petrobras.com.br (A.L.K.); atilasmeneses@petrobras.com.br (A.M.d.S.); epraxedes@petrobras.com.br (E.d.S.P.); vgorne@petrobras.com.br (V.G.S.)

* Correspondence: maria.reis@petrobras.com.br

Abstract: The capacity to predict the occurrence and quality of source rocks in a sedimentary basin is of great economic importance in the evaluation of conventional and non-conventional petroleum resources. Direct laboratory examinations of rock samples are the most accurate way to obtain their geochemical properties. However, rock information is usually sparse, and source rocks are often sampled at positions that may not be representative of the average organic content and quality of oil kitchens. This work proposes a work flow supported by machine learning methods (random forest, DBSCAN, and NGBoost) to automate the source rock characterization process to maximize the use of available data, expand data information, and reduce data analysis time. From the automated quality control of the input data through the extrapolation of laboratory measurements to continuous well logs of geochemical properties, culminating in the 3D estimation of these properties, we generate volumes of total organic carbon (TOC) by applying machine learning techniques. The proposed method provides more accurate predictions, reducing uncertainties in the characterization of source rocks and assisting in exploratory decision making. This methodology was applied in the presalt source rocks from Santos Basin (Brazil) and allowed us to quantify the TOC distribution, improving the interpretation of the main source rock interval top and base based only on seismic amplitude data. The result suggests higher TOC values in the northern and western grabens of the studied area and a higher charge risk in the eastern area.

Keywords: source rock; TOC; machine learning; Santos Basin; presalt

Citation: Reis, M.A.A.d.A.d.; Damasceno, A.C.; Roriz, C.E.D.; Korenchendler, A.L.; Silva, A.M.d.; Praxedes, E.d.S.; Silva, V.G. Source Rock Evaluation from Rock to Seismic Data: An Integrated Machine-Learning-Based Work Flow and Application in the Brazilian Presalt (Santos Basin). *Minerals* **2023**, *13*, 1179. https://doi.org/10.3390/min13091179

Academic Editor: Stavros Kalaitzidis

Received: 14 June 2023
Revised: 4 August 2023
Accepted: 6 August 2023
Published: 8 September 2023

Copyright: © 2023 by the authors. Licensee MDPI, Basel, Switzerland. This article is an open access article distributed under the terms and conditions of the Creative Commons Attribution (CC BY) license (https://creativecommons.org/licenses/by/4.0/).

1. Introduction

Risk assessment concerning the effectiveness of a petroleum system's elements and processes plays a major role in petroleum exploration. Source rocks are essential elements for the existence of unconventional resources (shale oil or shale gas) or conventional petroleum accumulations. Our major goal in this work is to assist a more accurate calculation of exploratory risks, particularly the assessment of hydrocarbon charge, by improving the quality and the vertical and lateral resolution of source rock characterization at the basin scale, applying an agile integrated approach using machine learning techniques.

The amount of total organic carbon (TOC) in source rocks depends on the balance between primary productivity, preservation, and mineral dilution controlled by the sedimentation rate. The kerogen quality is related to the type of organic matter deposited in the depositional substrate and the preservation degree, which is mainly regulated by the redox potential in the water column and within sediments. Source rocks under appropriate thermal evolution, as defined by temperature and time, can reach the process of generation and expulsion of petroleum [1].

Traditionally, source rocks are identified and characterized through geochemical analysis of rock and petroleum samples from wells. Rock-Eval pyrolysis is a fast method (approximately 30 min per analysis) and requires a small amount of pulverized rock. It is based on the selective detection and quantification of hydrocarbon and oxygenated compounds released by pyrolysis of organic matter on a predetermined heating schedule [2]. The amounts of hydrocarbons and CO_2 generated are measured as peaks as a function of time and recorded in the form of a pyrogram. Combined with TOC measurements, it is frequently used to measure the quantity, quality, and thermal maturity of organic matter in rock samples [1]. However, as the main objectives of exploration and production wells are reservoir rocks, source rocks samples are usually sparse, and are often sampled at positions that may not be representative of the average organic content or quality of the oil kitchens. To circumvent the limited core sample data, different methods have been proposed to obtain an estimate of the total organic carbon using geophysical well log data (e.g., [3,4]). However, source rock characterization involves the quantification of other geochemical properties, such as the hydrogen index (HI), hydrocarbon potential (S_2), and maturity (e.g., Tmax), which can also be obtained through Rock-Eval pyrolysis. We take all these properties into account in our approach.

Source rocks rich in organic matter tend to have lower density values than non-source rocks with the same mineralogy and burial, in association with higher gamma rays (GR), resistivity (RT), neutron porosity (NPHI), and slowness (DT) (e.g., [3,5,6]). Furthermore, their properties vary in relation to the type and thermal history of the organic matter [3]. Due to the vertical resolution of the logs (from 0.2 to 0.8 m in the cases of the logs used in this work), the source rocks can be identified even when their thickness is below the resolution of the curves used, although the quantification is imprecise.

Although well log data provide information with a relatively good vertical resolution, they are scattered in the basin and provide only local information. Seismic data can provide reliable information to spatially guide the identification of source rocks and the characterization of their geochemical properties, especially their organic content. Seismic inversion is an efficient technique to infer elastic properties of rocks that, in conjunction with geophysical logs and geochemical analysis, allows for an integrated characterization of the potential source rock interval with good vertical and lateral resolution (e.g., [7–10]).

Several authors have applied different methods to characterize the petrophysical and elastic properties of shales rich in organic matter and attempt to predict their occurrence and spatial variation using seismic data (e.g., [11–13]). The conventional seismic approach to source rock characterization is performed by calibrating linear regressions between acoustic impedance (P-impedance, the product of compressional wave velocity and density) and the TOC using well logs and extrapolating these relationships using seismic data to generate volumetric estimations of this property [8].

The presence of organically rich shales tends to reduce the seismic velocities and density and increase the anisotropy in comparison to organic lean shales of similar mineralogy and burial (e.g., [14–16]). Vernik and Nur (1992) [15] considered those changes to be relative to the kerogen content, microstructure, and maturity of the source rock. The physical–chemical interactions with the pore fluids [16] and the pore pressure produced by the conversion of the kerogen into oil [17] can also influence these parameters.

The ratio between compressional (Vp) and shear wave (Vs) velocities (Vp/Vs), which are sensitive to lithology and interstitial fluid, can increase or decrease with the organic content, which can be substantially affected by the variation in the mineral composition of the shales (e.g., [6,18–24]). Because of the lower density and velocity of the organic matter relative to the other minerals present in the rock, the acoustic impedance decreases in a non-linear way with the increase in the organic content [8]. The decrease in acoustic impedance and the increase in anisotropy result in characteristic seismic behavior [10]. The reflections of rich and thick source rocks have high amplitudes when compared to reflections of non-source rocks and are amplified with increased organic content [25], thermal maturation [11], and porosity [23]. The discrimination of each property that affects the seismic response is a big challenge and is usually achieved with rock physics numerical models (e.g., [26]).

The amplitudes of organically rich shales tend to vary with the offset or angle (AVO/AVA, respectively) in contrast to the host rocks (background) [7,10,11,23,25], which justifies the use of AVO analyses for the characterization of source rocks. Zhu et al. (2011) [19] observed that variation in lithology can significantly influence the AVO response of the source rock, since it is related to the VP/VS ratio. Numerical models of rock physics revealed that the mineralogy of the source rock can influence the class of seismically observed AVO anomaly. Silica-rich source rocks can result in AVO class 3, which occurs when there is a reduction in P-impedance and VP/VS (or Poisson's ratio) between the overlying (enclosing) rock and the source rock. On the other hand, clay-rich source rocks can result in AVO class 4, with reduced P-impedance but with little variation or even an increase in the Poisson ratio between the overlying and source rock.

The verification of the strong correlation between the reduction in the acoustic impedance and the increase in the TOC, combined with the availability of seismic stacks from different ranges of angles or offsets, allows for the quantitative evaluation of the source rock, increasing the reliability in the prediction of its occurrence and distribution [10]. Seismic attributes (any quantity calculated from the seismic data) can also provide relevant information and can highlight amplitude, phase, and frequency changes in seismic data.

Del Monte et al. (2018) [27] compared the signature of source rocks using different methods, including inversion, AVO analysis, and seismic attributes. Despite comparing the results of each methodology, the integration and interpretation of data were performed independently. As each elastic property has a different sensitivity to the petrophysical properties of the rock, the use of multiple properties for TOC prediction instead of just one (as is conventionally done) allows for a reduction in the intrinsic ambiguities between the effects of porosity, pore shape/fracture, and the amount and maturity of organic matter (OM). In this work, we use a broad set of elastic attributes as an input to a machine learning model that relates elastic and geochemical properties. The application of machine learning techniques provides some advantages compared with the aforementioned conventional approaches. For example, machine learning enables the review of large volumes of data to discover specific trends and patterns that would not be apparent to humans, and no human intervention is needed, allowing them to make predictions and improve the algorithms on their own. Moreover, machine learning techniques are good at learning non-linear representations from multidimensional data [28]. Thus, machine learning methods make it possible to obtain more accurate predictions of the properties, reducing ambiguities and enabling a better separation of the previously listed effects, which overlap in the elastic responses of the source rock [29]. After comparing numerous methods, random forest [30], DBSCAN [31], and NGBoost [32] were chosen in this work for their superior performances. A possible pitfall in the seismic estimation of geochemical properties is the ambiguity between the properties of the reservoir and the source rock intervals rich in organic matter. The behavior of elastic properties and electrical well logs may be similar in both situations. The use of multiple attributes tends to reduce this ambiguity. Another possible risk in applying this methodology is overfitting the machine learning model. To minimize these effects, we analyzed graphs of the variation of the metrics in the training, test, and validation data and applied a cross-validation process.

The main objective of this work was to develop an integrated work flow from rock samples to seismic data based on the use of machine learning algorithms for source rock evaluation. The strengths of the proposed methodology are the inclusion of automated quality control of the input data and the estimation of the quantity, quality, hydrocarbon potential, and maturity of the organic matter. The use of geochemical data from rock samples, well logs, and seismic data on a machine learning basis allows us to maximize the use of high-quality data, improving estimates of geochemical properties in terms of assertiveness, efficiency, and speed. The geochemical volumes (with adequate vertical and lateral resolution) support a more accurate calculation of exploratory risks, notably those related to the assessment of uncertainties regarding petroleum charge assessment.

2. Geologic Setting

The eastern Brazilian basins are classified as continental rift basins and are related to the rupture of the Gondwana Supercontinent and, consequently, the opening of the South Atlantic Ocean. They were formed during the Lower Cretaceous, when a thick succession of continental, fluvial and lacustrine sediments, siliciclastic and carbonate, were deposited in salty and freshwater lake environments controlled by the extensional stresses of the rift phase. Locally, intercalation of volcanic rock can be found. After the rift interval, the thermal subsidence phase began, with features of gravitational slip (e.g., [33]).

The Santos Basin, located in the southeastern Brazilian margin, is the biggest offshore Brazilian basin, with an area of more than 350 thousand square kilometers along the coasts of the sates of Rio de Janeiro, São Paulo, Paraná, and Santa Catarina [34] and is limited by the Cabo Frio structural high in the north and by the Pelotas structural high in the south (Figure 1). Recently, the Santos Basin became the largest producer of oil and natural gas in Brazil due to the discovery of large oil fields, such as Tupi and Búzios, after confirmation of the presalt play in 2006 through the drilling of the wildcat 1-BRSA-329D-RJS (Parati).

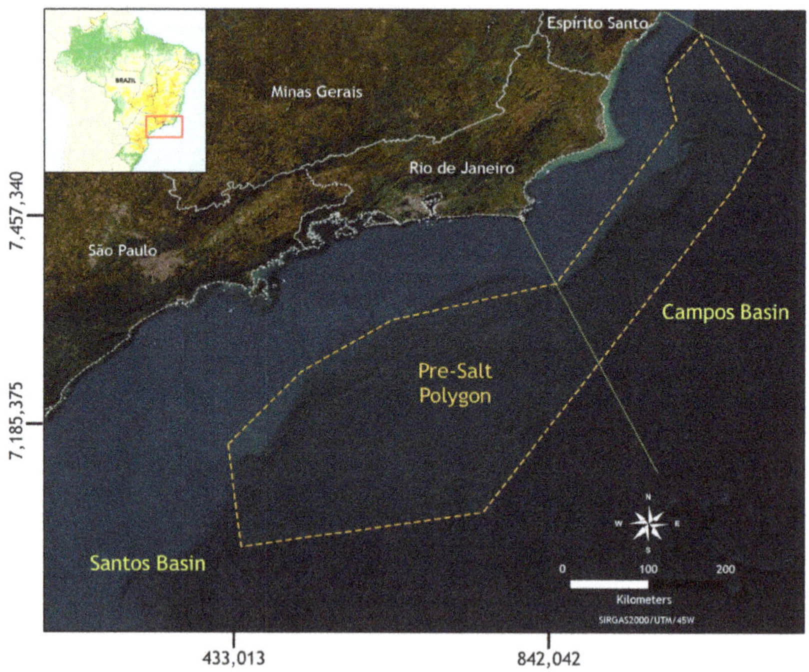

Figure 1. Location of the Santos Basin and the presalt province. The exact location of the research area cannot be shown for confidentiality reasons.

The presalt play is composed of two reservoirs: the main sag reservoir, which was deposited during the early and late Aptian, and the rift reservoir, which was deposited during the Late Barremian and early Aptian (Itapema Formation). The Itapema Formation, corresponding to the Jiquiá Brazilian Local Stage, is characterized by the intercalation of carbonates and black shales [34] and is the main source rock for the hydrocarbon accumulations in the basin. The wells that reached this interval have proven excellent characteristics for petroleum generation, with TOC up to 16% and hydrocarbon potential (S_2) up to 149 mg HC/g rock, possibly related to Ocean Anoxic Event 1a (Freitas et al., 2022 [35] and references therein).

3. Methods

To achieve the intent of this work, an area with a considerable number of wells that acquired data of the Jiquiá source rock (Itapema Formation) and covered by high-quality seismic data was selected. The work flow for source rock characterization is summarized in Figure 2. As presented by Damasceno et al. (2022) [36], the first steps consist of the application of automated quality controls for the measurements of geochemical properties in rock samples and well logs. Next, we fit a machine learning model relating the basic suite of well logs to the measurements of geochemical properties from the automatically validated rock samples. The final step is the fit of a model correlating elastic properties and the TOC well logs predicted in the previous step.

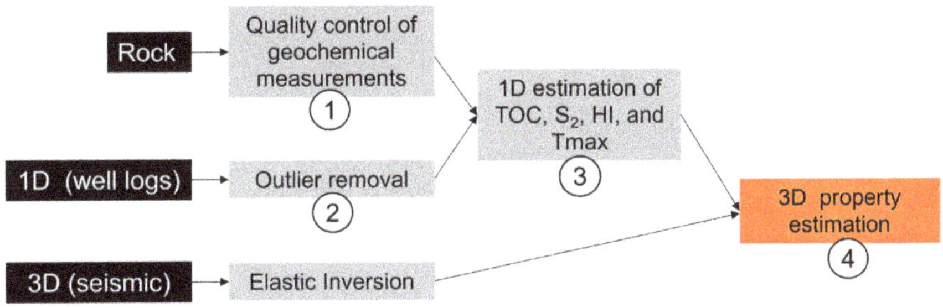

Figure 2. Work flow for automated source rock assessment. The numbers correspond to the machine-learning-based algorithms described in this work.

The methodology used in the application of each step of the proposed flow is described separately for each algorithm as follows.

3.1. *Algorithms 1 and 2: Quality Control of the Input Data*

The use of poor-quality data can substantially impact strategic decisions. To achieve satisfactory performance using machine learning techniques to predict the rock properties, it is imperative to ensure good data quality before training.

The wells used in this work were drilled with oil-based drilling fluid, which can contaminate the rock samples, considerably affecting the geochemical measurement results, as also reported by Freitas et al. (2022) [35] for presalt rocks. The evaluation of the degree of contamination is a costly task and is traditionally achieved by sample-by-sample evaluation of pyrograms, the results of Rock-Eval pyrolysis, coupled with other geochemical and geological data [37]. Therefore, Algorithm 1, which automatically qualifies the geochemical data and excludes contaminated rock samples, is the first stage of the work flow. Rock-Eval data from 167 rock samples from lacustrine and marine source rocks from different Brazilian sedimentary basins were selected and grouped into two classes (non-contaminated (reliable) or poorly contaminated and contaminated (unreliable)) based on the Rock-Eval product analysis of each rock sample.

As each pyrogram consists of around 1200 data points (one for each time step of the Pyrolysis analysis), we used principal component analysis (PCA) [38,39] to reduce the dimensions of the normalized data to seven components, guaranteeing 99% of the cumulative explained variance. Besides the seven components, the hydrocarbon potential (S_2), the production index (PI), and the number of free hydrocarbons normalized by the TOC (S_1/TOC) were chosen as attributes to train the model after the exploratory analysis (data visualization and analysis). A proportion of 33% of the data was used to evaluate the model's performance. Several machine learning methods were tested, and the random forest method was chosen, as it provided the best classification result on the validation set (Table 1).

Table 1. Metrics for quality control of automated classification of geochemical rock samples into reliable and unreliable.

	PRECISION	RECALL	F1 SCORE
RELIABLE	1.00	0.68	0.81
UNRELIABLE	0.83	1.00	0.91
ACCURACY			0.88
MACRO AVERAGE	0.91	0.84	0.86
WEIGHTED AVERAGE	0.90	0.88	0.87

With the same purpose, Algorithm 2 consists of the evaluation of the quality of the well log data, removing the outliers, mainly based on irregularities in the borehole. It is common for well log data to contain values affected by washouts, as well as other measurements that can be considered outliers for a given work flow. These problematic data need to be removed so that they do not result in misinterpretations in statistical analyses and machine learning flows. The traditional work flow is implemented manually by a geoscientist through the visual evaluation of two-dimensional curves and cross plots. This manual process is very time-consuming, often making it impossible to remove outliers from an extensive database. Accordingly, we developed an automatic outlier removal flow using the DBSCAN (density-based spatial clustering of applications with noise [29]) unsupervised clustering algorithm. The advantage of using DBSCAN for this purpose, compared to other clustering methods, is that in addition to classifying the data into clusters, it also allows for the identification of outliers, that is, points that do not belong to any cluster. The main idea of this application is to use the log data from several different wells so that the machine learning algorithm recognizes the existing patterns in the data and can indicate those measurements that are not part of the expected regular distribution.

3.2. Algorithm 3: 1D Property Estimation

The following step consists of the quantification of the quantity, quality, hydrocarbon potential, and maturity of organic matter using well logs coupled with geochemical data. Traditional methods propose the use of porosity and resistivity logs to estimate the TOC content (e.g., [3,4]). However, it is well known that the physical properties of source rocks allow them to be recognized in other well logs (e.g., [5,6]). Machine learning techniques can help to automatically find relations between those data to quantify not only the TOC but also other geochemical properties (e.g., [40,41]).

The dataset used to train the machine learning models to predict the TOC content and other properties comprises 92 wells. These wells were separated into two distinct datasets: a training dataset, with about 80% of the wells, and a test dataset, with the remainder of the data, acting as a blind test case. The test dataset was used to evaluate the model performance concerning new data, such as the newly drilled well case. The selection of the best model consists of choosing the one with the best performance on the test dataset. The following machine learning algorithms were evaluated: random forest [30], support vector machine (SVR) [42], XGBoost [43], NGBoost [32], and neural networks (multilayer perceptron (MLP)) [44]. TOC and Rock-Eval parameters from the non-contaminated and poorly contaminated cuttings, as well as sidewall and core samples from the Jiquiá source rock from Santos and Campos basins, were selected as targets to perform exploratory data analysis for log selection. At the end of this step, a total of 606 samples were selected, considering only sidewall and core samples due to the high inaccuracy of depths from the cuttings samples. The logs selected as input data to estimate TOC were gamma rays, density, neutron, deep resistivity, compressional sonic, and the burial depth of each well log measurement. These logs were selected due to their known correlation with TOC values [3–6]. The same logs were used for Tmax value estimation, as proposed by Tariq et al. (2020) [40] and Shalaby et al. (2020) [41].

To train these models, the scikit-learn python library was used [45]. Hyperparameter tunning of each algorithm was performed using scikit-learn's grid search with cross validation tool. This tool allows several parameters to be tested and compared. For each model, the best parameter combination is obtained by selecting the one with the lowest average error on the cross-validation dataset. In this study, 5 cross-validation groups were used. Finally, the best parameter of each algorithm was used to train a model with all the data available in the training dataset. To compare the performance of the models, we evaluated the correlation coefficient between the test data and the results obtained by each model, with Pearson and Spearman correlations explaining linear and non-linear relationships, respectively (Table 2). The F1 score was obtained by considering the TOC quality (poor when TOC < 0.5, fair when $0.5 \leq$ TOC ≤ 1, and good when TOC > 1 [46]) generated by the discretization of the TOC values. The models, in general, showed equivalent performance regarding the correlation between the estimates and the measured laboratory data. However, the NGBoost model showed better performance in terms of F1 score (Table 2). Additionally, besides predicting only a real value, NGBoost incorporates uncertainty estimation through probabilistic prediction, which is the approach whereby the model outputs a full probability distribution [32]. Accordingly, this was the model selected as the best model for TOC and Tmax estimation.

Table 2. Metrics for each applied method.

METHOD	PEARSON	SPEARMAN	F1-SCORE
RANDOM FOREST	0.72	0.77	0.68
SVR	0.66	0.76	0.67
MLP	0.74	0.81	0.68
XGBOOST	0.74	0.81	0.67
NGBOOST	0.72	0.80	0.74

A linear regression between the TOC and the hydrocarbon potential (S_2) measurements was used to obtain the S_2log, and the hydrogen index (HI) was derived from the following relation: $S_2\text{log}/\text{TOClog} \times 100$.

3.3. *Algorithm 4: 3D Property Estimation*

The low density and velocity characteristic of organic matter allow source rocks to be identified in the seismic data. Therefore, the following step is the 3D estimation of the geochemical properties. Although the application in this work was limited to obtaining a volume of TOC, it can also be applied to estimate other geochemical parameters. As shown in Figure 3, the set of elastic attributes used as an input for this study show a trend similar to that of the TOC well log obtained in the previous step. Hence, the focus of the methodology developed in this work is to capture these relationships according to the fit of a machine learning model relating the properties.

To train the model for 3D prediction of geochemical properties, we used a set of elastic attributes from the seismic inversion as features to obtain the TOC values, as represented by the TOC well log (target) described in Algorithm 3.

As in the 1D property estimation case, NGBoost with decision tree as the base learner was the chosen model. The training data were composed of the continuous TOC well logs (from Algorithm 3) from 3 wells (target), which represent a total amount of 626,372 samples for training, and the respective features were traces of each elastic attribute extracted at the location of the training wells. For validation, we used one well, which corresponds to 17,972 samples (16% of data). After the training step, we applied the trained model to the test data using one blind well with 26,758 samples (25% of data). All wells used in this step are located inside the limits of the 3D seismic data.

Before training, it was necessary to filter the TOC well logs to adapt the frequency of features and the target, since the well logs (target) have a wider-frequency bandwidth than the traces of the seismic inversion attributes (features).

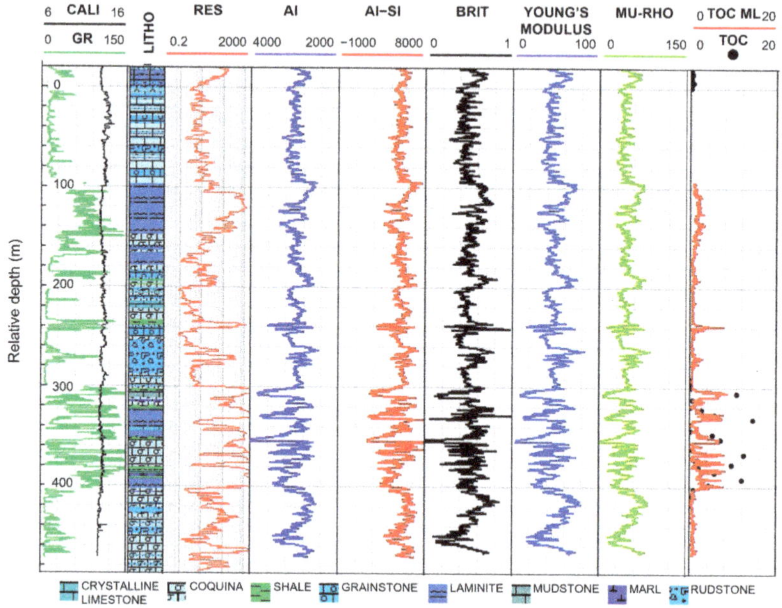

Figure 3. Set of well logs for well 3. From tracks 3 to 7, a set of elastic attributes is exhibited (acoustic impedance, difference between acoustic and shear impedances, brittleness, Young's modulus, and Mu-Rho). On track 8, the TOC log is overlaid with the laboratory measurements of this property. Note the good relation between TOC and all the elastic attributes (higher TOC values correspond to lower elastic attribute values).

4. Results and Discussion

As Algorithm 1 was not yet completed at the time of the development of Algorithm 3, validation of the rock measurements was performed manually. Based on the reliable sidewall and core data, the source rock corresponds predominantly to type I organic matter according to the Langford and Blanc-Valleron (1990) [47] diagram (Figure 4) and has excellent hydrocarbon potential, with an average and maximum TOC of 6% and 36%, respectively, and an average S_2 of 46 up to 358 mg HC/g rock.

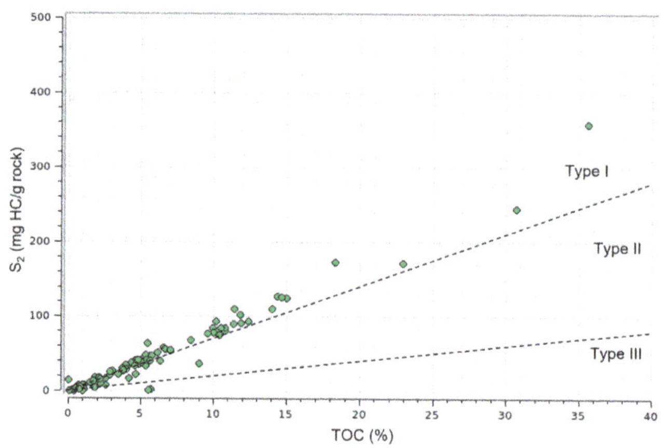

Figure 4. Classification of the source rock measurements using S_2 vs. TOC diagram based on Langford and Blanc-Valleron (1990) [47].

The use of the DBSCAN unsupervised clustering algorithm to recognize and remove outliers from the well-log curves used as features in the TOC prediction flow also presented consistent results. The inputs for the classification were P-sonic, S-sonic, and density well logs. Figures 5 and 6 show the results of the identification of outliers for a given well in the studied area. One can notice that the identified outliers correspond to the higher values of caliper, which suggests that well washout is the cause of the anomalous measurements.

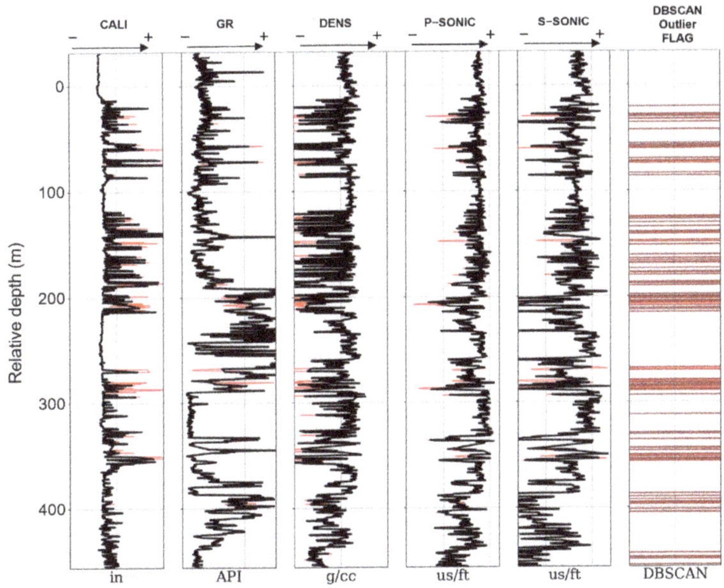

Figure 5. Identification of outliers (red) after performing the DBSCAN algorithm with the following well logs as features: density (DEN), compressional sonic (P-SONIC), and shear sonic (S-SONIC). The outlier flag is shown in the last track. The caliper well log was not used as a feature.

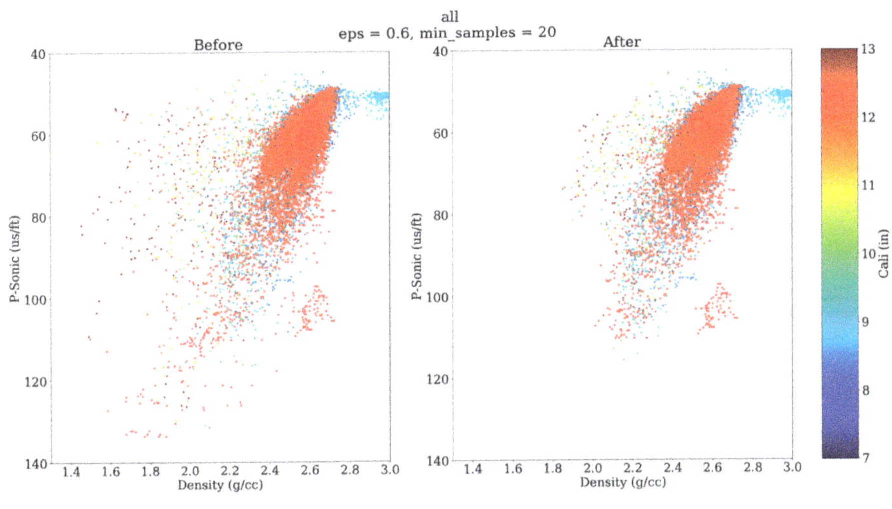

Figure 6. Cross plots of P-sonic as a function of density, colored by caliper before and after the removal of the outliers. Note that the data classified as outlier predominantly correspond to anomalously low densities (lower than 2 g/cc).

The validated rock measurements and well logs were used to train and test the 1D machine learning model, providing good estimations of the geochemical properties continuously in depth, as observed in the blind test result presented in Figure 7 and as indicated in Table 2. One can notice the trend of increasing maturity with increasing depth, as well as the intercalation between carbonates and black shales described by Moreira et al. (2007) [34], where TOC, HI, and S_2 contents tend to be higher in shales, laminites, and mudstones and lower in carbonate reservoir facies.

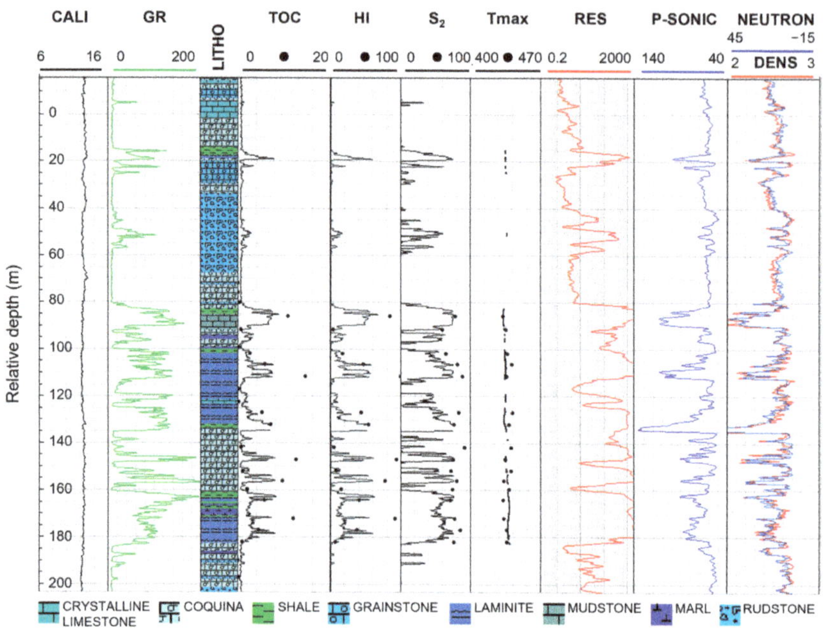

Figure 7. Well logs for well 3: caliper (CAL, in), interpreted lithology (LITHO), Gamma ray (GR, API), resistivity (RES, ohm.m), sonic compressional (P-SONIC, us/ft), density (DENS, g/cm^3), neutron (%); and geochemical logs: TOC (%), HI (mg HC/g TOC), S_2 (mg HC/g rock), and Tmax (°C) for one of the blind test wells, compared with the rock samples measurements (blind test data are represented as black dots, while continuous lines correspond to results obtained by our model).

Different combinations of elastic attributes were tested, aiming for greater accuracy in TOC prediction. Although the importance of the attributes for TOC prediction varies depending on the combination tested in the training, the acoustic impedance (IP) attribute remained with the highest index of importance (Figure 8). The TOC calculated from linear regression with the shear impedance (IS) was used as a benchmark for the prediction (Figure 9), as it is the most conventional approach to estimate TOC volumes from seismic data.

The peak frequency attribute was not used due to its low importance in TOC prediction. Also, we verified that removing the shear impedance (IS) attribute from the list of input attributes did not reduce the prediction accuracy, probably due to redundant attributes in the set generated from simple arithmetic combinations of IS with others. Figure 10 shows a blind test for the model whose metrics are indicated to be the best choice of parameters and attributes as features for TOC prediction. Note that the machine-learning-predicted TOC log (red) is very similar to that used as the target for the prediction (blue), whereas the regression with the IS attribute alone (green) does not provide a satisfactory prediction of this property. The values of Pearson's correlations between the predicted well logs and the target, also shown in Figure 10, validate the greater accuracy of NGBoost prediction. In addition to greater accuracy in prediction using machine learning compared to the benchmark, another advantage of using NGBoost is predicting parameters of a normal statistical distribution, allowing for the estimation and analysis of uncertainty in the predictions. The red line represented as the TOC prediction using NGBoost in Figure 10 is the P50 of the prediction, and the highlighted red space around the curve is the interval between P10 (pessimistic) and P90 (optimistic) predictions.

Figure 8. Feature importance of tested attributes. P-impedance always shows the highest value of importance. Among all nine tested attributes, peak frequency and S-impedance were discarded for the final training set.

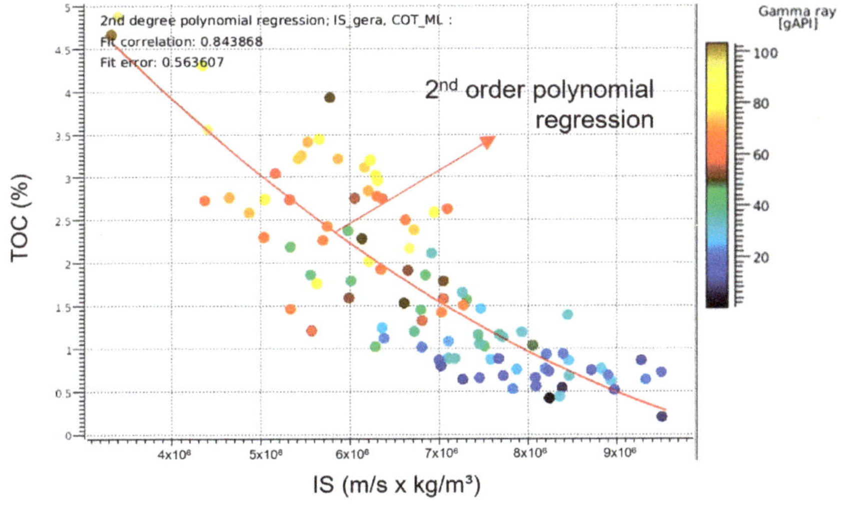

Figure 9. TOC as a function of S-impedance (IS). The linear regression is calculated as TOC = 8.57 − 1.37 × 10^{-6} × IS + 5.31 × 10^{-14} × IS^2.

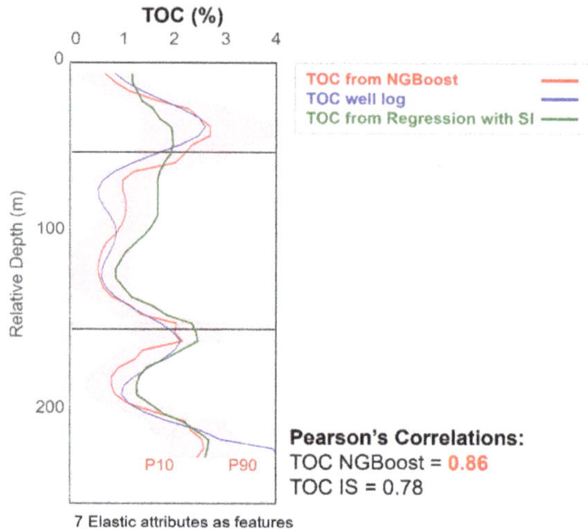

Figure 10. The plot above shows the results of the total organic carbon (TOC) prediction. The red strip is the range of possible TOC values between the P10 and P90 quantiles. The blue line is the well-log TOC (target), the red line is the NGBoost prediction (or P50), and the gray line is the TOC estimate from the linear regression with the single S-impedance attribute. The TOC calculated only with the S-impedance attribute (green line) shows a lower correlation with the target than that generated using machine learning (highlighted in red on the right side of the figure) in all tested scenarios.

Finally, based on the performed tests, the attributes chosen to generate the final trained model were P-impedance, brittleness, Young's modulus, Lambda-Rho, Mu-Rho, Poisson ratio, and the difference between P- and S-impedance. This trained model was used for the 3D prediction of TOC, as described below.

The vertical resolution of the input seismic data for this study is 105 m, and the inverted data (used as a feature for TOC prediction) have a resolution of 80 m. The inversion resolution gain is due to the deconvolution of the seismic pulse intrinsic to the inversion process. When the inversion results are used as input to machine learning models, which are complex and non-linear, this increase in vertical resolution is further enhanced, which translates into a better definition of the top and bottom of the richest layers in OM. Figure 11 shows a cross section from the P50 predicted TOC volume compared with the target well logs. One can observe a large vertical and lateral variability of TOC content on the source rock, alternating between rich and poor organic matter intervals, possibly due to the intercalation of carbonates and black shales described by Moreira et al. (2007) [34] and observed in the wells. The TOC well logs are shown in Figure 11 (black wiggle) superimposed on an arbitrary cross section of the TOC volume that crosses the wells. Note that the behavior of the TOC log agrees with that of the volumetric estimate of this property, evidencing the good fit between the logs (also generated via ML) and the predicted volume. The property contrasts observed in the wells can be traced laterally. Owing to the good lateral continuity of the estimated volume using ML techniques, the interpretation of the top and base of the main source rock intervals becomes a much easier task (Figure 12). After analyzing the TOC volumes generated in this study, the intervals with the highest TOC became more evident. Figure 13 depicts another cross section of the TOC volume where there is an indication of intervals rich in organic matter. Two of those intervals appear to be present only inside grabens (e.g., in the structural low in Figure 13, where the intervals highlighted by the yellow arrows seem to pinch out to the right-hand side of the figure, against the structural high) and possibly were not yet identified in any drilled wells. That is probably due to those intervals not being deposited over the adjacent structural highs (where there was a shallower paleobathymetry) or being eroded there. As well locations were selected with a focus on exploring better-quality carbonate reservoirs (thus, in shallower paleobathymetries), only the organically rich intervals also deposited on top of structural highs can be identified in wells. In our example, some of these organically rich intervals interpreted from our results were sampled by wells drilled in the area, but they do not appear as such in conventional seismic data. That is possibly due to the lower thickness of those potential source rock intervals relative to the original seismic data resolution, while our TOC volume has a higher resolution than that of the original seismic data, as mentioned above. Thus, TOC volumes can also be used to complement seismic and well data as indicators of potential source rock intervals.

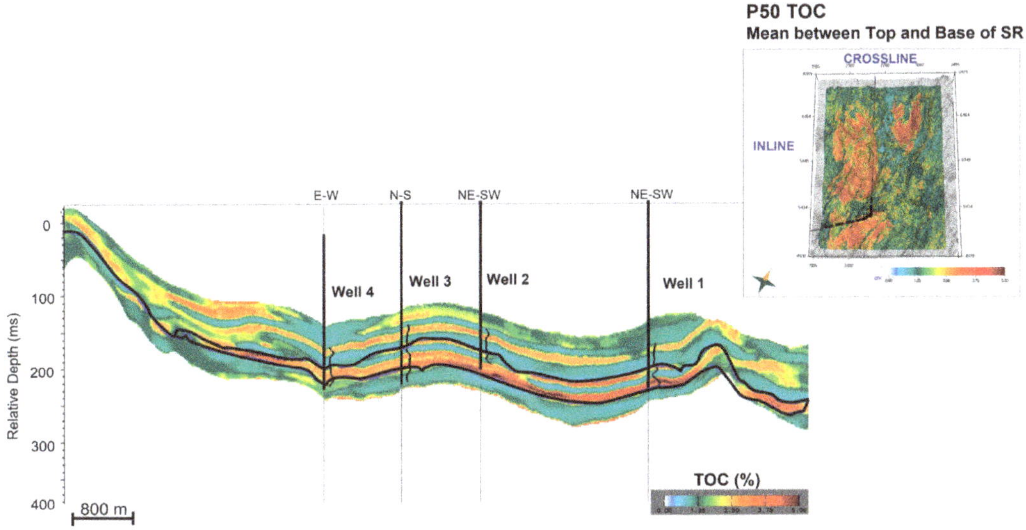

Figure 11. TOC well logs displayed on the same scale as seismic volumes. Note the good fit with the well logs and the vertical resolution gain provided by the ML application. The blue lines are the interpreted top and base of the main source rock (SR) interval.

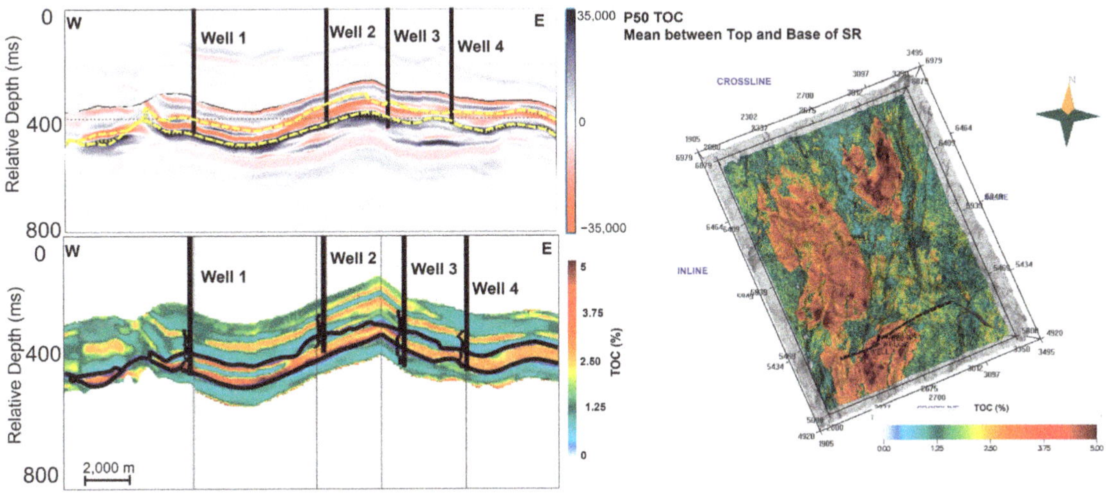

Figure 12. Seismic sections showing the seismic amplitude (above) and predicted TOC (below) with the interpreted top and base of the source rock (yellow dashed lines above and blue lines below).

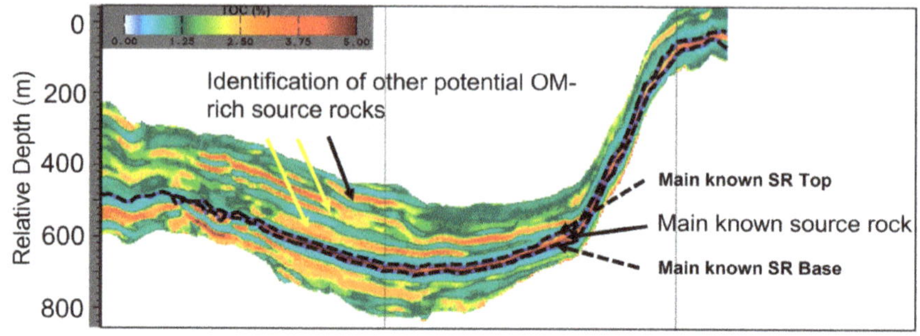

Figure 13. Seismic section showing the predicted TOC volume with the interpreted top and base of the main source rock (SR; black dashed lines). In addition to allowing for tracking of known organically rich source rock intervals, the TOC volume can also indicate new intervals presenting potential source rocks.

Figure 14 displays a map of the average TOC (P50) in the interval between the top and base of the source rock, showing the spatial variations of this property. It can be observed that TOC tends to increase from high structure blocks towards the lows in the northern and western portions of the studied area. However, in the eastern portion, the richness of the source rock is relatively lower. At shallower depths, the source rock was not deposited or was eroded. As the model provides a volumetric probabilistic result, it is possible to obtain a statistical distribution of the predicted property and to extract the information through maps and sections. Figure 15, for example, shows geobodies with high TOC values (higher than 2.5%) for optimistic (P90), pessimistic (P10), and realistic (P50) scenarios. One can observe that even in the pessimistic scenario, there are high values of TOC in the grabens on the north and west, and the eastern area represents the higher charge risk. All the information revealed from the application of this methodology is available for petroleum system analysis assessment and to be incorporated in the numerical basin modeling to reduce uncertainties in the quantification of exploration risks.

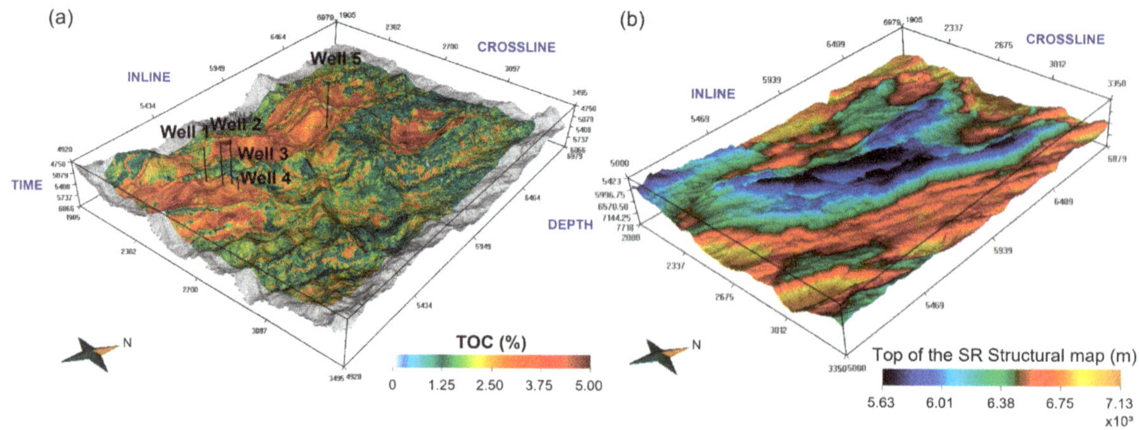

Figure 14. Average TOC map extracted between the top and the base of the source rock interval (**a**) and depth map of the base of the source rock (**b**).

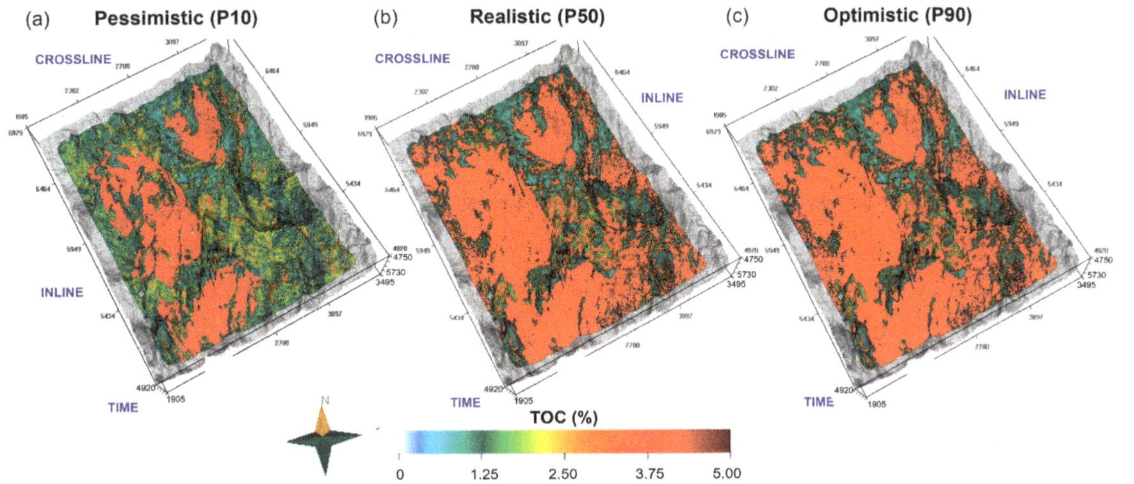

Figure 15. Geobodies of high TOC values (higher than 2.5%, in red) for pessimistic (**a**), realistic (**b**), and optimistic (**c**) scenarios.

5. Conclusions

We propose an integrated work flow based on machine learning methods to characterize source rocks from rock to seismic. Despite being a pilot project and the need to test it in other areas, as shown in the discussion of the results, the methodology provided satisfactory results. It improved the evaluation of input data quality and the estimation of geochemical properties when laboratory data are scarce or absent, ensuring the use of only reliable information and the integration of different datasets at different scales. The use of multiple attributes to estimate TOC combined with the application of machine learning techniques allowed for the estimation of TOC volumes with higher precision and resolution than the input seismic volumes. Once the machine learning models are trained, it is possible to predict 1D and 3D geochemical properties in real time. Therefore, it provided more robust results in a reduced time compared to traditional approaches.

The application of the methodology in the Jiquiá interval suggests that in the studied area, higher TOC values are located in the northern and western grabens, and the eastern area presents a higher charge risk. At the structural high, the source rock was not deposited or eroded. The reported results can be used as input in petroleum systems analysis and are important to mitigate charge risks and fluid assessments in exploration prospects. The interpretation of these results must be deeply allied to the geological knowledge of the area, since, even using sophisticated techniques of multiattribute predictions, ambiguities remain in shales poor in organic matter and reservoirs. We intend to continue this work by applying the methodology to study the Jiquiá source rock in similar areas, as well as to study other source rocks, possibly applying transfer learning techniques.

Author Contributions: Conceptualization, M.A.A.d.A.d.R. and A.C.D.; Methodology, M.A.A.d.A.d.R., A.C.D., C.E.D.R., A.L.K., A.M.d.S., E.d.S.P. and V.G.S.; Validation, M.A.A.d.A.d.R.; Data Curation, M.A.A.d.A.d.R., A.C.D., A.M.d.S. and V.G.S.; Formal Analysis, M.A.A.d.A.d.R., A.C.D., C.E.D.R., A.L.K., A.M.d.S., E.d.S.P. and V.G.S.; Software, C.E.D.R., A.L.K. and A.M.d.S.; Visualization, M.A.A.d.A.d.R., A.C.D., C.E.D.R., A.L.K., A.M.d.S., E.d.S.P. and V.G.S.; Writing—Original Draft Preparation, M.A.A.d.A.d.R., A.C.D., C.E.D.R., A.L.K., A.M.d.S., E.d.S.P. and V.G.S.; Writing—Review and Editing, M.A.A.d.A.d.R.; Supervision, M.A.A.d.A.d.R. and A.C.D.; Project Administration, M.A.A.d.A.d.R. All authors have read and agreed to the published version of the manuscript.

Funding: This research was funded by Petrobras.

Data Availability Statement: Data associated with this research are confidential and cannot be released.

Acknowledgments: The authors thank Petrobras for making this project possible and for allowing the results to be published. The authors also thank their colleagues, Luílson Leal and Sá, Sofia de Abreu e Lima Correia, Taíssa Rego Menezes, Regina Binotto, and Erick Costa e Silva Talarico for their input and support provided for the execution of this work.

Conflicts of Interest: The authors declare no conflict of interest.

References

1. Tissot, B.P.; Welte, D.H. *Petroleum Formation and Occurrence. A New Approach to Oil and Gas Exploration*; Springer: New York, NY, USA, 1978; 538p.
2. Espitalié, J.; Laporte, J.L.; Madec, M.; Marquis, F.; Leplat, P.; Paulet, J.; Boutefeu, A. Méthode rapide de caractérisation des roches mères, de leur potentiel pétrolier et de leur degré d'évolution. *Rev. De L'institut Français Du Pétrole* **1977**, *32*, 23–42. [CrossRef]
3. Passey, Q.R.; Creaney, S.; Kulla, J.B.; Moretti, F.J.; Stroud, J.D. A practical model for organic richness from porosity and resistivity logs. *AAPG Bull.* **1990**, *74*, 1777–1794.
4. Carpentier, B.; Huc, A.Y.; Bessereau, G. Wireline logging and source rocks estimation of organic carbon by the CARBOLOG method. *Log Anal.* **1991**, *32*, 279–297.
5. Fertl, W.H.; Chilingarian, G.V.; Yen, T. Organic Carbon Content and Source Rock Identification Based on Geophysical Well Logs. *Energy Sources* **1986**, *8*, 381–439. [CrossRef]
6. Sun, S.Z.; Sun, Y.; Sun, C.; Liu, Z.; Dong, N. Method of calculating total organic carbon from well logs and its application on rock's properties analysis. In Proceedings of the GeoConvention 2013: Integration, Calgary, AB, Canada, 6–10 May 2013.
7. Carcione, J.M. AVO effects of a hydrocarbon source-rock layer. *Geophysics* **2001**, *66*, 419–427. [CrossRef]
8. Løseth, H.; Wensaas, L.; Gading, M.; Duffaut, K.; Springer, M. Can hydrocarbon source rocks be identified on seismic data? *Geol. Soc. Am.* **2001**, *39*, 1167–1170. [CrossRef]
9. Jia, J.; Liu, Z.; Meng, Q.; Liu, R.; Sun, P.; Chen, Y. Quantitative evaluation of oil shale based on well log and 3-D seismic technique in the Songliao Basin, northeast China. *Oil Shale* **2012**, *29*, 128–150. [CrossRef]
10. Gading, M.; Wensaas, L.; Løseth, H. Source rocks from seismic, Part-2—Applications. In Proceedings of the Eage Conference & Exhibition Incorporating, Copenhagen, Denmark, 4–7 June 2012.
11. Yenugu, M.; Han, D. Seismic characterization of kerogen maturity: An example from Bakken shale. In Proceedings of the Society of Exploration Geophysicists International Exposition and Annual Meeting, Houston, TX, USA, 22–27 September 2013.
12. Sharma, R.K.; Chopra, S.; Vernengo, L.; Trinchero, E.; Sylwan, C. Reducing uncertainty in characterization of Vaca Muerta Formation Shale with poststack seismic data. *Lead. Edge* **2015**, *34*, 1462–1467. [CrossRef]
13. Ouadfeul, S.A.; Aliouane, L. Total organic carbon estimation in shale-gas reservoirs using seismic genetic inversion with an example from the Barnett Shale. *Lead. Edge* **2016**, *35*, 790–794. [CrossRef]
14. Mraz, T.; Dubow, J.; Rajeshwar, K. Acoustic wave propagation in oil shale: 1. Experiments. *Fuel* **1983**, *62*, 1215–1222. [CrossRef]
15. Vernik, L.; Nur, A. Ultrasonic velocity and anisotropy of hydrocarbon source rocks. *Geophysics* **1992**, *57*, 727–735.
16. Vernik, L.; Liu, X. Velocity anisotropy in shales: A petrophysical study. *Geophysics* **1997**, *62*, 521–532. [CrossRef]
17. Carcione, J.M. A model for seismic velocity and attenuation in petroleum source rocks. *Geophysics* **2000**, *65*, 1080–1092.
18. Eseme, E.; Urai, J.L.; Krooss, B.M.; Littke, R. Review of mechanical properties of oil shales: Implications for exploitation and basin modelling. *Oil Shale* **2007**, *24*, 159. [CrossRef]
19. Zhu, Y.; Liu, E.; Martinez, A.; Payne, M.A.; Harris, C.E. Understanding geophysical responses of shale-gas plays. *Lead. Edge* **2011**, *30*, 332–338. [CrossRef]
20. Vernik, L.; Milovac, J. Rock physics of organic shales. *Lead. Edge* **2011**, *30*, 318–323. [CrossRef]
21. Qin, X. Vp-Vs Relations of Organic-Rich Shales. Master's Thesis, Faculty of the Department of Earth and Atmospheric Sciences, University of Houston, Houston, TX, USA, 2013.
22. Qian, Z. Geophysical Responses of Organic-Rich Shale and the Effect of Mineralogy. Master's Thesis, Faculty of the Department of Earth and Atmospheric Sciences, University of Houston, Houston, TX, USA, 2013.
23. Li, W.; Zhang, Z.; Li, Y. Some aspects of excellent marine source rock formation: Implications on enrichment regularity of organic matter in continental margin basins. *Chin. J. Geochem.* **2015**, *34*, 47–54. [CrossRef]
24. Sayers, C.M.; Fisher, K.; Walsh, J.J. Sensitivity of P-and S-impedance to the presence of kerogen in the Eagle Ford Shale. *Lead. Edge* **2015**, *34*, 1482–1486. [CrossRef]
25. Sayers, C.M. The effect of kerogen on the AVO response of organic-rich shales. *Lead. Edge* **2013**, *32*, 1514–1519. [CrossRef]
26. Sengupta, M.; Jacobi, D.; Eichmann, S.; Wallet, B.; Altowairqi, Y.; Alsinan, S. Seismic assessment of maturity and richness in carbonate source rocks. In Proceedings of the International Meeting for Applied Geoscience & Energy, Houston, TX, USA, 28 August–2 September 2022.
27. Del Monte, A.A.; Antonielli, E.; De Tomasi, V.; Luchetti, G.; Paparozzi, E.; Gambacorta, G. Methods for source rock identification on seismic data: An example from the Tanezzuft Formation (Tunisia). *Mar. Pet. Geol.* **2018**, *91*, 108–124. [CrossRef]
28. Goodfellow, I.; Bengio, Y.; Courville, A. *Deep Learning*; MIT Press: Cambridge, MA, USA, 2016. Available online: https://www.deeplearningbook.org (accessed on 2 May 2021).

29. Ij, H. Statistics versus machine learning. *Nat. Methods* **2018**, *15*, 233.
30. Ho, T.K. Random Decision Forests. In Proceedings of the 3rd International Conference on Document Analysis and Recognition, Montreal, QC, Canada, 14–16 August 1995.
31. Ester, M.; Kriegel, H.; Sander, J.; Xu, X. A density-based algorithm for discovering clusters in large spatial databases with noise. In Proceedings of the Second International Conference on Knowledge Discovery and Data Mining, Portland, OR, USA, 2–4 August 1996.
32. Duan, T.; Avati, A.; Ding, D.Y.; Thai, K.K.; Basu, S.; Ng, A.Y.; Schuler, A. NGBoost: Natural Gradient Boosting for Probabilistic Prediction. *arXiv* **2019**, arXiv:1910.03225. Available online: https://arxiv.org/abs/1910.03225 (accessed on 2 May 2021).
33. Estrella, G.; Mello, M.R.; Gaglianone, P.C.; Azevedo, R.L.M.; Tsubone, K.; Rossetti, E.; Concha, J.; Bruning, I.M.R.A. The Espirito Santo Basin (Brazil) source rock characterization and petroleum habitat. *Am. Assoc. Pet. Geol. Bull.* **1984**, *35*, 253–271.
34. Moreira, J.L.P.; Madeira, C.V.; Gil, J.A.; Machado, M.A.P. Bacia de Santos. *Bol. De Geociências Da Petrobras* **2007**, *15*, 531–549.
35. Freitas, V.A.; Vital, J.C.S.; Rodrigues, B.R.; Rodrigues, R. Source rock potential, main depocenters, and CO_2 occurrence in the pre-salt section of Santos Basin, southeast Brazil. *J. S. Am. Earth Sci.* **2022**, *115*, 103760. [CrossRef]
36. Damasceno, A.C.; Korenchendler, A.L.; Da Silva, A.M.; Da Silva Praxedes, E.; De Almeida Dos Reis, M.A.A.; Silva, V.G. Source rock evaluation from rock to seismic: Integrated machine learning based workflow. In Proceedings of the IMAGE, Houston, TX, USA, 28 August–1 September 2022.
37. Peters, K.E. Guidelines for Evaluating Petroleum Source Rock Using Programmed Pyrolysis. *Am. Assoc. Pet. Geol. Bull.* **1986**, *70*, 318–329.
38. Pearson, K. On lines and planes of closest fit to systems of points in space. *Lond. Edinb. Dublin Philos. Mag. J. Sci.* **1901**, *2*, 559–572. [CrossRef]
39. Hotelling, H. Analysis of a complex of statistical variables into principal components. *J. Educ. Psychol.* **1933**, *24*, 417–441. [CrossRef]
40. Tariq, Z.; Mahmoud, M.; Abouelresh, M.; Abdulraheem, A. Data-Driven Approaches to Predict Thermal Maturity Indices of Organic Matter Using Artificial Neural Networks. *Am. Chem. Soc.* **2020**, *5*, 26169–26181. [CrossRef]
41. Shalaby, M.R.; Malik, O.A.; Lai, D.; Jumat, N.; Islam, M. Thermal maturity and TOC prediction using machine learning techniques: Case study from the Cretaceous–Paleocene source rock, Taranaki Basin, New Zealand. *J. Pet. Explor. Prod. Technol.* **2020**, *10*, 2175–2193. [CrossRef]
42. Cortes, C.; Vapnik, V. Support vector networks. *Mach. Learn.* **1995**, *20*, 273–297. [CrossRef]
43. Chen, T.; Guestring, C. XGBoost: A Scalable Tree Boosting System. In Proceedings of the 22nd ACM SIGKDD International Conference on Knowledge Discovery and Data Mining, San Francisco, CA, USA, 13–17 August 2016; pp. 785–794.
44. Rumelhart, D.E.; Hinton, G.E.; Williams, R.J. Learning representations by back-propagating errors. *Nature* **1986**, *323*, 533–536. [CrossRef]
45. Pedregosa, F.; Varoquaux, G.; Gramfort, A.; Michel, V.; Thirion, B.; Grisel, O.; Blondel, M.; Prettenhofer, P.; Weiss, R.; Dubourg, V.; et al. Scikit-learn: Machine Learning in Python. *J. Mach. Learn. Res.* **2011**, *12*, 2825–2830.
46. Peters, K.E.; Cassa, M.R. Applied source rock geochemistry. In *The Petroleum System—From Source to Trap*; Magoon, L.B., Dow, W.G., Eds.; American Association of Petroleum Geologists: Tulsa, OK, USA, 1994; Volume 60, pp. 93–120. [CrossRef]
47. Langford, F.F.; Blanc-Valleron, M.M. Interpreting Rock-Eval pyrolysis data using graphs of pyrolyzable hydrocarbons versus total organic carbon. *Am. Assoc. Pet. Geol. Bull.* **1990**, *74*, 799–804. [CrossRef]

Disclaimer/Publisher's Note: The statements, opinions and data contained in all publications are solely those of the individual author(s) and contributor(s) and not of MDPI and/or the editor(s). MDPI and/or the editor(s) disclaim responsibility for any injury to people or property resulting from any ideas, methods, instructions or products referred to in the content.

minerals

MDPI

Article

Prediction of Reflection Seismic Low-Frequency Components of Acoustic Impedance Using Deep Learning

Lian Jiang [1], John P. Castagna [1,*], Zhao Zhang [2] and Brian Russell [3]

1 Department of Earth and Atmospheric Sciences, University of Houston, Houston, TX 77204, USA; jiang2015leon@gmail.com
2 Chevron, Houston, TX 77002, USA; zhaozhang@chevron.com
3 GeoSoftware, Calgary, AB T2R 0C6, Canada; brian.russell@geosoftware.com
* Correspondence: jpcastagna@uh.edu; Tel.: +1-(713)-743-8699

Abstract: The unreliable prediction of the low-frequency components from inverted acoustic impedance causes uncertainty in quantitative seismic interpretation. To address this issue, we first calculate various seismic and geological attributes that contain low-frequency information, such as relative geological age, interval velocity, and integrated instantaneous amplitude. Then, we develop a method to predict the low-frequency content of seismic data using these attributes, their high-frequency components, and recurrent neural networks. Next, we test how to predict the low-frequency components using stacking velocity obtained from velocity analysis. Using all the attributes and seismic data, we propose a supervised deep learning method to predict the low-frequency components of the inverted acoustic impedance. The results obtained in both synthetic and real data cases show that the proposed method can improve the prediction accuracy of the low-frequency components of the inverted acoustic impedance, with the best improvement in a real data example of 57.7% compared with the impedance predicted using well-log interpolation.

Keywords: low frequencies; acoustic impedance; deep learning; GRU; seismic attributes; seismic inversion

Citation: Jiang, L.; Castagna, J.P.; Zhang, Z.; Russell, B. Prediction of Reflection Seismic Low-Frequency Components of Acoustic Impedance Using Deep Learning. *Minerals* **2023**, *13*, 1187. https://doi.org/10.3390/min13091187

Academic Editors: Paulo T. L. Menezes and Xiaomin Zhu

Received: 2 August 2023
Revised: 3 September 2023
Accepted: 5 September 2023
Published: 10 September 2023

Copyright: © 2023 by the authors. Licensee MDPI, Basel, Switzerland. This article is an open access article distributed under the terms and conditions of the Creative Commons Attribution (CC BY) license (https://creativecommons.org/licenses/by/4.0/).

1. Introduction

The low frequencies in inverted acoustic impedance (AI) profiles are crucial in quantitative seismic interpretation since the impedance can be directly related to reservoir parameters such as porosity and water saturation.

One way to build the low-frequency model is to extract the low frequencies from well logs (e.g., P-impedance, the product of the density and sonic logs) and interpolate the low frequencies from different wells laterally along interpreted horizons. Both the picked horizons and the interpolation can be incorrect. This can cause problems when the geological setting becomes complex or if the number of wells is limited. For instance, the rock properties (e.g., velocity and density) can change laterally abruptly due to faulting or changes in the depositional environment. Another way of predicting the low-frequency components of AI is based on L1-norm regularized sparse spike inversion [1–6]. These methods assume a sparse set of reflection coefficients. Therefore, we can make a prediction of the reflectivity by adjusting the regularization parameter based on the L1 norm. However, we are still not able to fully recover the low-frequency components due to the absence of the low frequencies in seismic data or low signal-noise ratio (SNR), although this method may accurately predict the location of major reflection coefficients. Bianchin et al. [7] proposed a method to predict the broadband acoustic impedance using harmonic reconstruction and interval velocity, which improves the prediction accuracy of the low frequencies in the AI. However, one of the issues in this method is that the interval velocity derived from raw seismic data can be unreliable, and this will make the prediction unstable if the data quality is low.

Another family of low frequencies prediction is to predict the low frequencies of seismic data directly using their high frequencies in the seismic data processing stage. For example, Wu et al. [8] and Hu and Wu [9] proposed direct envelope inversion to retrieve the low wavenumber components based on the instantaneous amplitude and phase of the seismic data. These authors demonstrated that this method could effectively recover the low wavenumber components, improving the full waveform inversion (FWI) of the velocity model. Hu et al. [10] proposed a beat tone inversion to predict the low wavenumber components using the high-frequency components of seismic data. The main idea in their work is that the interference between two signals with slightly different frequencies can be used to demodulate a low-frequency signal using its envelope. Then, they developed a deep transfer learning approach to predict the low frequency based on the beat tone concept [10]. Li and Demanet [11,12] proposed a phase and amplitude tracking method to extrapolate the low-frequency components by decomposing seismic signals into atomic events, which are then parameterized by solving a nonconvex least-squares optimization problem. Sun and Demanet [13] use a deep learning method to directly predict the low frequencies of the raw band-limited shot records using their high-frequency components (e.g., 5.0–20.0 Hz). These works show promise in estimating the low frequencies of seismic data directly using their high-frequency components. In this paper, we use the high frequencies of seismic data to predict their low frequencies using recurrent neural networks (RNNs), which helps to improve the prediction accuracy of AI inversion. Due to the success of the application of deep learning methods in many different fields [14–20], deep learning methods have gained attention in geophysics, both in the industry and academia [7,10,21]. The inverse problems in geophysics are similar to problems in many other fields, such as face recognition, natural language processing, and self-driving cars, and involve building a predictive model to predict the future using preexisting data. Hampson et al. [22] proposed the use of a probabilistic neural network (PNN) to predict well logs directly using seismic attributes. Li and Castagna [23] used the support vector machine (SVM) to classify seismic attributes. Saggaf et al. [24] investigated how to classify and identify reservoir facies from seismic data using a competitive neural network.

Recently, convolutional neural networks (CNN) and recurrent neural networks (RNN) have been successfully applied in many fields [25–27]. These methods have been used on seismic data for a variety of tasks, such as fault detection and facies classification [28–31]. Das et al. [21] predicted acoustic impedance and petrophysical properties using CNNs. Alfarraj and AlRegib [32] proposed a semi-supervised sequence model to estimate the elastic impedance from prestack seismic data.

The prediction of the low-frequency component for inverted acoustic impedance remains challenging. In this paper, we attempt to address this challenge using seismic attributes that may contain low-frequency information. These include the relative geological age attribute and the apparent time thickness attribute. We introduce a deep learning architecture and demonstrate how to predict the low frequencies of AI using various combinations of attributes and methods, first using synthetic seismic data. Finally, we apply the proposed methods to field seismic data from the Midland basin and evaluate the result.

The objective of this paper is to predict and substitute for the missing low-frequency content of the seismic data without the use of well logs or possibly incorrect seismic interpretations of, for example, lithology or mineralogy away from well control. Although we have mineralogy information from the interpretation of the well log curves, such as the volumetric percentage of quartz and limestone, we do not directly use this information to predict the low frequencies between wells because we do not have accurate mineralogy information away from wells. On the other hand, we do not use mineralogy as our output because the relationship between seismic data and mineralogy is highly nonunique and underdetermined and thus beyond the scope of this work.

2. Workflow and Method

In this section, we assume that the seismic signal $s(t)$ can be represented as the convolution of a wavelet $w(t)$ and reflectivity $r(t)$ plus random noise $\delta(t)$:

$$s(t) = w(t) * r(t) + \delta(t), \tag{1}$$

where the reflectivity is defined by

$$r(t) = \frac{I_p(t) - I_p(t-1)}{I_p(t) + I_p(t-1)}, \tag{2}$$

where I_p is the AI at time t. Seismic processing attempts to create this ideal model on the seismic data, but this goal is often not achieved due to problems such as multiples and incorrect scaling. Furthermore, the temporal and spatial variation of the seismic wavelet is only approximately known, even at well locations.

One of the main goals of seismic inversion is to invert Equations (1) and (2) to predict the acoustic impedance using seismic data using various inversion techniques. In conventional seismic inversion [5,6], we first need to extract a wavelet using seismic and well-log data, then construct the objective function based on the physical relationship between the wavelet and seismic data. We can invert the seismic data to AI using a variety of optimization methods [1,3,4].

In contrast, machine/deep learning algorithms try to build the linear/non-linear mapping function between the target inversion parameters and seismic data by training a model using feature-label data, where the "feature" and "label" refer to the seismic data and target inversion parameters, respectively [33].

Potential issues for the conventional inversion methods include: (1) uncertainties caused by inaccurate wavelets; and (2) the oversimplified physics of the forward model in Equation (1). Since the learning-based inversion methods do not require extracting wavelets or making physical assumptions, they offer an alternative approach that may have advantages in certain cases. However, learning-based inversion methods have a high requirement for training data quantity and quality, which might not be met easily in a real case. This can result in poor generalization for predicting new data. The question remains: Can such methods outperform physics-based methods in predicting absolute acoustic impedance?

2.1. Convolutional Neural Networks

Convolutional neural networks (CNNs) have achieved success in many kinds of applications [15,16,18], so we will apply this method here. The CNN architecture we used contains three convolution layers at the beginning and one dense layer at the end of this network (see Figure 1). Each convolution layer is followed by batch normalization and a ReLU activation function. The batch normalization helps reduce the covariate shift and smooths the objective function, which improves the model performance. The ReLU activation function enhances the learning of a non-linear relationship and has been shown to improve the performance over the sigmoidal functions used in older networks. In our case, the inputs are poststack seismic data and attributes after being properly processed, and the target log is the AI or reflectivity. Note that the poststack seismic data contain AVO effects, which could decrease the prediction accuracy of rock properties, here calculated assuming normal incidence. With sufficient training data, a neural network could learn the proper relationship between stacked amplitude and acoustic impedance in the presence of such AVO effects. A better approach would be to use the prestack data or pseudo-P and S sections derived from the prestack data, but we restrict the problem to the normal incidence in this study.

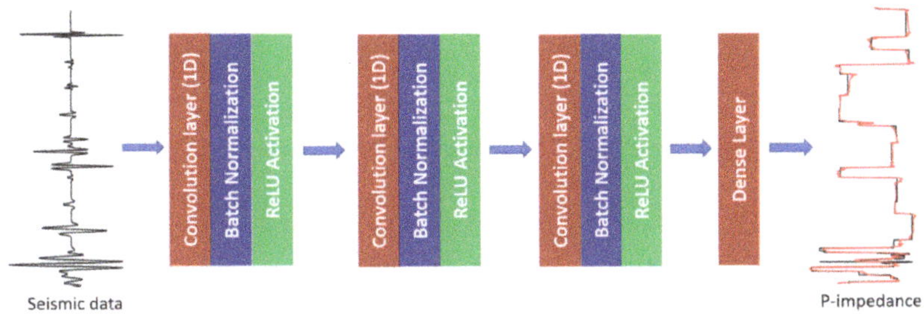

Figure 1. Architecture of CNNs used in this study.

The hyperparameters in this network include the channel, kernel size, and learning rate, which need to be set case by case. Since the theory of deep learning offers insufficient prior insight, the number of convolutional layers used in a real work needs to be tested for each situation.

2.2. Recurrent Neural Networks

Recurrent neural networks (RNNs) are deep learning methods that have been successfully applied to process time series data [14,34]. In this work, we used three bidirectional gated recurrent unit (GRU) stacked layers and one dense layer at the end of this network (see Figure 2). Each stacked layer has three stacked GRUs. The input, in our case, is post-stack seismic data after being properly processed, and the target log is the AI or reflectivity. The number of stacked layers and the number of single GRUs are subject to change in each case. The hyperparameters in this network include the hidden layer size and learning rate, which need to be tested.

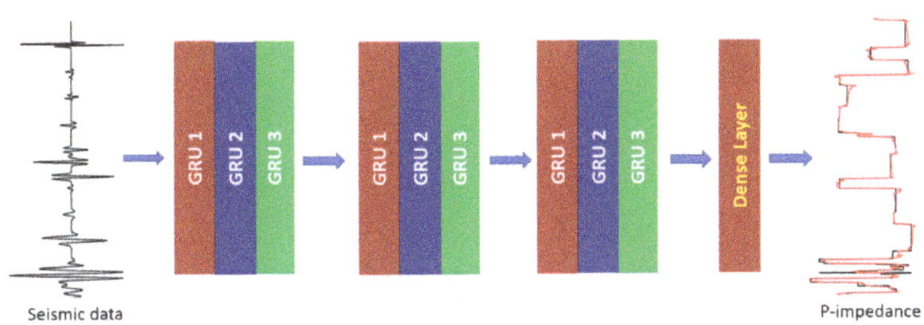

Figure 2. Architecture of RNNs.

3. Results

3.1. Synthetic Data Test

3.1.1. Synthetic Data Test Based on Rock Physics Modeling

The first synthetic data we use to test the proposed methods are produced using rock physics modeling (RPM) and a convolutional model. The data are purely synthetic data, and there is no added noise in this dataset. Therefore, we can use this dataset to evaluate how well each machine learning model performs with perfect data.

We use 20 pseudo wells to train the predictive model, 5 wells for validation, and 5 wells for the blind test. Figure 3 shows the reflectivity predicted from the seismic data using the CNN and RNN methods. The black curve shows the true reflectivity and the red curve shows the predicted reflectivity. The input we use in this example is original synthetic seismic data (seismic amplitudes, not attributes). As seen in this figure, both CNN and RNN successfully predict the reflectivity with high accuracy, with an R^2 score of about 0.86 from the CNN and 0.87 from the RNN and a correlation coefficient (CC) of 0.93

from the CNN and RNN for the blind test. Therefore, both the CNN and RNN algorithms successfully map the physical relationship between seismic data and reflectivity.

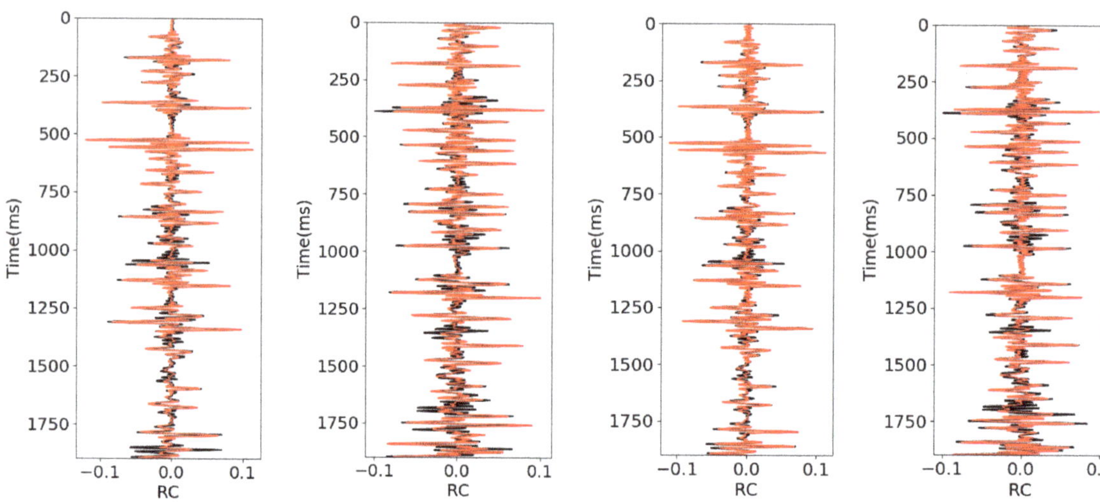

(a) (b) (c) (d)

Figure 3. The reflectivity is predicted from seismic data using different methods. (**a**) Training result from CNN. (**b**) Test result from CNN. (**c**) Training result from RNN. (**d**) Test result from RNN. Red curve: Predicted; Black curve: True. The hyperparameters used in both CNNs and RNNs are: batch size = 2, maximum epoch = 300, and learning rate = 0.005. The activation function used in this test is the tanh function.

We then use the same synthetic data to test if we can predict the AI using the reflectivity. Figure 4 is the predicted AI from reflectivity using the CNN and RNN algorithms. The black curve in the figures is the true AI, and the red curve is the predicted AI. The results show that the CNN does not predict well in this case, with an R^2 score of 0.34 and a CC of 0.70 for the blind test. In comparison, the results from RNN show that it performs well in this task, with an R^2 score of 0.95 and a CC of 0.98 for the blind test.

We then predict the AI directly using seismic data as input. Figure 5 shows the predicted AI from seismic data using the same CNN and RNN approaches used in the last section. The results show that the CNN gives a poor prediction of the AI well in this case, with an R^2 score of 0.20 and CC of 0.77 for the blind test. In contrast, the results from RNN show that it performs well in this task, with an R^2 score of 0.67 and CC of 0.85 for the blind test.

From these three tests, we find that: (1) the RNN is suitable for all three tasks, but the performance deteriorates when it comes to predicting the AI using seismic data directly; (2) the CNN architecture used here is suitable for predicting the reflectivity from seismic data, but not for predicting the AI using either reflectivity or seismic data.

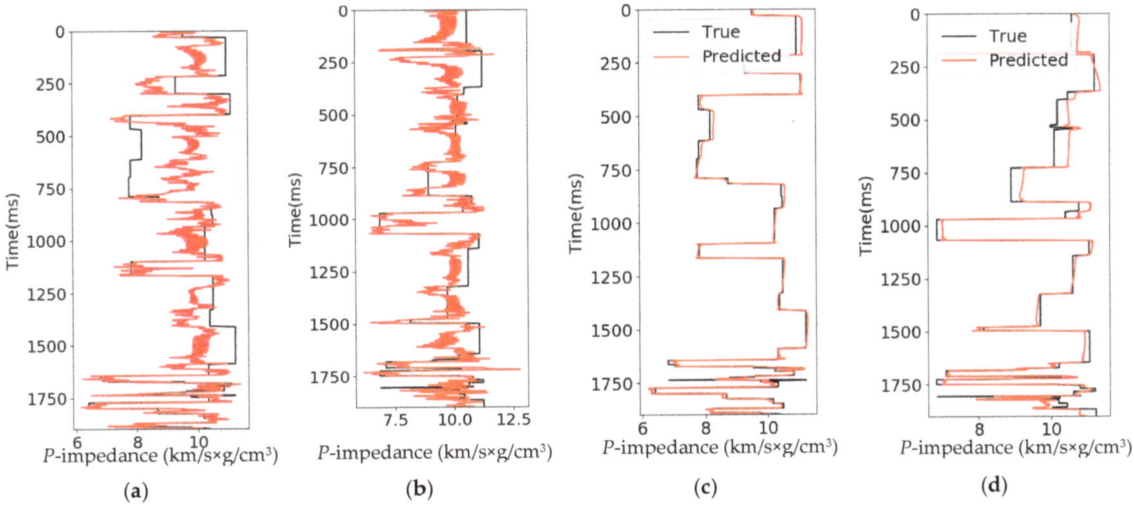

Figure 4. The AI predicted from reflectivity using different methods. (**a**) Training result from CNN. (**b**) Test result from CNN. The CC is 0.70. (**c**) Training result from RNN. (**d**) Test result from RNN.

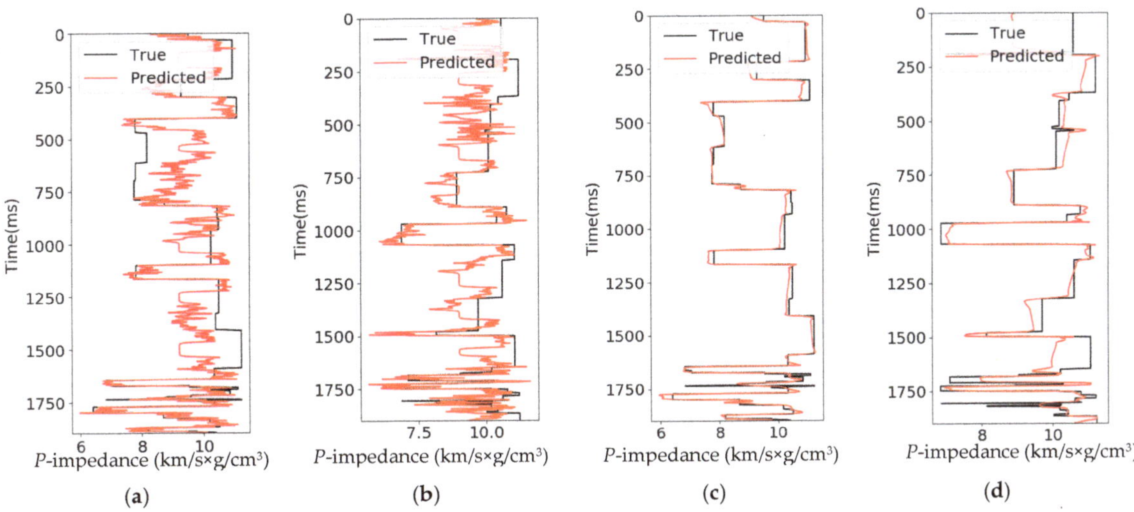

Figure 5. The AI predicted directly using seismic data. (**a**) Training result from CNN. (**b**) Test result from CNN. (**c**) Training result from RNN. (**d**) Test result from RNN.

3.1.2. Marmousi Model 2 Test

To further test how well the proposed methods work, we used the Marmousi model 2 [35]. This model is an extension of the original acoustic Marmousi model created by the Institute Français du Pétrole (IFP) [36], which has the same structure and velocity as Marmousi model 2, but the model size of Marmousi model 2 is wider and deeper and it is fully elastic [35]. Marmousi model 2 is 17.0 km in width and 3.5 km in depth. More details about this model are given by Martin [35]. The model has velocity and density volumes only in the depth domain, so we convert them into the time domain with a time interval of 4.0 ms using the true velocity model. We also reduced the size of the model to a maximum time of 2700.0 ms and a CDP number range of 400 to 2350. We create synthetic seismic data using a 30 Hz Ricker wavelet based on a convolution model to test the proposed methods. Figure 6a is the AI, and Figure 6b is the poststack seismic data after prestack time migration (PSTM) processing.

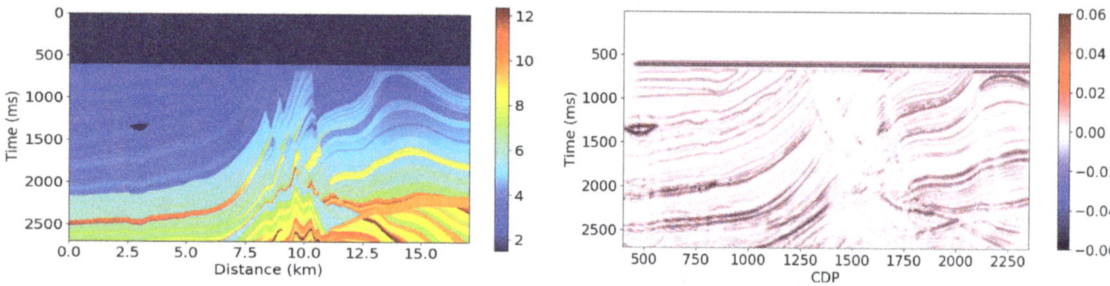

(a) (b)

Figure 6. (**a**) AI. (**b**) PSTM seismic data.

3.1.3. Low-Frequency Attributes

The geological age of a formation at a given location is associated with the depositional environment and various rock properties, such as velocity, density, and porosity. The geological age can, therefore, contribute to the prediction of the low-frequency components of AI. We first choose the horizons and faults from the Marmousi model 2 (see Figure 7a). Then, we use these faults and horizons to construct the relative geological age attribute as an index with different values for each formation (see Figure 7b). The index is approximately proportional to geologic age, but not exactly since it ignores the fluctuation in sedimentation rate and erosion.

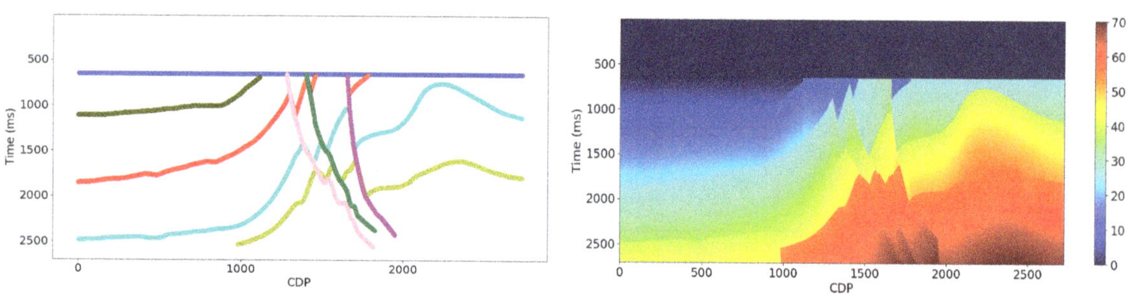

(a) (b)

Figure 7. (**a**) The horizons and faults extracted from the Marmousi model 2 and marked with different colors. (**b**) Relative geological age attribute created using the horizons and faults.

The rock properties are not just controlled by geological age but also governed by burial depth. For instance, rocks deposited at the same geological time but buried in different depths can have different velocities. Therefore, one can use depth to constrain the prediction of the low frequencies in AI. We use the time-depth relationship along the horizons at pseudo-well locations to convert the time to depth. Then, we interpolate the depth vertically to obtain the depth volume. The depth volume here is an attribute that we use to calculate the interval velocity. It is calculated using very limited information (e.g., 5 or 10 pseudo wells). Compared with the true depth and the predicted depth in Figure 8, we find that they give a good match, which means that we can predict the depth attribute using this method accurately. Note that there are some mismatches at around CDP of 1500. This is because of the complexity of the geological structure near this location, which means that the velocity changes rapidly both horizontally and vertically. Since the method we used to calculate the depth attribute uses only a few pseudo wells and horizons, it fails to capture these velocity variations.

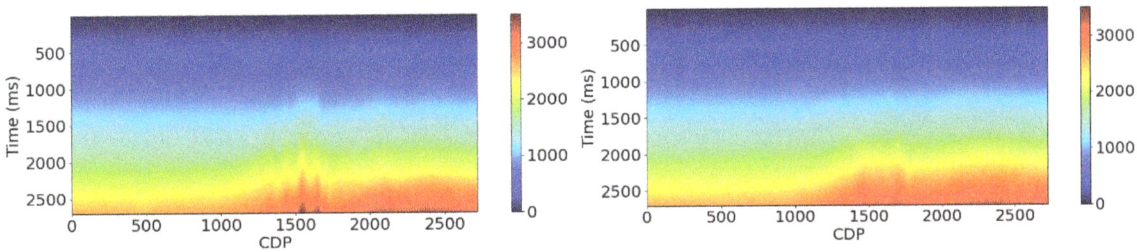

(**a**) (**b**)

Figure 8. (**a**) True depth. (**b**) The depth attribute estimated using the time-depth relationship at pseudo-well locations along the horizons.

With the depth information at each time location, we can then calculate the interval velocity. Figure 9a is the true interval velocity, and Figure 9b is the interval velocity calculated using the depth attribute we create in Figure 8b. Although the details between these two figures are different, the velocity structure looks similar, which means that the interval velocity we create can provide low-frequency information. Note that we only use four horizons in our example. Increasing the number of horizons to calculate the interval velocity would improve the vertical resolution and the prediction of low frequencies.

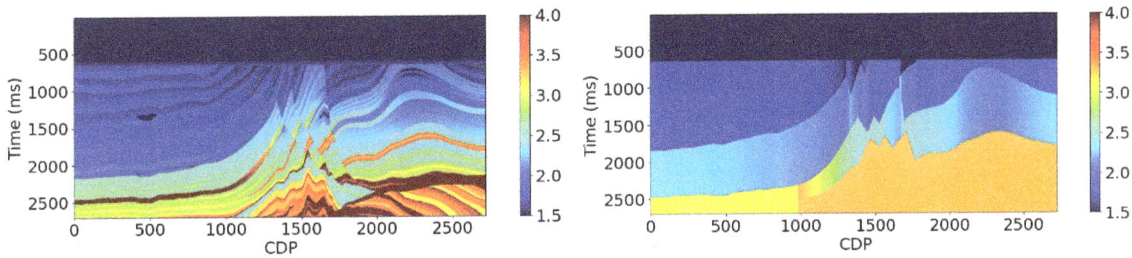

(**a**) (**b**)

Figure 9. (**a**) True interval velocity. (**b**) Interval velocity calculated using the depth attribute.

Instantaneous amplitude has been used to recover the ultra-low frequencies of the seismic signal (i.e., frequencies below the lowest frequency in the source spectrum). Wu et al. [8] proposed an envelope inversion method using a non-linear seismic signal model to recover the low frequencies using instantaneous amplitude. This is based on an empirical relationship between the instantaneous envelope and the low frequencies at a given locality.

In contrast, we use instantaneous amplitude as one input for deep learning models and let the model learn the relationship. We also produce two more attributes (Figure 10) using the instantaneous amplitude: integrated instantaneous amplitude and apparent time thickness (time thickness between adjacent peaks on the instantaneous amplitude curve). Figure 11 shows the 2D sections for each of the attributes showing the apparent geological structures, which could, in turn, be related to the low frequencies. Together, these attributes make a positive contribution to predicting the low frequencies of AI, although it is difficult to evaluate them separately.

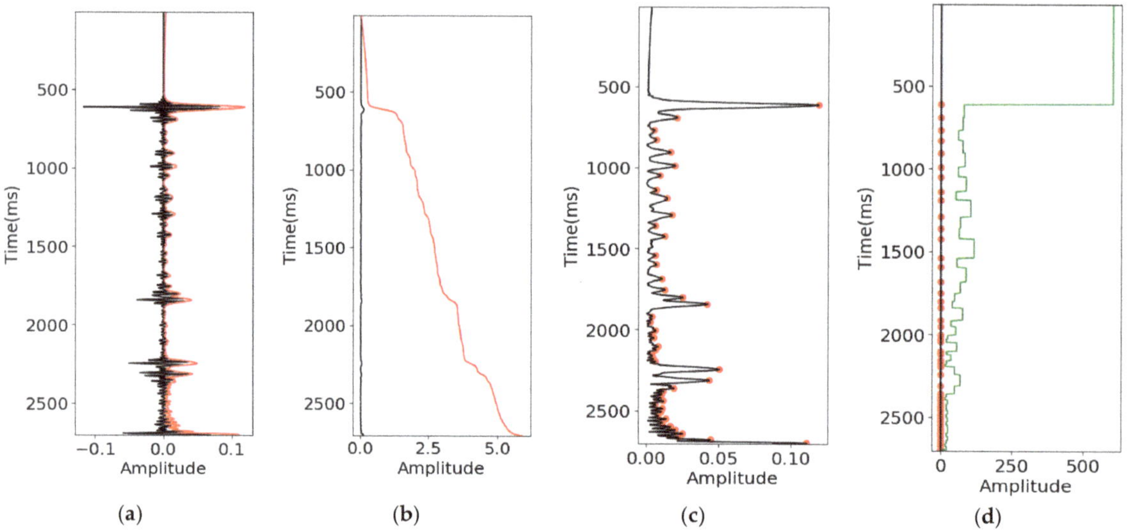

Figure 10. (**a**) Seismic data (black) and instantaneous amplitude (red). (**b**) Instantaneous amplitude (black) (same as the red curve in (**a**)) and integrated instantaneous amplitude (red). (**c**) Instantaneous amplitude (black) and peaks (red dot). (**d**) Instantaneous amplitude (black) and peaks (red dot) and apparent time thickness (green).

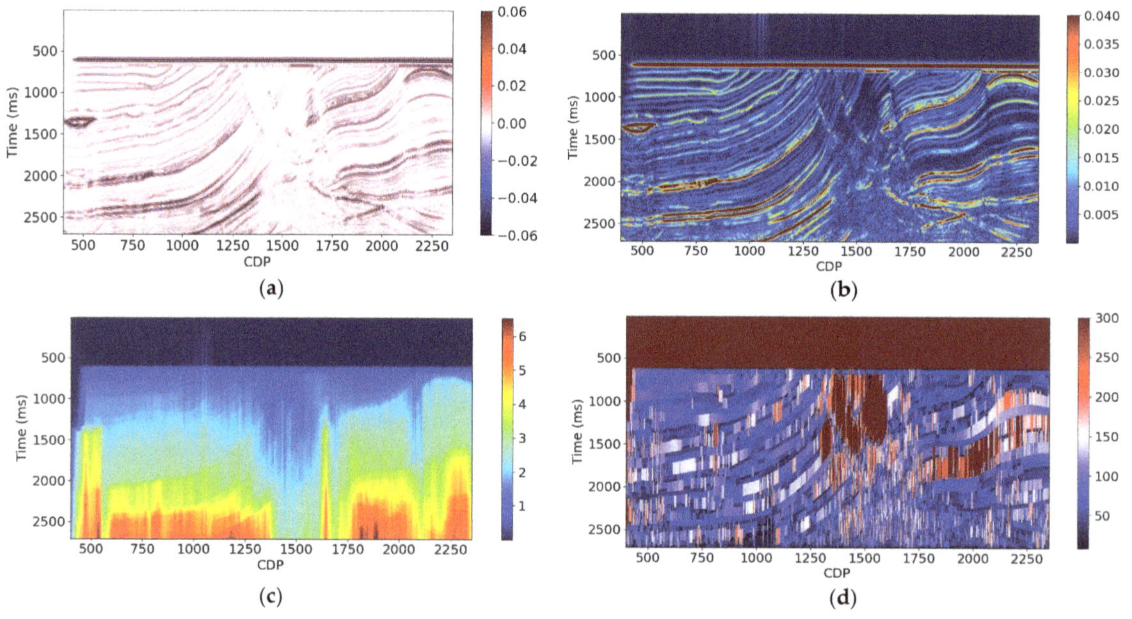

Figure 11. (**a**) PSTM seismic data (black). (**b**) Instantaneous amplitude. (**c**) Integrated instantaneous amplitude. (**d**) Apparent time thickness.

3.1.4. Predicting the Low-Frequency Components of Seismic Data

Seismic data often lacks frequencies below 5.0 Hz and this limits our ability to invert seismic data for absolute rock properties. This section shows how we predict the low-frequency components of seismic data using higher frequencies that are present (e.g., 10.0–40.0 Hz).

We will use the Marmousi model for illustration. Figure 12a shows the simulated PSTM data with a 5–40 Hz pass band. We create 0–5 Hz synthetic seismic data using a convolution model with a 0–1–4–5 Hz bandpass filter (see Figure 12b). This is the low-frequency model we wish to recover from the simulated PSTM data. Next, we choose 10 traces of data from the 5.0–40.0 Hz PSTM seismic data (Figure 12a) and 0.0–5.0 Hz synthetic seismic data (Figure 12b) as the training set. The 10 pseudo wells are generated randomly, and they are marked using black lines in Figure 12a. Finally, we predict the whole section using the model we trained with these 10 traces. We tested various CNN and RNN architectures and found that the RNN achieved better performance.

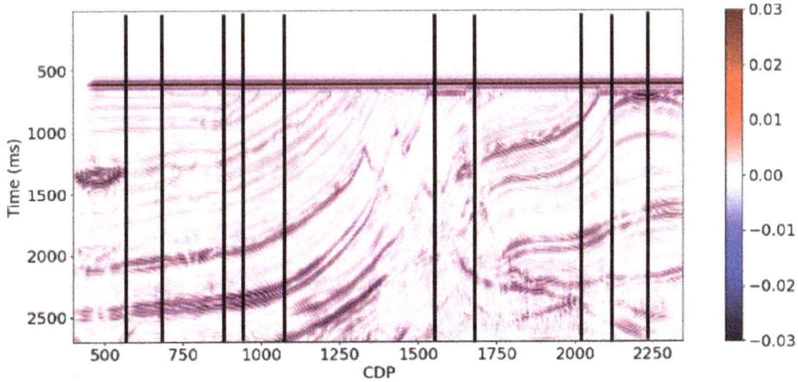

(a)

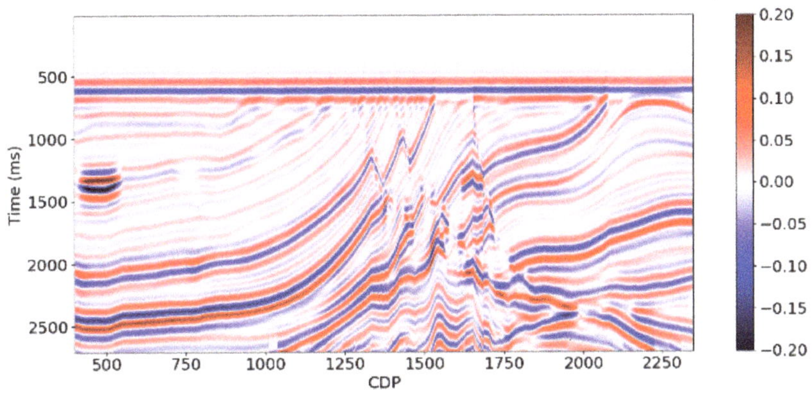

(b)

Figure 12. *Cont.*

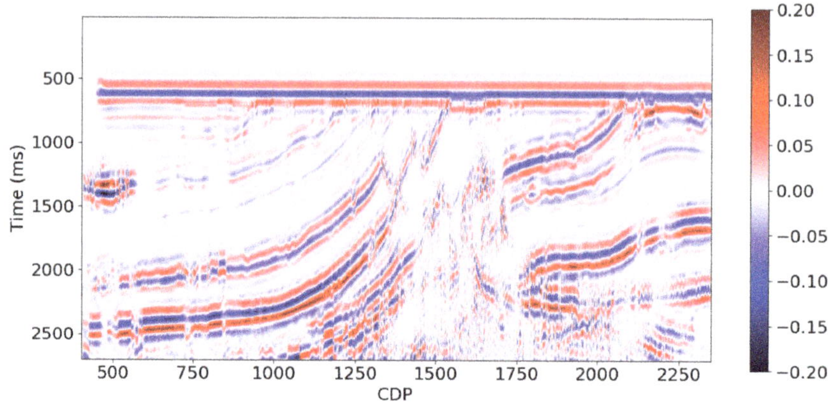

(c)

Figure 12. (**a**) 5.0–40.0 Hz PSTM seismic data. (**b**) Synthetic seismic data created using the convolutional model (0.0–5.0 Hz). (**c**) 0.0–5.0 Hz seismic data predicted using the data in (**a**).

The results from the RNN are demonstrated in Figure 12c. The seismic section (including predicted low frequencies) in Figure 12c looks promising when compared with the seismic section (including true low frequencies) in Figure 12b. Figure 13 shows the training and blind test results taken from the 2D line, with CDP = 2000 and 900, respectively. The black curve is the true synthetic seismic data (0–5.0 Hz), and the red curve is the predicted one. It shows that the R^2 score and CC are 0.92 and 0.96 for training results and 0.74 and 0.86 for blind test results. Considering the original PSTM seismic data are noisy, the prediction accuracy is acceptable. However, as seen in Figure 12c, the results are unstable and produce non-geological variations.

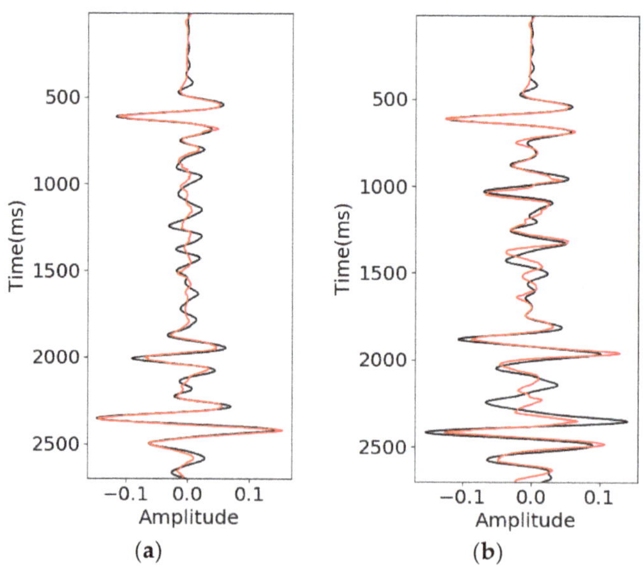

Figure 13. (**a**) Training results from RNN at CDP 2000. The CC is 0.96, and R^2 score is 0.92. (**b**) Validation results from RNN at CDP 900. The CC is 0.86, and R^2 score is 0.74. The black curve is the true synthetic seismic data (0–5.0 Hz), and the red curve is the predicted one.

3.1.5. Predicting the Low Frequencies of AI

With all the attributes we created in addition to the conventional seismic attributes, we can test how each method works using different combinations of seismic attributes as input on the 5 -40 Hz Marmousi-simulated PSTM.

We first compute a range of conventional seismic attributes. Then, we apply the stepwise regression method to optimize both the conventional attributes and the attributes proposed in this paper. In this method, we first select the attribute that has the strongest correlation with the AI; then, we select the second attribute from the remaining attributes list that best fits the regression model with the AI by combining the first attribute selected in the previous step; we keep adding the remaining attributes to the regression model until the model performance improvement cannot meet the criteria we set beforehand; at last, we check the model performance using a validation dataset to ensure there is no overfitting. Finally, we obtain the following 11 best attributes:

- Depth attribute
- Interval velocity
- Average frequency
- Time
- Instantaneous amplitude
- Apparent polarity
- Apparent time thickness
- Amplitude weighted frequency
- Integrated instantaneous amplitude
- Relative geological age
- Filter 5/10–15/20

The additional attributes we created are on this list, so they are helpful for improving the prediction of the low frequencies. We then use these optimized attributes to test different methods with the same 10 traces for training. In this paper, we compared the RNN to two of the most common algorithms. One is the Probabilistic Neural Network (PNN) [22], and the other one is the multi-linear regression of the attributes. The PNN is trained by minimizing the validation error for each training data point by searching for the optimal smoothing parameter sigma (σ). The new target log is then computed by combining the distance of the training and target log attributes and the target logs of the training set (see more details in [22]). Figure 14a,b is the 0.0–5.0 Hz AI predicted using the new and conventional seismic attributes using PNN. Figure 15c,d is the AI (0.0–5.0 Hz) predicted using the predicted low-frequency seismic data and new attributes using RNN. The results from PNN show that the R^2 score and CC are 0.99 and 1.0 for training and 0.95 and 0.98 for the blind test. The results from RNN show that the R^2 score and CC are 0.99 and 1.0 for training and 0.95 and 0.97 for the blind test, respectively. As observed from Figure 14b, the blind test results from PNN have more high-frequency noise than the blind test in Figure 14d from RNN, although the prediction accuracy is close. Figure 15 shows the results predicted using a different combination of attributes and methods, which reveals that the result predicted using the predicted low-frequency seismic data and the new attributes produced in this work using RNN are the best. However, the performance could be different in different parts of this 2D section for different methods. For example, the results predicted from the new attributes and conventional attributes using PNN at around a CDP of 500 are better than the results in Figure 15e at the same location, although the overall prediction accuracy from this result is better than the result from Figure 15b.

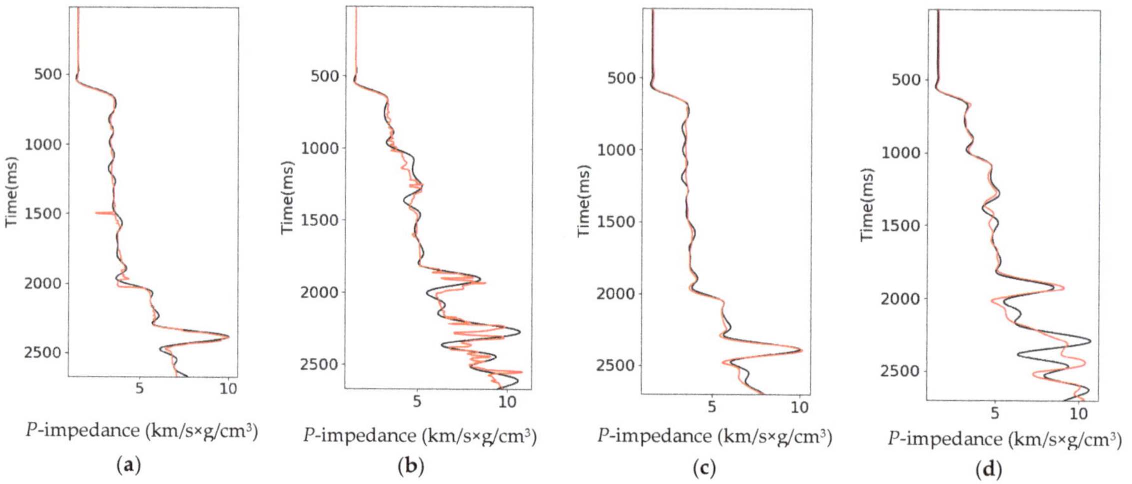

Figure 14. (**a**) Training result from PNN. The CC is 1.0 and the R^2 score is 0.99. (**b**) Validation result from PNN. The CC is 0.98 and the R^2 score is 0.95. (**c**) Training result from RNN. The CC is 1.0 and the R^2 score is 0.99. (**d**) Validation result from RNN. The CC is 0.97 and the R^2 score is 0.95. The black curve is the true synthetic seismic data (0–5.0 Hz), and the red curve is the predicted one.

Table 1 is the validation summary for the AI (0.0–5.0 Hz) predicted from various combinations of attributes and methods. The R^2 score and CC in this table are the average values calculated using the entire 2D line. In comparison, the AI (0.0–5.0 Hz) predicted from the conventional attributes using PNN has the lowest prediction accuracy, with an average R^2 score of 0.84 and CC of 0.94. The result predicted from the proposed new attributes and RNN method achieves the highest prediction accuracy, with an average R^2 score of 0.93 and CC of 0.97. The improvement is 10.7% for the R^2 score.

Table 1. Validation summary for the AI (0.0–5.0 Hz) predicted using various kinds of combinations of attributes and methods. The R^2 score and CC in this table are the average values calculated using the whole 2-D line.

Attributes and Methods	**AI (0.0–5.0 Hz)**	
	R^2 Score	**CC**
Conventional attributes + PNN	0.84	0.94
New and conventional attributes + Linear regression	0.87	0.94
New and conventional attributes + PNN	0.91	0.96
New and conventional attributes + RNN	0.90	0.95
New attributes and predicted low-frequency seismic data + RNN	0.93	0.97

Table 2 shows that the predicted low-frequency components of AI improve the prediction accuracy of the absolute AI, with an improvement of 11.0%, which indicates that the low-frequency components are critical in the AI inversion. RNN shows some benefit over PNN, and the newly calculated attributes make a major contribution to the overall improvement in the prediction of low frequencies in this example.

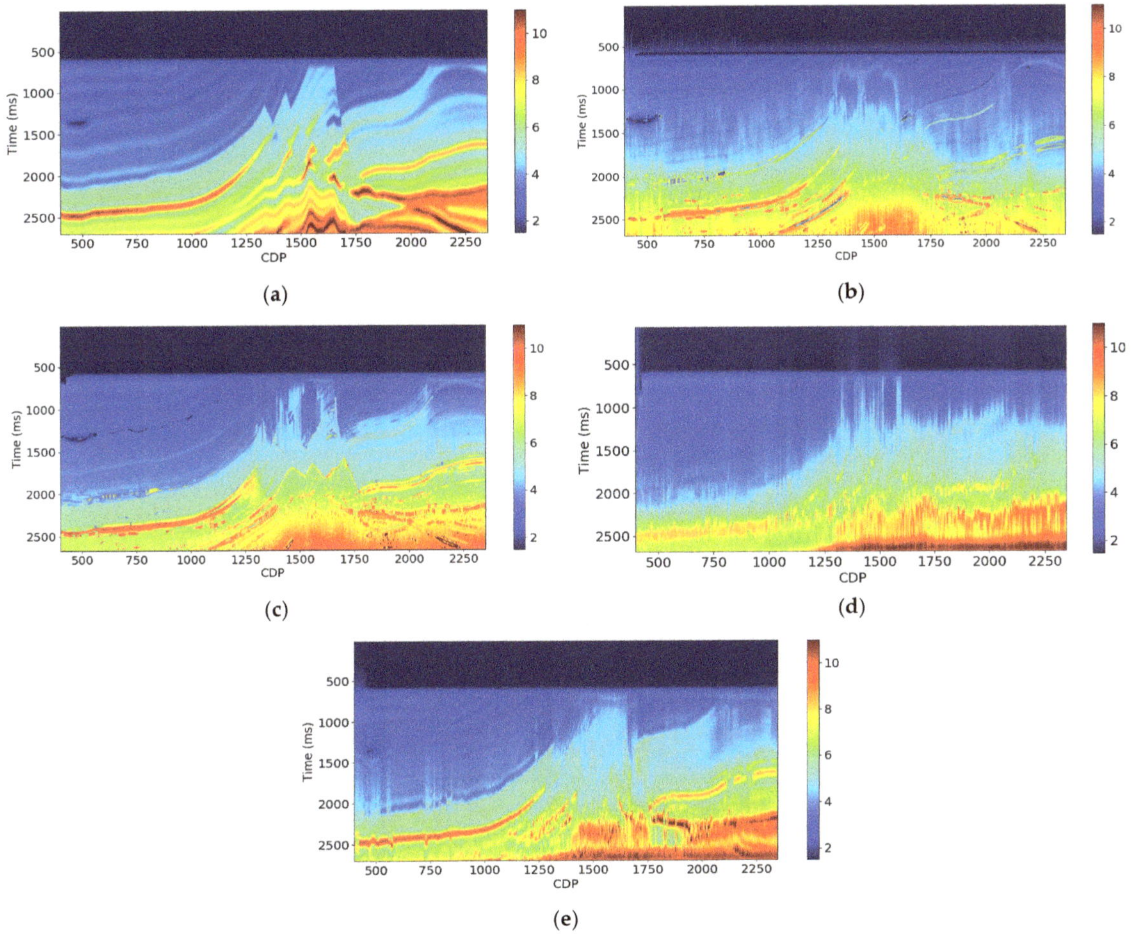

Figure 15. Prediction results from various kinds of combinations of attributes and methods. (**a**) True AI (0.0–5.0 Hz). (**b**) AI (0.0–5.0 Hz) predicted from the conventional seismic attributes using PNN. (**c**) AI (0.0–5.0 Hz) predicted from the new and conventional attributes using PNN. (**d**) AI (0.0–5.0 Hz) predicted from the new and conventional attributes using RNN. (**e**) AI (0.0–5.0 Hz) predicted from the predicted low-frequency seismic data and new attributes using RNN.

3.1.6. Predicting the Low Frequencies Using RMS Velocity

Another important low-frequency source is the travel time information at different offsets recorded in seismic data. Bianchin et al. [7] discussed how to improve AI estimation using interval velocity calculated from tomography based on autoregressive models. In contrast, we attempt to directly predict the low frequencies of AI using deep neural networks from the RMS stacking velocity obtained from the velocity analysis on common-midpoint (CMP) gathers containing nonzero offset recordings.

Table 2. Validation summary for the AI (full band) predicted using various kinds of combinations of attributes and methods. The R^2 score and CC in this table are the average values calculated using the whole 2-D line.

Attributes and Methods	AI (Absolute)	
	R^2 Score	CC
Conventional attributes + PNN	0.82	0.93
Raw seismic + RNN	0.82	0.93
Predicted low-frequency components and conventional attributes + PNN	0.88	0.94
Predicted low-frequency components+ Raw seismic + RNN	0.90	0.95
Predicted low-frequency components from RNN + Raw seismic + RNN	0.91	0.96

The Marmousi interval and RMS velocities are shown in Figure 16. We chose 10 traces of these data as the training set and trained a predictive model using RNN. Finally, we predicted the whole AI volume from the RMS velocity volume using this predictive model. The result is displayed in Figure 17b. As one can see in this figure, the predicted AI (0.0–5.0 Hz) is very close to the true AI (0.0–5.0 Hz) in Figure 17a, with an R^2 score of 0.98 and CC of 0.99. We then try to predict the AI (0.0–10.0 Hz) using the RMS velocity using the same workflow, and the results indicate that the prediction accuracy is decreased (see Figure 18). Next, we add the PSTM seismic data as an input in the training and the prediction result is displayed in Figure 19b, which shows that the true and predicted AI (0.0–10.0 Hz) match each other very well, with an R^2 score of 0.98 and CC of 0.99. Finally, we try to predict the AI (0.0–10.0 Hz) from the RMS velocity and PSTM seismic data using RNN using only five pseudo wells. The result in Figure 19c closely matches the true AI (0.0–10.0 Hz), with an R^2 score of 0.97 and CC of 0.99, although it is not as good as the result predicted using 10 pseudo wells in Figure 19b. It reveals that the proposed method is not sensitive to the number of traces in the training set for this synthetic example.

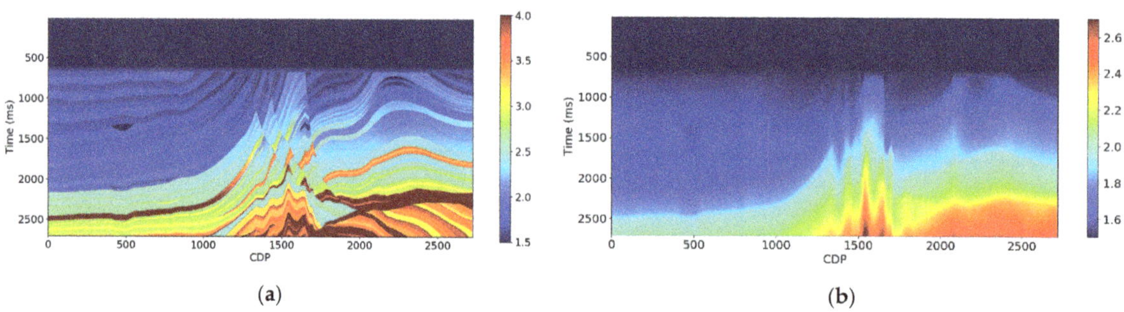

Figure 16. (**a**) Interval velocity from Marmousi model 2. (**b**) RMS velocity calculated using the Dix equation.

Although we can theoretically predict the low frequencies accurately using the RMS velocity, there can be problems in a real case. First, the field data could be very noisy, which could cause many errors in the picking of the RMS stack velocity. Second, even a small, localized mis-picking can lead to a physically implausible interval velocity value [37]. Third, the static error from the inaccurate near-surface velocity could also cause an error in the picking of the RMS stack velocity. Fourth, the multiples in the raw seismic data can cause picking errors. Therefore, we need to be cautious in using the RMS stack velocity as an input to predict the low-frequency components of AI.

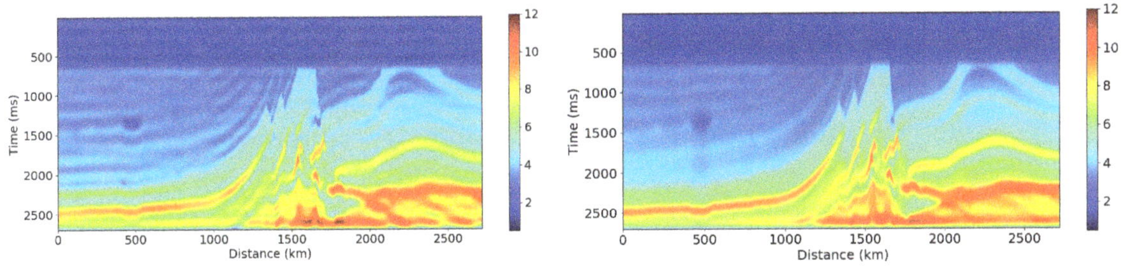

(**a**) (**b**)

Figure 17. (**a**) 0.0–5.0 Hz AI (True). (**b**) AI (0.0–5.0 Hz) predicted using RMS velocity.

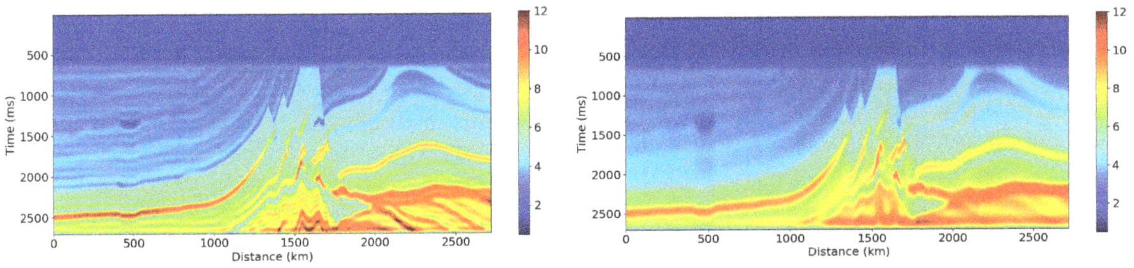

(**a**) (**b**)

Figure 18. (**a**) 0.0–10.0 Hz AI (True). (**b**) AI (0–10.0 Hz) predicted using RMS velocity.

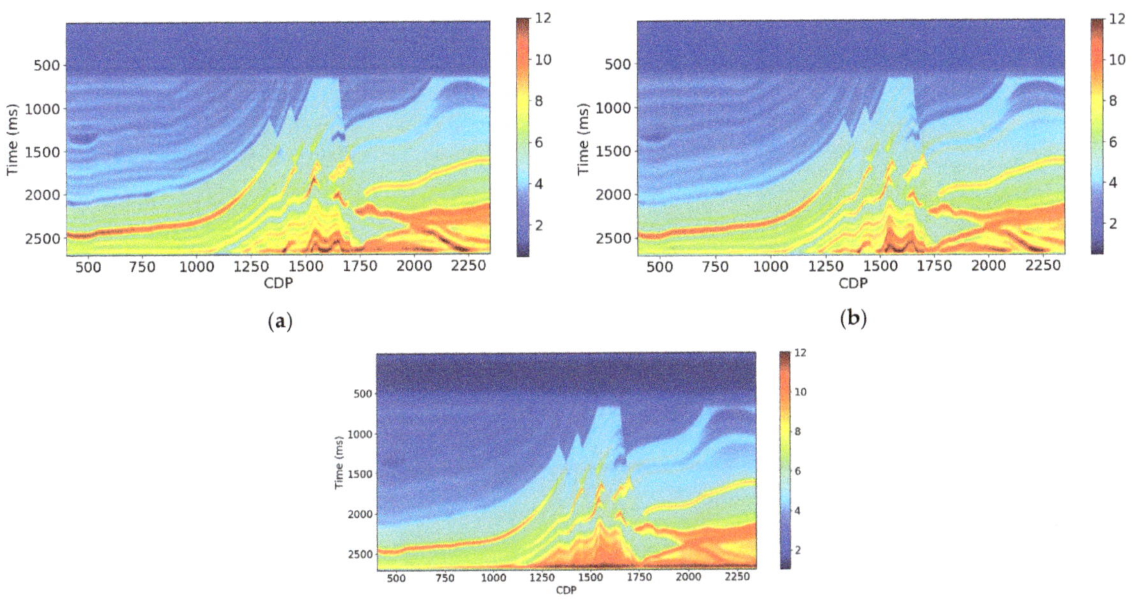

(**a**) (**b**)

(**c**)

Figure 19. (**a**) 0–10.0 Hz AI (True). (**b**) AI (0–10.0 Hz) predicted using the RMS velocity and 5.0–10.0 Hz PSTM seismic data using RNN (10 wells). (**c**) AI (0–10.0 Hz) predicted using the RMS velocity and 5–10 Hz PSTM seismic data using RNN (5 wells).

3.2. Real Data Example

Next, we apply our techniques to a dataset from the Midland basin, which is an eastern sub-basin of the Permian basin, one of the largest Hercynian (Middle Devonian-Middle Triassic) structural basins in North America (Figure 20). Due to intense structural deformation and orogenic movement, the depositional environment of the Midland basin is very complex from late Mississippian to early Permian time [38]. The sediments include coarse clastic sediments deposited near the basin shorelines, limestone, and extensive reef developed toward the ocean.

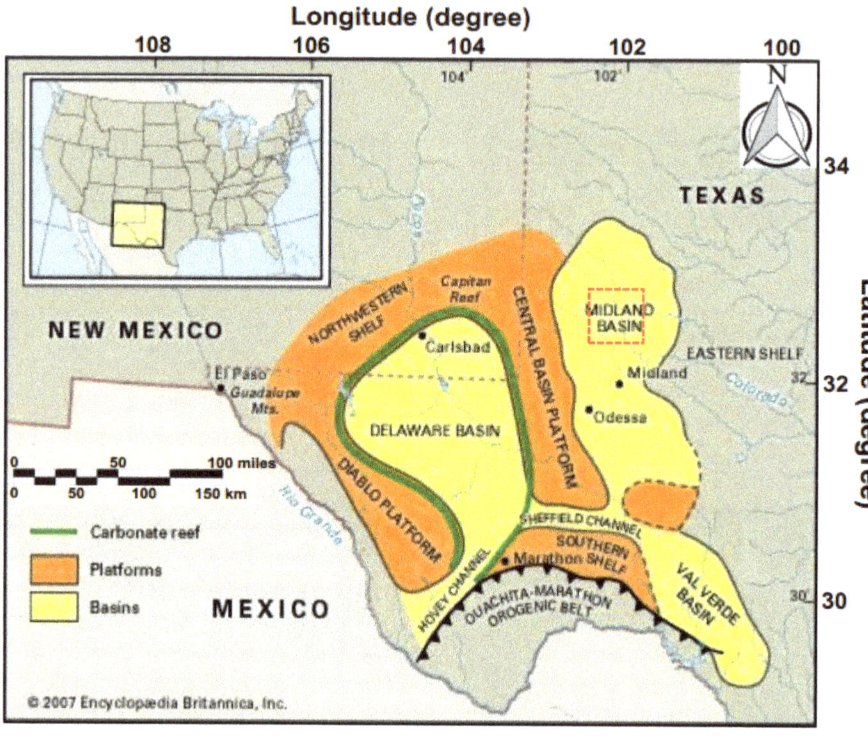

Figure 20. Geological map of the Permian Basin. The red dashed box is the general location of the 3D seismic data.

The target formations of interest in the field studied include Wolfcamp, Dean, and Spraberry, with depths varying from 6000 to 8000 ft (1.0 to 1.5 s two-way seismic travel time). The lithology includes sandstone, shale, limestone, and dolomite and shows dramatic lateral changes, making the AI prediction very difficult. The 3D seismic data for the study area covers approximately 50.0 mile^2. The sample interval is 2.0 ms, and the central frequency is approximately 50.0 Hz. The data contain prestack CDP gather and poststack PSTM seismic data (see Figure 21). In general, the seismic data quality is good. There are nine wells with high-quality logs in this area, and they are well-documented and edited. The log curves include density (RHOB), P-wave slowness (DT), S-wave slowness (DTS), and neutron porosity. The RMS velocity is stacking velocity picked manually by experienced seismic processing engineers. The black lines in Figure 21 are the training wells used in our deep learning training, and the red lines are the wells we used for testing.

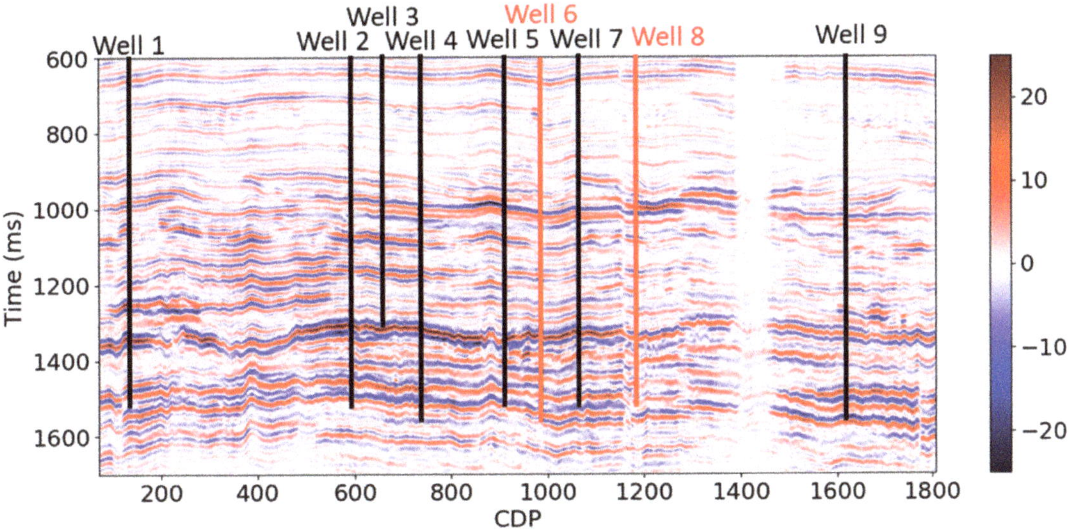

Figure 21. Seismic arbitrary line going through all the wells (PSTM).

We created the relative geological age, instantaneous amplitude, integrated instantaneous amplitude, and apparent time thickness attributes from the PSTM seismic data using the same procedure described above (see Figures 22–24). We did not produce the depth and interval velocity attributes due to the flat horizons in this area.

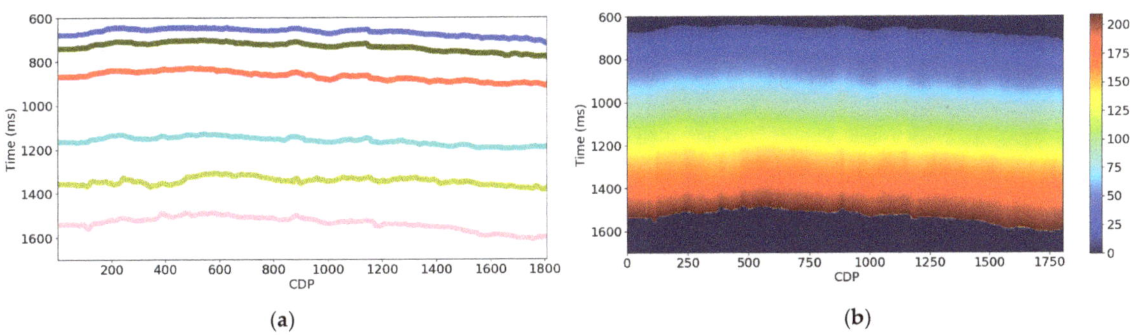

Figure 22. (**a**) The horizons interpreted on the PSTM seismic data from Figure 21 (marked with different colors). (**b**) Relative geological age attribute calculated from the horizons in (**a**).

Figure 25 is the low-frequency seismic data (0.0–5.0 Hz) predicted from the high frequencies of the PSTM seismic data. We use seven wells as the training and validation set and two wells as the blind test set in this example (see the well locations in Figure 21). The average CC for these two wells is 0.67. Since this is real data, we could not evaluate the whole 2D section; however, the prediction results displayed in Figure 25 indicate that the predicted low-frequency seismic section is geologically reasonable, with instability at a few traces that can be removed by median filtering.

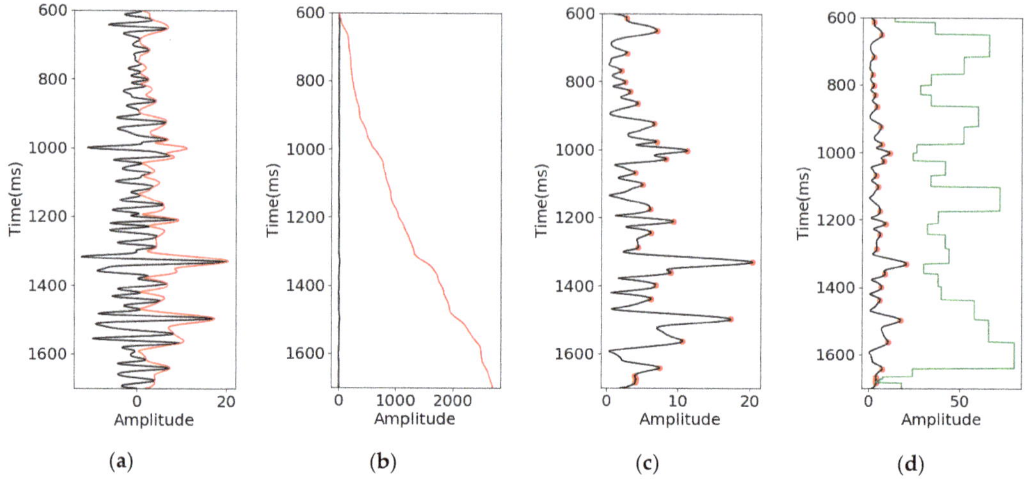

Figure 23. (**a**) Seismic data (black) and instantaneous amplitude (red). (**b**) Instantaneous amplitude (black) and integrated instantaneous amplitude (red). (**c**) Instantaneous amplitude (black) and peaks (red dot). (**d**) Instantaneous amplitude (black) and peaks (red dot) and apparent time thickness (green).

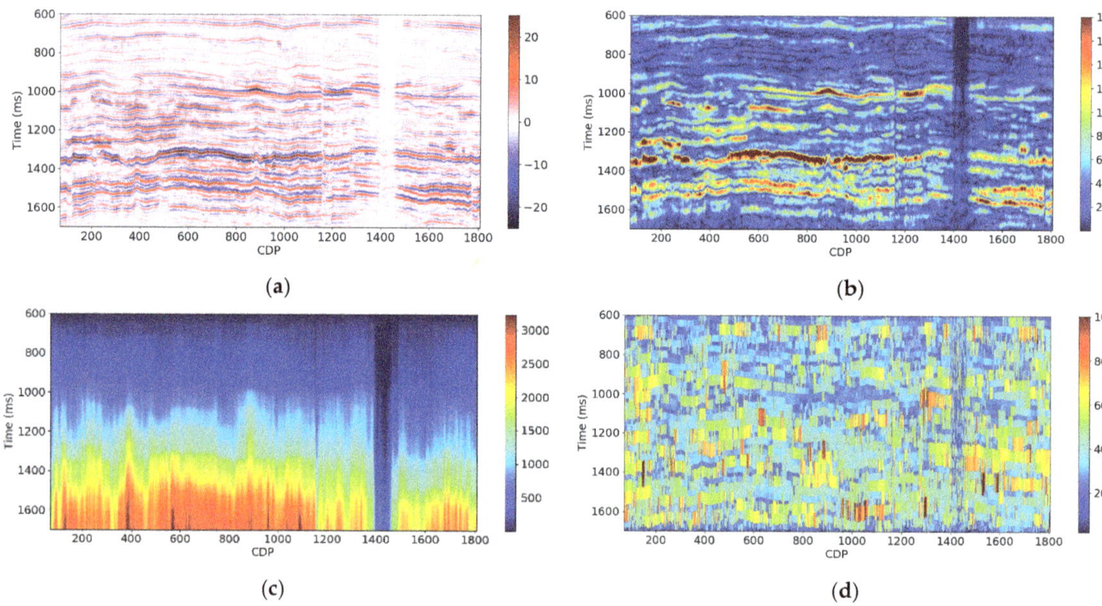

Figure 24. (**a**) PSTM seismic data. (**b**) Instantaneous amplitude. (**c**) Integrated instantaneous amplitude. (**d**) Apparent time thickness.

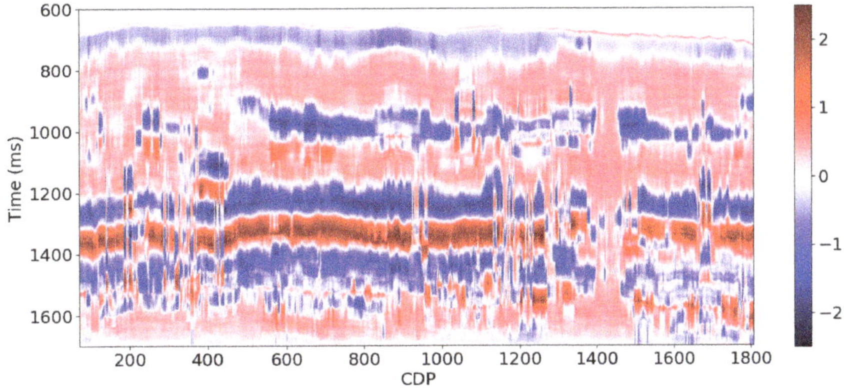

Figure 25. 0.0–5.0 Hz seismic data predicted using band-limited PSTM seismic data.

The analysis of synthetic and real data tests demonstrates that the 0.0–2.0 Hz AI interpolated using well logs along the horizons could potentially provide important constraints in the prediction of 0.0–5.0 Hz AI; therefore, we use it as one of the attributes to test the proposed method (see Figure 26).

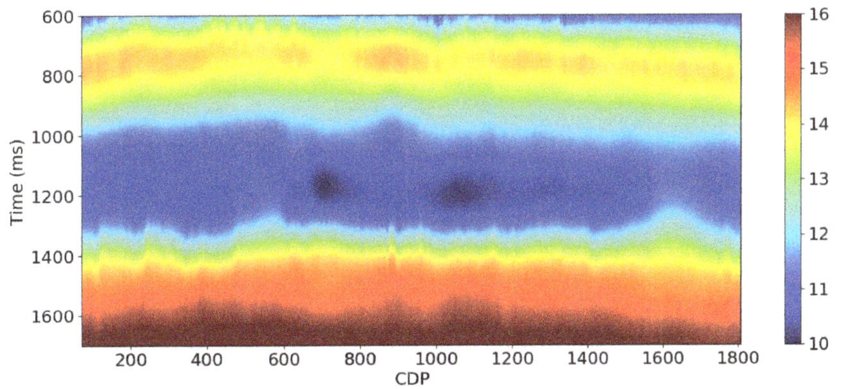

Figure 26. 0.0–2.0 Hz AI interpolated using well logs along the horizons.

We predict the AI (0.0–5.0 Hz) using a different combination of attributes and methods using RNN for this field dataset (See Table 3 and Figure 27). All the results are displayed in Figure 27, and we also summarize them in Table 3. We found that the relative geological age, interpolated AI (0.0–2.0 Hz), instantaneous amplitude, integrated instantaneous amplitude, predicted low-frequency seismic data, and RMS velocity were the best attributes needed for the prediction of AI (0.0–5.0 Hz). The baseline prediction in this example is from the interpolated AI (0.0–5.0 Hz), with an average R^2 score of 0.52 and CC of 0.91. Note that the prediction accuracy is very different between the two blind test wells. The R^2 score is 0.13, and the CC is 0.83 for the blind test well 1. The R^2 score is 0.91, and the CC is 0.98 for the blind test well 2. This occurs very often in a real case because the interpolation of wells could be accurate when there is no lateral geological variation and vice versa. This is also why we need to introduce new attributes to help improve the prediction accuracy and stability of the low frequencies of AI.

Table 3. Validation summary for the AI (0.0–5.0 Hz) predicted using various kinds of combinations of attributes and methods. Test 1 is the result of blind test well 1, and test 2 is the result of blind test well 2.

Attributes/Methods	AI (0.0–5.0 Hz)					
	Test 1 R^2 Score	Test 1 CC	Test 2 R^2 Score	Test 2 CC	Average R^2 Score	Average CC
Well-log interpolation along horizons	0.13	0.83	0.91	0.98	0.52	0.91
Relative geological age + Instantaneous amplitude + Integrated instantaneous amplitude + Apparent thickness	0.47	0.89	0.14	0.49	0.30	0.69
Relative geological age + Predicted low-frequency seismic data	0.19	0.82	0.93	0.98	0.56	0.90
Relative age + Predicted low-frequency seismic data + Instantaneous amplitude	0.23	0.86	0.92	0.98	0.57	0.92
Relative geological age + RMS stack velocity + Predicted low-frequency seismic data + Instantaneous amplitude	0.41	0.90	0.82	0.93	0.61	0.91
AI (0–2.0 Hz) (interpolated using well logs) + Instantaneous amplitude	0.38	0.89	0.94	0.98	0.66	0.94
AI (0–2.0 Hz) (interpolated using well logs) + RMS stack velocity	0.16	0.83	0.93	0.97	0.55	0.90
AI (0–2.0 Hz) + Predicted low-frequency seismic data	0.42	0.92	0.94	0.98	0.68	0.95
AI (0–2.0 Hz) + Predicted low-frequency seismic data + RMS stack velocity	0.43	0.93	0.95	0.98	0.69	0.95
AI (0–2.0 Hz) + Predicted low-frequency seismic data + RMS stack velocity + Instantaneous amplitude	0.39	0.93	0.91	0.98	0.65	0.95
AI (0–2.0 Hz) + RMS stack velocity + Predicted low-frequency seismic data + Integrated instantaneous amplitude	0.63	0.88	0.80	0.92	0.71	0.90
AI (0–2.0 Hz) + Integrated instantaneous amplitude	0.59	0.91	0.45	0.68	0.52	0.80
RMS stack velocity	0.85	0.93	0.71	0.94	0.79	0.94
RMS stack velocity + Predicted low-frequency seismic data	0.81	0.91	0.82	0.97	0.82	0.94

The combination of the relative geological age, instantaneous amplitude, and predicted low-frequency seismic data achieves a better performance than the baseline prediction, with an average R^2 score of 0.57 and CC of 0.92. The improvement is 9.6% for the R^2 score. The benefit of this combination is that the predicted results are relatively clean. There is almost no noise in this prediction. This feature could make the prediction results stable for a variety of datasets. In addition, we could improve the prediction accuracy and stability by adding more horizons in the calculation of the relative geological age attribute.

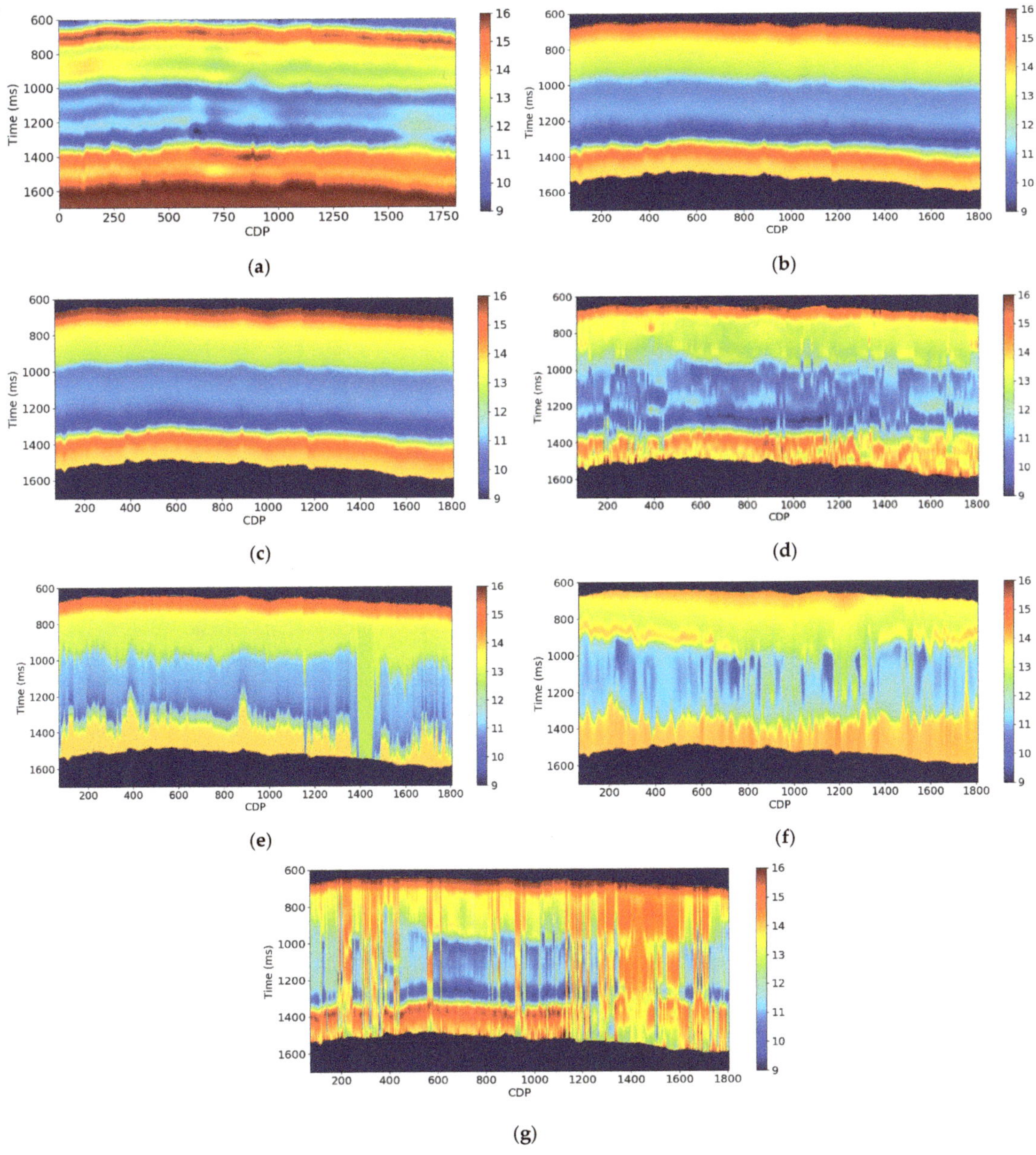

Figure 27. (**a**) AI (0.0–5.0 Hz) calculated using the well-log interpolation along the horizons. (**b**) AI (0.0–5.0 Hz) predicted using the relative geological age and predicted low-frequency seismic data. (**c**) AI (0.0–5.0 Hz) predicted using the relative geological age, predicted low-frequency seismic data, and instantaneous amplitude. (**d**) AI (0.0–5.0 Hz) predicted using the AI (0–2.0 Hz), predicted low-frequency seismic data, and RMS velocity. (**e**) AI (0.0–5.0 Hz) predicted using the AI (0–2.0 Hz), predicted low-frequency seismic data, RMS velocity, and integrated instantaneous amplitude. (**f**) AI (0.0–5.0 Hz) using the RMS stack velocity. (**g**) AI (0.0–5.0 Hz) using the RMS velocity and predicted low-frequency seismic data.

The combination of the AI (0–2.0 Hz), predicted low-frequency seismic data, and RMS stack velocity also achieves a better performance than the baseline prediction, with an average R^2 score of 0.69 and CC of 0.95. The improvement is 32.7% for the average R^2 score and 4.4% for the average CC. When we add the integrated instantaneous amplitude, the R^2 score improves slightly, but the CC decreases from 0.95 to 0.90. Thus, we need to decide whether to use this attribute on a case-by-case basis. The benefit of this combination is that the noise level and accuracy are intermediate compared with other cases. This combination of attributes could give a relatively stable and accurate prediction result.

The combination of the RMS stack velocity and predicted low-frequency seismic data achieves the best performance in all cases, with an average R^2 score of 0.82 and CC of 0.94. The improvement is 57.7% for the R^2 score and 3.3% for CC. However, if we compare the prediction results in Figure 27g with Figure 27b,d, we find that the results are noisy, with many stripes in the section. This can be explained by the unreliable velocity analysis. We could alleviate this issue by smoothing the RMS stack velocity and improving the RMS stack velocity accuracy by performing a more careful velocity analysis. We could also try to invert the interval velocity using tomography inversion, which could improve the velocity estimation accuracy.

4. Discussion and Conclusions

Here, we showed how to predict the low frequencies of AI from various combinations of seismic attributes and geological information using deep learning. We first tested various deep learning methods to compare their performance in predicting seismic reflectivity and AI using synthetic seismic data produced by rock physics modeling. We found that CNNs are suitable for predicting the reflectivity using seismic data but not for the impedance using reflectivity. However, we found that RNNs are suitable for both prediction tasks. In addition, RNNs also performed well in predicting the AI using seismic data directly.

We then tested how to predict the low frequencies of AI using various combinations of attributes and methods using the Marmousi model 2. The results show that the attributes we created in this work, including the relative geological age, interval velocity, integrated instantaneous amplitude, etc., helped supply low frequencies for AI prediction. Further, we proposed a method to predict the 0.0–5.0 Hz low frequencies of seismic data using their high-frequency components. With these attributes, we investigated how each combination of the attributes and methods performed in predicting the low frequencies of AI. We found that we can achieve good performance in predicting the low frequencies of AI by combining the new attributes, predicted low-frequency seismic data, and the RNN, with an R^2 score of 0.93 and CC of 0.97. The biggest improvement compared with other combinations is 10.7% for the R^2 score. We also found that we can perform better in absolute AI inversion by adding the predicted low-frequency components of AI, with the best improvement of 11.0% compared with other cases. The RNN showed some benefits over the PNN, but the improvement is not significant.

Next, we studied how each new attribute contributed to the prediction of low frequencies using convolutional synthetic seismic data created using the Marmousi model 2. The results reveal that: (1) the interval velocity, instantaneous amplitude, integrated instantaneous amplitude, predicted low-frequency seismic data, and RMS velocity made a positive contribution to the prediction of the low frequencies of AI; (2) the integrated instantaneous amplitude and apparent time thickness played a smaller role in improving the prediction accuracy; and (3) the results predicted from the combinations of the interval velocity, instantaneous amplitude, predicted low-frequency seismic data, and RMS velocity were better than the well-log interpolation.

Finally, we tested the proposed methods and attributes using field data from the Midland basin. We found that the relative geological age, interpolated AI (0.0–2.0 Hz), instantaneous amplitude, integrated instantaneous amplitude, predicted low-frequency seismic data, and RMS stack velocity can help improve the prediction accuracy of the low frequencies. The combination of the relative geological age, instantaneous amplitude, and predicted low-frequency seismic data made a better performance than the baseline prediction, with an average R^2 score of 0.52 and CC of 0.92. The average improvement for the R^2 score is 9.6%. One of the advantages of this combination is that the predicted result is relatively clean, with only a little noise. In other words, this method could make a robust prediction compared with other methods when the signal-noise ratio of the data is low. This method could be enhanced by adding more horizons to the attributes' calculation.

The combination of the AI (0.0–2.0 Hz), predicted low-frequency seismic data, and RMS stack velocity made a good prediction for the low frequencies, with an average R^2 score of 0.69 and CC of 0.95. The improvement is 32.7% for the R^2 score and 4.4% for CC. The noise level and accuracy are intermediate compared with other cases; therefore, it could provide a relatively robust and accurate prediction result.

The combination of the RMS stack velocity and predicted low-frequency seismic data achieved the best performance in all cases, with an average R^2 score of 0.82 and CC of 0.94, with an improvement of 57.7% for the R^2 score and 3.3% for CC compared with the baseline prediction. One of the drawbacks of this method is that the result is noisy, with many stripes in the section caused by the low data quality and unstable velocity analysis process. Therefore, we need to be very cautious when using this method to predict the low frequencies of AI.

The real seismic data shown in this case are processed with a high standard, and the geological structure is relatively simple in our real case study. Therefore, the equivalent prediction accuracy might not be able to be achieved if the seismic data are not processed well and/or the geological conditions become more complicated. How well the model will perform in these cases is still not clear to us. We will be investigating this issue further in our future work.

Author Contributions: Conceptualization, J.P.C. and L.J.; methodology, L.J. and Z.Z.; software, L.J.; validation, J.P.C., B.R. and Z.Z.; formal analysis, L.J. and Z.Z.; investigation, L.J. and Z.Z.; resources, J.P.C., B.R. and Z.Z.; data curation, L.J.; writing—original draft preparation, L.J.; writing—review and editing, J.P.C. and B.R.; visualization, L.J. and Z.Z.; supervision, J.P.C., B.R. and Z.Z.; project administration, J.P.C.; funding acquisition, J.P.C. All authors have read and agreed to the published version of the manuscript.

Funding: This research received no external funding.

Data Availability Statement: The real seismic data and well data is unavailable due to privacy restrictions.

Acknowledgments: We thank Leon Thomsen, Evgeni Chesnokov, Yingcai Zheng, Hao Hu, Hongyu Sun, and many others for their discussion, suggestions, and help. Thanks to the research computing data core center and EAS from the University of Houston for providing computing resources. We also want to thank the anonymous reviewers for their thoughtful suggestions and comments, which greatly helped to improve this paper.

Conflicts of Interest: The authors declare no conflict of interest.

References

1. Levy, S.; Fullagar, P.K. Reconstruction of a sparse spike train from a portion of its spectrum and application to high-resolution deconvolution. *Geophysics* **1981**, *46*, 1235–1243. [CrossRef]
2. Oldenburg, D.W.; Scheuer, T.; Levy, S. Recovery of the acoustic impedance from reflection seismograms. *Geophysics* **1983**, *48*, 1318–1337.
3. Sacchi, M.D.; Velis, D.R.; Comínguez, A.H. Minimum entropy deconvolution with frequency-domain constraints. *Geophysics* **1994**, *59*, 938–945. [CrossRef]
4. Velis, D.R. Stochastic sparse-spike deconvolution. *Geophysics* **2008**, *73*, R1–R9. [CrossRef]
5. Zhang, R. Seismic reflection inversion by basis pursuit. *Geophysics* **2010**, *76*, Z27. [CrossRef]
6. Liang, C. Seismic Spectral Bandwidth Extension and Reflectivity Decomposition. Ph.D. Thesis, University of Houston, Houston, TX, USA, 2018.

7. Bianchin, L.; Forte, E.; Pipan, M. Acoustic impedance estimation from combined harmonic reconstruction and interval velocity. *Geophysics* **2019**, *84*, 385–400.
8. Wu, R.-S.; Luo, J.; Wu, B. Seismic envelope inversion and modulation signal model. *Geophysics* **2014**, *79*, WA13–WA24.
9. Hu, Y.; Wu, R.-S. Instantaneous-phase encoded direct envelope inversion in the time-frequency domain and the application to subsalt inversion. In *SEG Technical Program Expanded Abstracts*; Society of Exploration Geophysicists: Houston, TX, USA, 2020; pp. 775–779.
10. Hu, W.; Jin, Y.; Wu, X.; Chen, J. A progressive deep transfer learning approach to cycle-skipping mitigation in FWI. In *SEG International Exposition and Annual Meeting*; SEG: Houston, TX, USA, 2019; pp. 2348–2352.
11. Li, Y.E.; Demanet, L. A short note on phase and amplitude tracking for seismic event separation. *Geophysics* **2015**, *80*, WD59–WD72. [CrossRef]
12. Li, Y.E.; Demanet, L. Full Waveform Inversion with Extrapolated Low Frequency Data. *Geophysics* **2016**, *81*, R339–R348. [CrossRef]
13. Sun, H.; Demanet, L. Extrapolated full-waveform inversion with deep learning. *Geophysics* **2020**, *85*, R275–R288. [CrossRef]
14. Rumelhart, D.E.; Hinton, G.E.; Williams, R.J. Learning representations by back-propagating errors. *Nature* **1986**, *323*, 533–536. [CrossRef]
15. LeCun, Y.; Bengio, Y.; Hinton, G. Deep learning. *Nature* **2015**, *521*, 436–444.
16. Jordan, M.I.; Mitchell, T.M. Machine learning: Trends, perspectives, and prospects. *Science* **2015**, *349*, 255–260. [CrossRef] [PubMed]
17. Dehaene, S.; Lau, H.; Kouider, S. What is consciousness, and could machines have it? *Science* **2017**, *358*, 486–492. [CrossRef] [PubMed]
18. Fernando, C.; Banarse, D.; Blundell, C.; Zwols, Y.; Ha, D.; Rusu, A.A.; Pritzel, A.; Wierstra, D. PathNet: Evolution Channels Gradient Descent in Super Neural Networks. *arXiv* **2017**, arXiv:1701.08734v1.
19. George, D.; Lehrach, W.; Kansky, K.; Lázaro-Gredilla, M.; Laan, C.; Marthi, B.; Lou, X.; Meng, Z.; Liu, Y.; Wang, H.; et al. A generative vision model that trains with high data efficiency and breaks text-based CAPTCHAs. *Science* **2017**, *358*, eaag2612. [CrossRef]
20. Sanchez-Lengeling, B.; Aspuru-Guzik, A. Inverse molecular design using machine learning: Generative models for matter engineering. *Science* **2019**, *361*, 360–365. [CrossRef]
21. Das, V.; Pollack, A.; Wollner, U.; Mukerji, T. Convolutional neural network for seismic impedance inversion. *Geophysics* **2019**, *84*, R869–R880.
22. Hampson, D.; James, S.S.; John, A.Q. Use of multiattribute transforms to predict log properties from seismic data. *Geophysics* **2001**, *66*, 220–236.
23. Li, J.; Castagna, J. Support Vector Machine (SVM) pattern recognition to AVO classification. *Geophys. Res. Lett.* **2004**, 31. [CrossRef]
24. Saggaf, M.M.; Toksöz, M.N.; Marhoon, M.I. Seismic facies classification and identification by competitive neural networks. *Geophysics* **2003**, *68*, 1984–1999. [CrossRef]
25. Xie, S.; Girshick, R.; Dollár, P.; Tu, Z.; He, K. Aggregated residual transformations for deep neural networks. In Proceedings of the IEEE conference on computer vision and pattern recognition, Honolulu, HI, USA, 21–26 July 2017; pp. 1492–1500.
26. Haris, M.; Shakhnarovich, G.; Ukita, N. Deep back-projection networks for super-resolution. In Proceedings of the 2018 IEEE/CVF Conference on Computer Vision and Pattern Recognition, Salt Lake City, UT, USA, 18–23 June 2018; pp. 1664–1673. [CrossRef]
27. Anwar, S.; Barnes, N. Densely Residual Laplacian Super-Resolution. *IEEE Trans. Pattern Anal. Mach. Intell.* **2020**, *44*, 1192–1204. [CrossRef]
28. Xiong, W.; Ji, X.; Ma, Y.; Wang, Y.; AlBinHassan, N.M.; Ali, M.N.; Luo, Y. Seismic fault detection with convolutional neural network. *Geophysics* **2018**, *83*, O97–O103. [CrossRef]
29. Wrona, T.; Pan, I.; Gawthorpe, R.L.; Fossen, H. Seismic facies analysis using machine learning. *Geophysics* **2018**, *83*, O83–O95. [CrossRef]
30. Chopra, S.; Marfurt, K.J. Seismic facies classification using some unsupervised machine-learning methods. In *SEG Technical Program Expanded Abstracts*; SEG: Houston, TX, USA, 2018; pp. 2056–2059.
31. Zhang, Z.; Halpert, A.D.; Bandura, L.; Coumont, A.D. Machine-learning based technique for lithology and fluid content prediction: Case study from offshore West Africa. In *SEG Technical Program Expanded Abstracts*; SEG: Houston, TX, USA, 2018; pp. 2271–2276.
32. Alfarraj, M.; AlRegib, G. Semi-supervised Sequence Modeling for Elastic Impedance Inversion. *Interpretation* **2020**, *7*, SE237–SE249. [CrossRef]
33. Russell, B. Machine learning and geophysical inversion—A numerical study. *Lead. Edge* **2019**, *38*, 498–576. [CrossRef]
34. Hopfield, J.J. Neural networks and physical systems with emergent collective computational abilities. *Proc. Natl. Acad. Sci. USA* **1982**, *79*, 2554–2558. [CrossRef] [PubMed]
35. Martin, G.S. The Marmousi2 Model, Elastic Synthetic Data, and an Analysis of Imaging and AVO in a Structurally Complex Environment. Master's Thesis, University of Houston, Houston, TX, USA, 2004.
36. Versteeg, R. The Marmousi experience: Velocity model determination on a synthetic complex data set. *Lead. Edge* **1994**, *13*, 927–936. [CrossRef]

37. Yilmaz, O. *Seismic Data Analysis: Processing, Inversion, and Interpretation of Seismic Data*; Society of Exploration Geophysicists: Tulsa, OK, USA, 2001; 1028p.
38. Robinson, K. *Petroleum Geology and Hydrocarbon Plays of the Permain Basin Petroleum Province West Texas and Southeast New Mexico*; US Geological Survey: Reston, VA, USA, 1988.

Disclaimer/Publisher's Note: The statements, opinions and data contained in all publications are solely those of the individual author(s) and contributor(s) and not of MDPI and/or the editor(s). MDPI and/or the editor(s) disclaim responsibility for any injury to people or property resulting from any ideas, methods, instructions or products referred to in the content.

minerals

MDPI

Article

Application of Electromagnetic Methods for Reservoir Monitoring with Emphasis on Carbon Capture, Utilization, and Storage

César Barajas-Olalde [1,*], Donald C. Adams [1], Ana Curcio [2], Sofia Davydycheva [3], Ryan J. Klapperich [1], Yardenia Martinez [3], Andri Y. Paembonan [3,4], Wesley D. Peck [1], Kurt Strack [3,*] and Pantelis Soupios [5]

1 Energy & Environmental Research Center, 15 North 23rd Street, Stop 9018, Grand Forks, ND 58202, USA; dadams@undeerc.org (D.C.A.); wpeck@undeerc.org (W.D.P.)
2 Proingeo SA, Agüero 1995 6A, Ciudad Autónoma de Buenos Aires 1425, Argentina; anitacurcio@gmail.com
3 KMS Technologies, Inc., 11999 Katy Freeway, Suite 160, Houston, TX 77079, USA; sdavyd@gmail.com (S.D.); yardenia@kmstechnologies.com (Y.M.); andriyadip@gmail.com (A.Y.P.)
4 Sumatera Institute of Technology, Selatan 35365, Indonesia
5 Department of Geosciences, College of Petroleum Engineering and Geosciences, King Fahd University of Petroleum and Minerals, Dhahran 31261, Saudi Arabia; p.soupios@gmail.com
* Correspondence: cbarajas@undeerc.org (C.B.-O.); kurt@kmstechnologies.com (K.S.); Tel.: +1-701-777-5414 (C.B.-O.); +1-832-630-8144 (K.S.)

Abstract: The Controlled-Source ElectroMagnetic (CSEM) method provides crucial information about reservoir fluids and their spatial distribution. Carbon dioxide (CO_2) storage, enhanced oil recovery (EOR), geothermal exploration, and lithium exploration are ideal applications for the CSEM method. The versatility of CSEM permits its customization to specific reservoir objectives by selecting the appropriate components of a multi-component system. To effectively tailor the CSEM approach, it is essential to determine whether the primary target reservoir is resistive or conductive. This task is relatively straightforward in CO_2 monitoring, where the injected fluid is resistive. However, for scenarios involving brine-saturated (water-wet) or oil-wet (carbon capture, utilization, and storage—CCUS) reservoirs, consideration must also be given to conductive reservoir components. The optimization of data acquisition before the survey involves analyzing target parameters and the sensitivity of multi-component CSEM. This optimization process typically includes on-site noise measurements and 3D anisotropic modeling. Based on our experience, subsequent surveys tend to proceed smoothly, yielding robust measurements that align with scientific objectives. Other critical aspects to be considered are using magnetotelluric (MT) measurements to define the overall background resistivities and integrating real-time quality assurance during data acquisition with 3D modeling. This integration allows the fine tuning of acquisition parameters such as acquisition time and necessary repeats. As a result, data can be examined in real-time to assess subsurface information content while the acquisition is ongoing. Consequently, high-quality data sets are usually obtained for subsequent processing and initial interpretation with minimal user intervention. The implementation of sensitivity analysis during the inversion process plays a pivotal role in ensuring that the acquired data accurately respond to the target reservoirs' expected depth range. To elucidate these concepts, we present an illustrative example from a CO_2 storage site in North Dakota, USA, wherein the long-offset transient electromagnetic method (LOTEM), a variation of the CSEM method, and the MT method were utilized. This example showcases how surface measurements attain appropriately upscaled log-scale sensitivity. Furthermore, the sensitivity of the CSEM and MT methods was examined in other case histories, where the target reservoirs exhibited conductive properties, such as those encountered in enhanced oil recovery (EOR), geothermal, and lithium exploration applications. The same equipment specifications were utilized for CSEM and MT surveys across all case studies.

Keywords: CO_2 storage monitoring; fluid imaging; controlled-source electromagnetics (CSEM); magnetotelluric (MT) survey; CCUS; EOR; geothermal monitoring; lithium exploration

Citation: Barajas-Olalde, C.; Adams, D.C.; Curcio, A.; Davydycheva, S.; Klapperich, R.J.; Martinez, Y.; Paembonan, A.Y.; Peck, W.D.; Strack, K.; Soupios, P. Application of Electromagnetic Methods for Reservoir Monitoring with Emphasis on Carbon Capture, Utilization, and Storage. *Minerals* **2023**, *13*, 1308. https://doi.org/10.3390/min13101308

Academic Editor: Paulo T. L. Menezes

Received: 4 August 2023
Revised: 27 September 2023
Accepted: 28 September 2023
Published: 10 October 2023

Copyright: © 2023 by the authors. Licensee MDPI, Basel, Switzerland. This article is an open access article distributed under the terms and conditions of the Creative Commons Attribution (CC BY) license (https://creativecommons.org/licenses/by/4.0/).

1. Introduction

Clean energy technology deployment must accelerate rapidly to meet climate goals [1]. Carbon capture, utilization, and storage (CCUS) technology for capturing CO_2 emissions to use them sustainably or store them is crucial for reaching net zero emissions [2]. As part of the CCUS green gas reduction strategies, injecting CO_2 into saline aquifers for storage and oil reservoirs to enhance oil recovery (CO_2-EOR) addresses the challenge of reducing atmospheric concentrations of CO_2 and satisfying the worldwide energy demand [3]. This technology permits us to store CO_2 within deep geological formations permanently. These formations must be deeper than 800 m deep, have a thick and extensive seal, sufficient porosity for large volumes, and be permeable enough to allow high flow rates without requiring excessive pressure [4]. It is worth noting that at depths of around 800 m and below, the natural temperature and fluid pressure are higher than the critical point of CO_2 (temperature 31.1 °C and pressure 72.9 atm) in most parts of the world. Supercritical CO_2 is dense like a liquid but viscous like a gas. Therefore, injecting supercritical CO_2 at this depth or lower makes storage more space-efficient [5].

Transitioning from coal and oil to geothermal energy is another priority envisioned in the Net Zero Emissions (NZE) scenario, as heat flux from the Earth's core provides a reliable and abundant green source of energy for decarbonization of the energy system [1,6]. In recent years, research efforts have been focused on assessing the advantages and efficiency of using CO_2, an alternative to water, as heat transmission fluid for geothermal energy recovery from enhanced geothermal systems (EGS) where the permeability of the underground source is enhanced by hydrofracturing. In a CO_2-EGS scenario, CO_2 geologic storage is an ancillary objective of geothermal operations ([7] and the literature herein). There are several challenges associated with transitioning to geothermal energy. One of the significant historic challenges is its dependence on specific geographical locations. Geothermal fluids cannot be transported far from their source, which limits investment in this field by developed countries. Drilling activities and the disposal of geothermal fluid can have environmental impacts, such as emitting greenhouse gases and inducing seismic activity. However, these challenges are being addressed by bringing lower enthalpy geothermal reservoirs online, developing more efficient power generation and storage technology, using efficient heat pumps, and improving geophysical technology to measure near urban areas.

CO_2-Plume Geothermal (CPG) [8,9] is another emerging technology to harvest geothermal energy by circulating CO_2 through geothermal reservoirs with naturally sufficiently porous and permeable formations overlain by caprock. In CPG systems, the key objective is to provide simultaneous CO_2-injection-induced pressure relief by producing hot brine with consumptive beneficial uses [7,10]. The CO_2-sequestration potential of CPG systems is more significant than that of CO_2-EGS systems as they utilize natural large high-permeability reservoirs such as the Williston Basin, USA [8].

The deployment of these clean energies has generated a higher demand for critical minerals [1,6]. Defining critical minerals depends on the context [11]. While lithium, nickel, cobalt, manganese, and graphite are required for developing efficient lithium-ion batteries, rare earth elements are needed for wind turbines and electric vehicles, and copper and aluminum for solar photovoltaic technology and electricity networks [2].

Geophysical methods are essential for monitoring reservoirs for CO_2 sequestration, geothermal, and EOR activities, as well as for exploring critical minerals. These methods are vital for ensuring the permanent containment of CO_2 within reservoirs and increasing the success of finding critical mineral resources. Time-lapse reflection seismics is the standard geophysical method used to monitor CO_2 plume migration due to its high horizontal resolution [12].

When CO_2 is injected into a geologic formation, the existing fluid (brine or oil) is replaced, creating an electrical resistivity contrast. Forward modeling is needed to understand whether the injected CO_2 in a brine reservoir will increase or decrease the resistivity. Figure 1 shows some models for the fluids in a CO_2 reservoir (top) [13] and the structure in a hydrocarbon reservoir (bottom) [14]. In the CO_2 injection case, the CO_2 combines with the water in the brine and builds larger molecules [13] that are neutrally charged (colored purple in the figure). As a result, fewer electrons are available to provide current flow, and the reservoir is more resistive. In the figure on the bottom, a rock model made up of grains and pore space [14] is considered. The pores are filled with fluid, and when strain/stress are impressed upon it, the brine fluid's mobility and electron flow increase, causing a resistivity reduction. Once the pressure on the grain-to-grain contact increases, the reservoir seal breaks, generating a microseismic signal. Similar to the fluid model described above, any water saturation in the reservoir fluid contains free electrons. When the plume front moves through the reservoir, it exerts pressure on existing fluid (and even causes microseismic signals when the seal breaks). This pressure will cause the free electrons to flow and strong resistivity reduction right at the plume front, followed by the resistive plume. In the case of CCUS, this is caused by a fluid pressure wave due to the injection. Thus, CO_2 injection causes a resistive plume in a brine reservoir, and resistivity reduction at the edges can be observed.

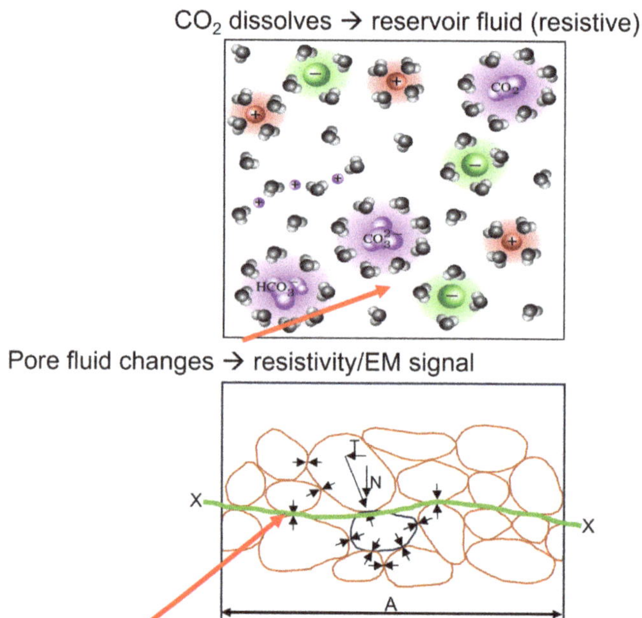

Figure 1. Rock models illustrate the pore fluid response to CO_2 injection and flooding (modified after [13,14]). The rock models focus on the fluids (top) and depict brine molecules (pore water: grey; cations: red; and anions: green) and CO_2-saturated ones (purple). Below, a model of a clastic rock with grains and pore space is displayed. The grain-to-grain contacts (green line) build the fluid boundary.

Electromagnetic (EM) methods are sensitive to resistive contrasts [15]; therefore, they are ideal for estimating the location and extent of an injected CO_2 plume. Multiple time-lapse surveys can be used to observe the change in resistivity because of the injection of CO_2 and its subsequent movement. Measurements taken before CO_2 injection can provide a baseline against which future changes can be evaluated. Time-lapse Controlled-Source ElectroMagnetic (CSEM) surveys combined with magnetotelluric (MT) surveys are a valuable tool for monitoring injected industrial CO_2 as a part of CCUS processes. Even if the changes caused by the CO_2 are small, it should be confirmed with confidence that changes seen in the data are only because of the movement of the CO_2 and not noise sources such as naturally occurring electric fields, surveying mistakes, or local anthropogenic features.

When surface CSEM measurements are tied to 3D anisotropic models derived from the available logs and lithology from the area under investigation as part of an initial 3D feasibility workflow, baseline measurements can be verified within the context of the borehole information, representing a significant risk reduction.

An initial 3D feasibility modeling workflow includes field noise measurements to determine the noise level and, thus, the best experimental source-sensor geometry and acquisition parameters based on the expected fluid substitution models. Careful instrument calibration and verification of all acquisition parameters are essential for data acquisition. Concurrent with the acquisition, near-real-time quality assurance and control (QA/QC) are carried out. The feedback verification loop results positively influence the acquisition's data quality by optimizing operation times/equipment moves.

This process defines the proper length of acquisition time and consequently yields good data quality that can then be used for unsupervised inversion. The only influencing components in the inversion are the data weights derived from the repeated measurements (stacking weights). CSEM and MT data acquisition can benefit from this process and achieve high sensitivities for deep reservoir targets.

Achieving log scale resolution from surface measurements is a significant breakthrough; thus, baseline measurements can be verified before acquiring the monitoring measurements. Based on the results, the time-lapse (monitoring) survey can be further fine-tuned to reduce the success risk by taking the data integration from a subjective interpretation to an automated mere data display. This step requires hardware control and operation parameters (frequently repeated calibration) to obtain higher accuracy than is normal for exploration purposes.

Applying EM data acquisition systems to rare minerals exploration has demonstrated the need for careful hardware design to achieve high amplifier fidelity. This results in the ability to record subsurface responses requiring a high dynamic range from high resistivity contrast. Lithium prospecting represents such a challenge as lithium has a resistivity comparable to seawater at several hundred meters' depth. Electric and EM methods are suited for lithium prospecting due to their sensitivity to very low electrical resistivity (typically 0.1 to 0.5 ohm-m) associated with the brine-saturated multilayers present and with a frequency/time window. These methods require specific equipment characteristics (as used for time domain CSEM) with instantaneous dynamic range.

In the next sections, we present CCUS, geothermal, EOR, and critical mineral case histories using EM methods.

2. Case Histories

The initial case study exemplifies the application of a feasibility workflow to a baseline data set obtained through the long-offset transient electromagnetic method (LOTEM) [16]. This method is a variation of the CSEM method, which was integrated with the MT method for a time-lapse survey in Center, North Dakota, USA, as a part of a CCUS project. This example describes rigorous quality control and assessment methods, including noise removal and data validation with 1D and 3D inversion to tie results to borehole logs. The results are an accurate and representative CSEM data set and information that stakeholders can use to inform future time-lapse survey costs and designs.

The second case history is a geothermal example from Saudi Arabia focusing on exploration. The EOR case history examples include a modeling and feasibility study and a field test from the Middle East.

Finally, lithium exploration's critical minerals case histories in Argentina and Saudi Arabia are described. The first is from an area where lithium reserves are known, and the second is from an area where they are suspected based on our findings.

All case studies were conducted utilizing the same equipment for CSEM and MT surveys. The equipment was designed for monitoring purposes, ensuring high repeatability, long-term stability, and minimum influence of system bandwidth limitations on signal bandwidth. The transmitter used was either a 100 (EOR example) or 150 (CCUS and Saudi Arabia examples) kVA transmitter, as described in [17]. The array acquisition system for the CSEM (LOTEM) and MT configurations is described in [18]. For more information on the system architecture and survey design, see [15]. Furthermore, additional references for data processing, data interpretation, and inversion activities considered in the case studies can be found in Appendix A.

2.1. Monitoring of CO_2 in a Saline Reservoir in North Dakota, USA

Measurements taken before CO_2 injection can provide a baseline against which future changes can be evaluated. Multiple time-lapse surveys can be used to observe the change in resistivity because of the injection of CO_2 and subsequent movement.

Baseline EM methods were applied to assess their performance in CO_2 monitoring in the lower Permian Broom Creek and Cambrian and lower Ordovician Deadwood formations in the Williston Basin near Center, North Dakota, USA. Approximately 4 million metric tons of CO_2 annually are expected to be injected into the Broom Creek and Deadwood formations at depths of approximately 1450 and 2835 m and average thicknesses of 70 m and 85 m, respectively. Figure 2 shows a simplified cross-section of North Dakota with the studied reservoirs (yellow ellipses) and a simplified stratigraphic column, including the formations overlying and underlying the studied reservoirs. The vertical black line represents the approximated location of the studied area.

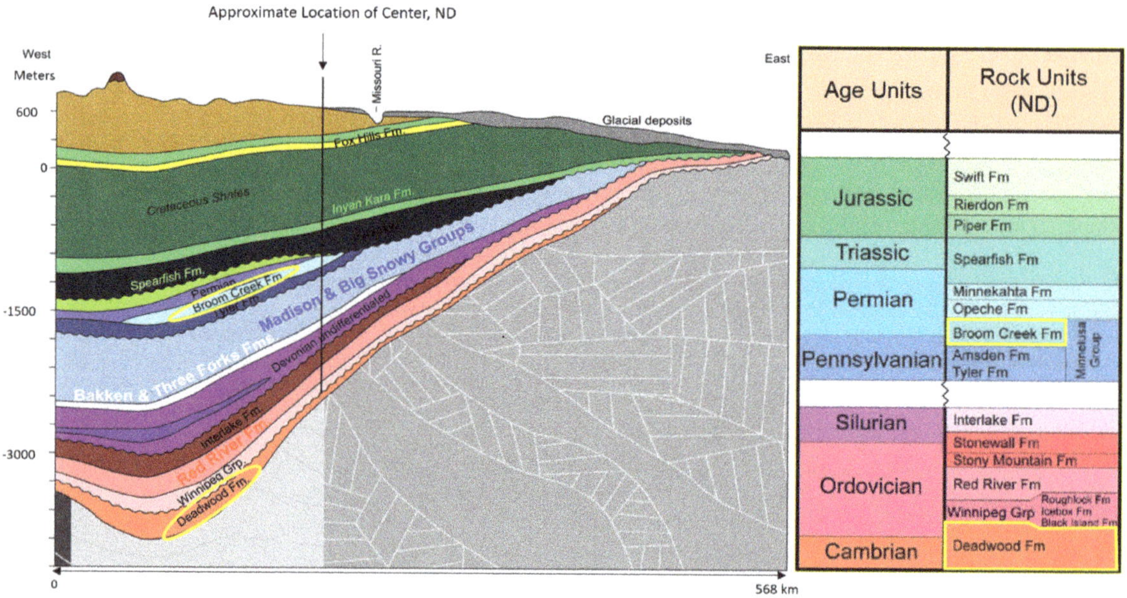

Figure 2. Simplified geologic cross-section (W–E) through the Williston basin, North Dakota (left), and simplified stratigraphic column (right). Potential CO_2 injection reservoirs are lower Permian Broom Creek and Cambrian to lower Ordovician Deadwood. (Modified after North Dakota Geological Survey—ndstudies.gov). The yellow outlines highlight the potential carbon capture and sequestration formations.

The major Broom Creek lithofacies are eolian sandstone, nearshore marine sandstone, marine carbonate, and anhydrite. The Broom Creek Formation in the study area can be divided into upper, middle, and lower sandstone-dominated intervals, with an average porosity of 23% and median permeability of 100 mD. Mudstones, siltstones, and interbedded evaporites of the undifferentiated middle Permian Opeche and upper Permian–Triassic Spearfish formations unconformably overlie the Broom Creek Formation. Mudstones and siltstones of the lower Jurassic lower Piper Formation (Picard Member and lower) overlie the Opeche–Spearfish formations. The lower Piper and Opeche–Spearfish formations serve as the primary confining zone for the CO_2 storage reservoir, with an average thickness of 47 m. The upper Pennsylvanian Amsden Formation (dolostone, limestone, and anhydrite) unconformably underlies the Broom Creek Formation and serves as the lower confining zone, with an average thickness of 82 m [19]. The base of the Broom Creek Formation is approximately 1500 m above the Precambrian basement.

The Deadwood Formation unconformably overlies the Precambrian of the Williston Basin and consists of siliciclastics, carbonates, and evaporites. The Deadwood can be divided into six members, A–F [20]. The earliest Member A is Cambrian, composed of alluvially deposited conglomerates and sandstones. Member B consists of glauconitic shallow marine sandstones and siltstones. Members C–F consist of a lower Ordovician succession of three regressive–transgressive sequences containing sandstones, siltstones, mudstones, and carbonates. Members A–E are present in the study area. The Winnipeg Group unconformably overlies the Deadwood and consists of three formations: Black Island, Icebox, and Roughlock. The Black Island Formation is a mixture of sandstone and shale deposited in a fluvial–deltaic to shallow marine environment [21]. The Icebox Formation conformably overlies the Black Island Formation. The Icebox is a marine shale that serves as the primary upper confining zone, with an average thickness of 36 m. The Roughlock is a calcareous shale to argillaceous limestone. The continuous shales of the Deadwood Formation B member serve as the lower confining zone, with an average thickness of 10 m. In addition to the Icebox Formation, there are 174 m of impermeable rock formations between the Black Island Formation and the next overlying porous zone, the Red River Formation [19].

2.1.1. CSEM Feasibility Study

Injection of supercritical CO_2 into brine-filled reservoir rocks reduces the electrical conductivity and potentially produces a 4D anomaly that can be measured from the surface. We selected the CSEM method for CO_2 monitoring because it showed the strongest coupling to resistive and conductive formations [22,23]. A feasibility study was performed to determine the effectiveness of CSEM monitoring CO_2 injected into the Broom Creek and Deadwood formations [24]. Although the ultimate proof will be the three-dimensional time-lapse image from potential repeat surveys, the feasibility study's prediction was verified with the initial field data by comparing the model derived from the well logs with inversion results for the one-, two- and three-dimensional EM methods used in the project plus the three-dimensional anisotropic model from fitting the measurements of all CSEM components.

The goal of the 3D modeling [25] was to estimate the expected surface EM field response level caused by an increase in CO_2 saturation and determine whether signals of that magnitude could be detected in the field in the presence of observed noise levels. Moreover, the feasibility study defined survey parameters, such as station spacing along survey lines, to maximize signal levels from target formations.

Integrating surface and borehole data is an essential requirement derived from 3D modeling. Data integration is achieved by measuring between surface and borehole, calibrating the information with conventional well logs, and considering resistivity anisotropy [23]. This process reduces the risk of imaging false anomalies [26–29]. Hence, combining advances in acquisition hardware, imaging methods, 3D modeling, and workflow to integrate surface models with borehole measurements in a CO_2-monitoring scenario is paramount. High measurement accuracy and overall repeatability (better than 0.5%) are required in this scenario [15]. This includes everything from instruments, operations, location repeatability, and environmental issues (atmospheric, cultural, and geologic noise). The target reservoir signal band must be well within the center of the entire system signal band to avoid deconvolution of potential system responses [22].

Figure 3 shows the CSEM feasibility workflow. The workflow's input data are seismic horizons from two 3D seismic data sets and well logs from the four wells (A, B, C, and D) in the study area. Borehole resistivity logs are considered ground truth. Using an equivalencing process [22,30] and 3D modeling based on physics, we built a 31-layer equivalent model (3D anisotropic) that honors the lithological boundary and the log [31].

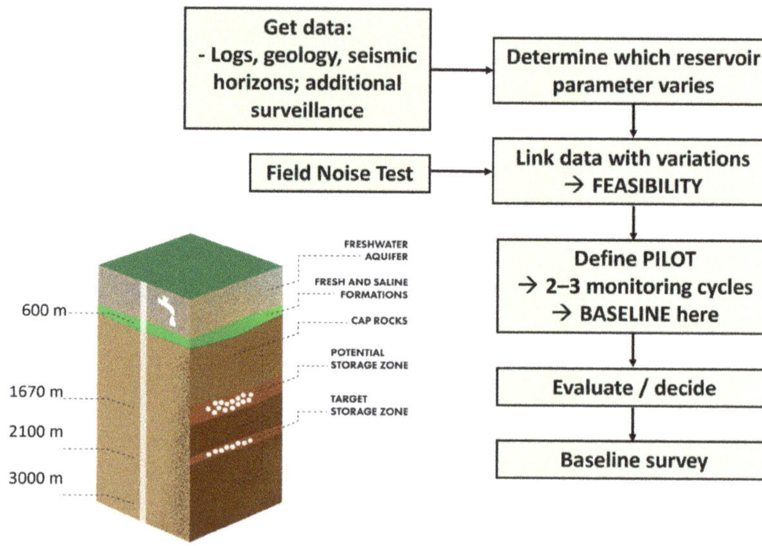

Figure 3. Feasibility workflow for CSEM fluid monitoring. The greatest modeling effort is in the feasibility step, which includes fluid substitution for various CO_2 and water saturations.

Rock physics is an essential element of the feasibility workflow, which provides the means to connect rock properties and fluid saturation determined from geology or petrophysics with measurements from geophysics. Rock physics can be used to predict how changes to rock properties and fluid saturation will affect the geophysical measurements and, thus, the feasibility of using geophysics to monitor the storage and sequestration of CO_2.

The geological data used in the rock physics analysis consist of regional geology, core descriptions from whole cores and thin sections, and core plug analysis results, including X-ray diffraction (XRD), X-ray fluorescence (XRF), and measurements of porosity, permeability, and grain density. Petrophysical data were corrected for tool and borehole effects and checked for consistency between wells. The volumes of the dominant minerals were used to determine the rock properties used in rock physics modeling. Lithology volume logs in the Broom Creek and Deadwood formations at each well were generated from the well logs using the TechLog Quanti.Elan module. The normalized solid volume results are shown in Figures 4 and 5. The results were quality-controlled, comparing the predicted volume fractions and porosities with the weight fractions from core XRD and core porosities.

The results show that the Broom Creek Formation's dominant lithologies are quartz, dolomite, and anhydrite. The Deadwood Formation's dominant lithologies are quartz and carbonate, with smaller volumes of clay and feldspar.

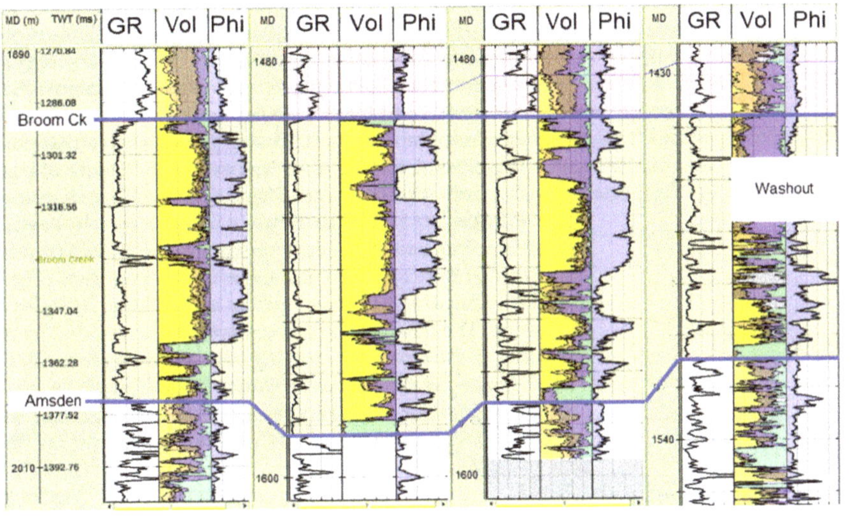

Figure 4. Q-ELAN lithology logs from the Broom Creek Formation. The wells from left to right are A, B, C, and D. The logs for each well, left to right, are environmentally corrected gamma ray, normalized lithologic volume (0–1), and porosity (0–0.4). The volumes were normalized and color-coded as anhydrite (green), dolomite (dark blue), calcite (light blue), quartz (yellow), clay (pink), and feldspar (light brown).

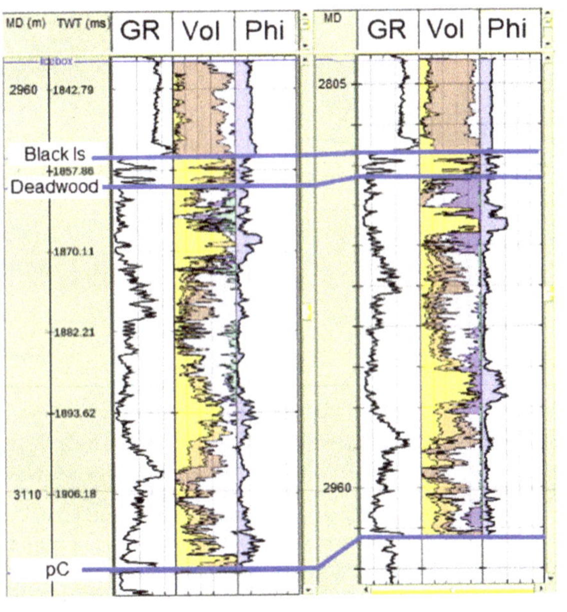

Figure 5. Q-ELAN lithology logs from the Deadwood Formation. The wells from left to right are B and D. The logs for each well, left to right, are environmentally corrected gamma ray, normalized lithologic volume (0–1), and porosity (0–0.4). The volumes are normalized and color-coded as anhydrite (green), dolomite (dark blue), calcite (light blue), quartz (yellow), clay (pink), and feldspar (light brown).

Table 2. Deadwood Formation elastic and electrical fluid properties.

Fluid	T (°C)	P (MPa)	Density (kg/m^3)	Vp (m/s)	K (GPa)	σ (S/m)	Boerner Coefficient	Elastic Reference	Electrical Reference
Brine	83	31.7	1168	1816	3.852	57.14	NA *	[33] [1]	[44]
CO_2	83	31.7	750	488	0.179	0	NA	[34] [2]	Assumed
CO_2-Saturated Brine	83	31.7	1176	1811	2.86	52.98	0.9272	[36]	[13]

[1] RokDoc FLAG fluid property calculator [33]. [2] National Institute of Standards and Technology CO_2 property calculator [34]. * Not applicable.

CSEM 3D Modeling

The CSEM setup includes a transmitter and receiver moved along profile lines, as shown in Figure 12. The survey layout consists of three lines of receivers and two separate CSEM transmitter sources to the north and south of these lines. We chose two transmitters to see and account for larger directional structural changes and anisotropy effects. More details on the hardware can be found in Strack et al. [15]. Each transmitter had two dipoles in perpendicular directions to account for local anisotropy variations. The log was scaled to 32 layers with anisotropic resistivities shown superimposed on the log on the right of Figure 12. The upscaling was verified with 3D modeling within 1%, and seismic and lithologic boundaries were maintained. The upscaled log represents the ground truth for CSEM based on the well logs.

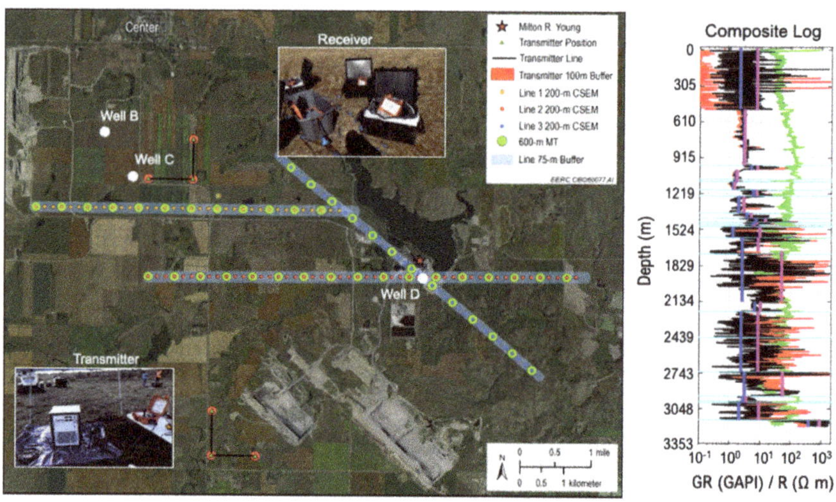

Figure 12. CSEM and MT survey layout and composite log used in the feasibility study. CSEM receivers were placed at 200 m intervals (small red, orange, and blue dots) along the three shaded lines, and sources were placed at two transmitter sites. MT sites were deployed at 600 m intervals along the receiver lines, indicated by the large green dots. The white dots represent the location of wells B, C, and D (the map does not cover the location of well A). The big red circles and black lines represent the transmitters. Exemplary equipment images are also shown. To the right is the composite resistivity log, and superimposed is the upscaled log with 31 layers shown with vertical (magenta) and horizontal (blue) resistivities. The light blue lines represent seismic reflection horizons that were kept fixed along with lithological boundaries.

For the 3D modeling, one transmitter and fourteen receivers along the three-receiver lines were used to build various 3D modeling tests. In the field, 13 receivers were moved along the profile lines. The receivers were at the injection well and represent the most distant receivers from Transmitter Location 1 (north location in the study area) and serve as a reference. All electromagnetic field responses and the time changes were modeled. An exemplary display is shown in Figure 13 for the Broom Creek Formation. These models were also used to estimate the optimum parameter range. Here, the results are displayed in the context of defining the optimum station spacing by displaying the 100 m, 200 m, and 300 m site spacing. The curve colors correspond to the site locations shown at the bottom of the figure, along with the chosen injection plume radius. Note that the plume radius can be seen on all curves.

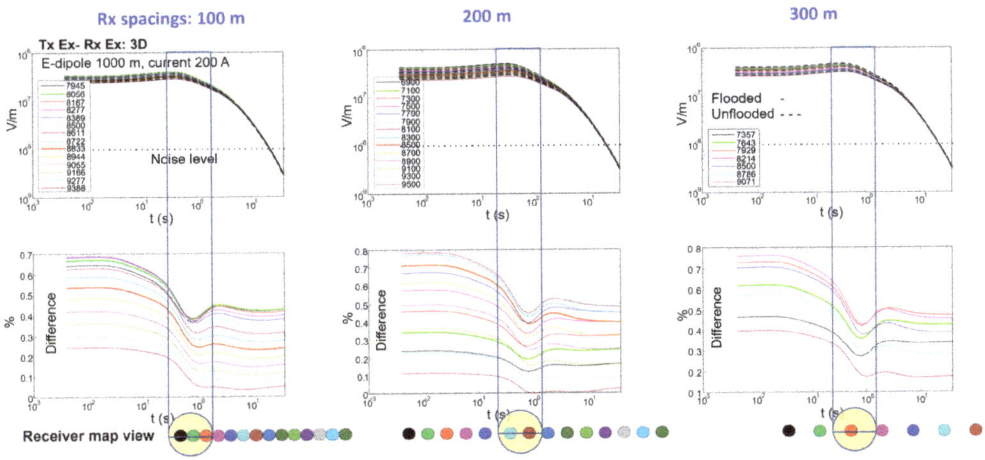

Figure 13. Example of 3D modeling results for various offsets (meters) between transmitter and receiver for different receiver station spacings.

Furthermore, the curves can be interpolated from the left to the right. The largest total amplitude in the electric field values can be seen for the 200 m spacing, which is why this was chosen for the survey. All other components behaved similarly. The benchmark models covering most field scenarios were based on petrophysical analysis. Then, the equipment/sensor choice was added to the 3D modeling results. The result is a set of models including the expected anomaly within the measurable time window.

Twenty-one months of CO_2 injection in the Broom Creek Formation was simulated using a 60% average CO_2 saturation and an injection radius of 500 m. The simulation showed that 15–18 months of injection produced a strong anomaly. Next, an injection radius of 150 m was used, and the required receiver spacing was estimated. Figure 13 depicts the 3D modeling results for the Broom Creek Formation for Ex (Ey and dBz/dt are not shown here). The reference noise from the noise test is shown at each component (horizontal dotted line). The response of all the components is above the noise level. The curve variations between the three spacings are smooth; therefore, the CO_2 anomaly can be reconstructed for 100 m, 200 m, and 300 m receiver spacings.

As the Deadwood Formation has lower porosity and is significantly deeper than the Broom Creek Formation, a conservative 30% CO_2 saturation after injection (representing a 150 m flood zone radius) was considered. The responses and differences corresponding to reservoir conditions before and after CO_2 injection were between 1% and 5% for a 1D model and below 1% for 3D models. Based on these results, monitoring injected CO_2 in the Deadwood Formation under the assumed field and survey parameters will be challenging, and novel anomaly-enhancing methods will be needed.

Field Noise Test

Concurrent with the 3D modeling, field measurements were conducted near the potential injection plant to assess noise conditions. This allows optimization of the survey design and estimation of the data's signal-to-noise ratios. Magnetic field sensors, electric sensors, and recording units were used in the test. An example of the amplitude spectra of the noise measurements is shown in Figure 14. The amplitude of the noise recorded by the induction coils was higher than the air coil's noise. This difference suggests that the areal averaging of the air coil reduces some of the localized noise (we did not see the same at different survey sites). The 60 Hz noise and its harmonics observed in the raw data were attenuated as part of the data acquisition/processing. Subsequently, the receiver data were used to simulate noise combined with the transmission cycle and signal processing. A CSEM transmitter's response was modeled for four transmitter-to-receiver offsets used in the field using the 31-layer anisotropic model. The analysis shown yields stacked data as an excellent noise-level estimator on the right of Figure 14. The resulting estimated optimum recording for CSEM data acquisition is 3:40 h. This proved very useful during the 24-rotation acquisition (a minimum of 4 h of recording time was used depending on local noise conditions and real-time quality assurance).

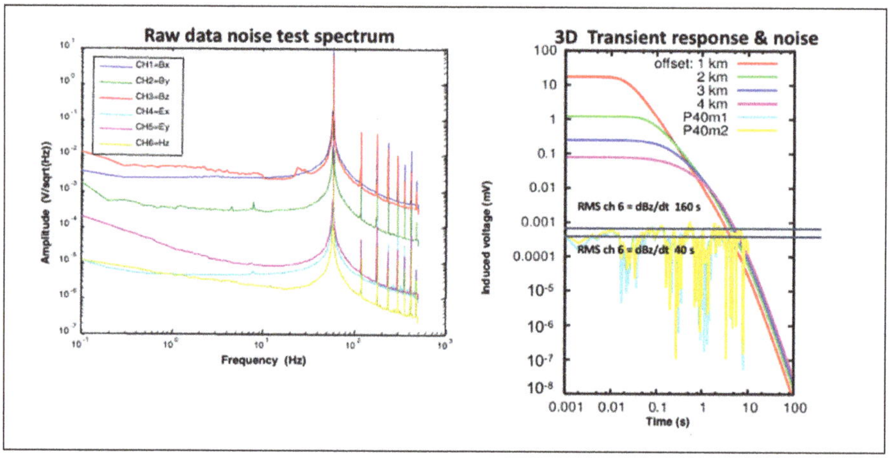

Figure 14. Summary of the noise test and results compared with the 3D modeling responses. The top left shows the power spectra. The power spectra for two different acquisition settings, m1 and m2, and the 3D modeling responses for four offsets are displayed on the right. Two power spectra use different recording times. The black horizontal lines mark the average noise level for magnetic field sensors.

For the MT acquisition, the skin depth formula and the estimated lowest frequency at approximately 3000 m deep were used to derive the recording time for the MT data acquisition. The high frequency (HF) range includes power line noise and uses HF data processing, and the low frequency (LF) range is below the power line noise and uses LF data processing (all robust processing). This range was used for overnight data acquisition. The duration variation depended on when the station setup was finished and when the station was moved the next day.

2.1.2. CSEM and MT Data Acquisition

Based on these feasibility study results, a time-lapse CSEM monitoring project was designed. The survey layout and design are shown in Figure 12. The baseline survey was taken before the CO_2 injection into the formations. Furthermore, an MT survey was performed in conjunction with the baseline CSEM survey to measure field site background resistivity. The survey was carefully designed, and special attention was paid to noise levels in the field to ensure that any time-lapse differences observed would be solely due to changes in CO_2 concentration within the reservoir. All acquisition hardware was carefully controlled for variations and calibrations were conducted before, during, and after the survey. Long-term stability was shown to be better than 0.5% in the worst case (1 out of 14 instruments).

The initial EM surveying lines were designed to overlap the 2D seismic lines recorded in the study area. Protected areas were avoided when setting out the lines, and most stations were located at least 100 m away from power lines to avoid disturbance from EM noise. The survey layout consisted of three lines of receivers and two separate CSEM transmitter sources to the north and south, each with two dipoles. Given the optimal 200 m station spacing as determined from the feasibility study, 125 CSEM receiver locations were used. The receivers consisted of 100 m long dipoles oriented north–south (Ex) and east–west (Ey). An air loop or buried induction coil was laid out on the northwest quadrant of each receiver site to measure the vertical magnetic field (Hz). Contact resistance was logged at each site for quality assurance during acquisition, processing, and interpretation.

CSEM in the time domain was selected for this project because of its high sensitivity in onshore applications (e.g., [15,46–48]). CSEM with grounded electric dipole excitation is better suited for reservoir analysis since the grounded transmitter excites horizontal and vertical currents in the formation, making the method sensitive to thin anisotropic resistors. However, in CO_2 monitoring, sensitivity to both resistors and conductors is needed; namely, a full 3D anisotropic model [23].

Each source consisted of buried electrodes connected by a 1 km cable to the transmitter. The transmitter was set to transmit between 160 and 250 amperes of current. The actual current was monitored and recorded during transmission to normalize the data. The recording times for each receiver station were between 4 and 5 h. Extended long recording times (total of 7 h per transmitter) were conducted at overlapping sites for use as reference locations should later time-lapse processing require this.

The total frequency range of MT data can be from 40 kHz to less than 0.0001 Hz, depending on the equipment used (here, we mainly used up to 1 kHz and sometimes up to 20 kHz). The ratio of the electric (Ex, Ey) and magnetic (Hx and Hy) recorded values gives an estimate of the apparent resistivity of the Earth at different depths. This estimate can be used to differentiate between rocks with contrasting resistivities, such as sandstones saturated with brine, and those where CO_2 has been injected.

Forty-two MT site locations were deployed, as depicted by large green dots in Figure 12, with a spacing of 600 m between each MT recording site. Three planned stations near the power plant were skipped during the survey because of noise/access issues. Six additional sites were located close to a noise source; the recorded data were reviewed in these cases, and the measurements were repeated because of unsatisfactory quality. A primary remote reference site was in Grand Forks, North Dakota, and a backup site was operated in Austin, Texas. MT data quality was greatly improved by using the remote reference during data processing.

Quality Assurance/Quality Control

Quality assurance/quality control (QA/QC) procedures were performed both in the field during collection and after data processing. Additional measures were considered when the receiver sites were in relatively high EM noise areas, including carefully selecting the appropriate electric line length when laying out the site, extending data acquisition time, and conducting necessary repetitions based on closely monitoring daily operations.

For MT measurements, the data were uploaded to the Cloud, and results were returned the following day, including remote reference processing where available. An example is given in Figure 15. The single site processing on the left still shows a mismatch between data (circles) and the inversion model response (line). This difference was caused by local (discharge enthalpy), and 183 mW/M^2 (heat flow) to provide long-term electricity to the Al-Lith area. Due to limited geophysical measurements in this area, these claims require further verification. We report here only the initial findings of the MT measurements, which are used to plan a more detailed CSEM survey later to drive future new drilling targets. Notice the primary targets for geothermal energy are conductive, while for CO_2 monitoring, the CO_2 plume is resistive. Therefore, MT plays a more critical role in geothermal energy than in CCUS. However, both cases require all EM components from the MT and CSEM methods to differentiate the fluid signature from the surrounding host rock.

Thirteen MT stations were collected using non-polarizable electrodes, and low-frequency induction coils. This data acquisition system used in a 40 °C environment in Saudi Arabia is identical to the equipment used in the North Dakota CCUS example mentioned above at −20 °C. The data were acquired for about 7 h at three different sampling frequencies of 4 kHz, 1 kHz, and 40 Hz. The raw data were processed in detail to ensure high-quality results. This stage involved a spectrum analysis and coherence to examine the relationship between two orthogonal and parallel fields. The impedance tensor was estimated using a robust statistical technique. The multiple impedance tensor results were further processed to extract the final transfer function data. Finally, the impedance tensors were inverted using a smoothness constraint algorithm for better image and interpretation.

The final 2D inverted section along the N-S profile (blue sites) in the study area is shown in Figure 20. The selected subset shows the flow and the reservoir best. The remaining data acquisition is ongoing and will be subject to a later complete analysis. The preliminary results of this MT study are the following:

- The geophysical survey detects and clearly maps the heat source, a low-resistivity hot body below a depth of 7 km.
- The geothermal flow cell observed at the bottom of the figure (hot fluids—green area) consists of the fractured basement acting as the pathway for the geothermal fluids to reach the surface.
- The heat exchanger is represented by the whole system, where the heat transfer consists of conduction through the hot body and convection via the flow cell.

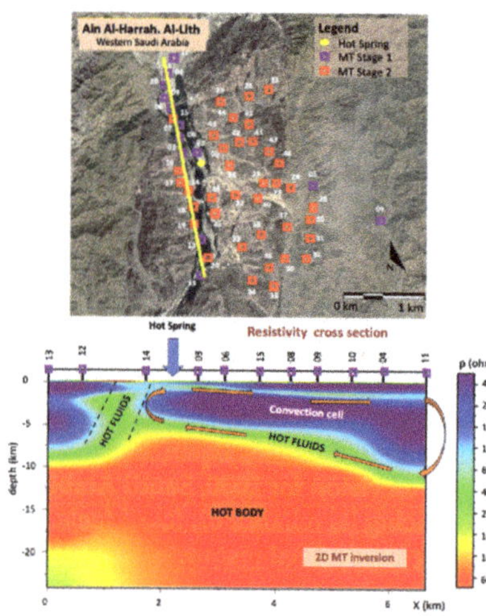

Figure 20. Two-dimensional geoelectrical model presenting the geothermal system in Al-Lith area, Saudi Arabia. In the low-resistivity hot body, the hot fluids form the heat exchanger of a convection cell. The fracture zone depicted with dashed lines acts as the pathway for the hot fluids to reach the surface. The arrow shows the hot spring at the top section. Above the section is a map showing the location of MT sites. The purple rectangles along the hot spring (yellow line) represent the sites displayed on the profile.

A fluid cell of lower resistivity and a deeper heat source can be identified from the MT data. The blue arrow in the figure indicates the location of a known hot spring as a reference.

Future activities in this project include other geophysical measurements (gravity, passive seismic, CSEM) with denser station spacings and data integration.

2.3. *Enhanced Oil Recovery (EOR)*

Imaging a flood front can often increase the EOR recovery factor by 30%–50% [23], which means less carbon footprint per barrel produced. When using CO_2 as a driving gas, one can obtain further carbon credits and move toward CO_2-neutral operations. From an EM plume imaging sensitivity viewpoint, EOR lies between CCUS and geothermal energy, as in most cases an unbiased resistivity of the fluids at the boundary of the flood front is needed. The first example is a heavy oil application where oil reservoirs are often shallow and expensive, and steam is used to drive the oil. This scenario represents a resistivity contrast at the flood front that would be replaced in case of CO_2 flooding with a short conductive front followed by a resistive CO_2 plume. The second key issue in the EOR case is the ability to record EM signals through the casing. This issue can be overcome by using a surface-to-borehole configuration. Another typical EOR case is waterflood, where a strong contrast between the resistive oil and the conductive water is present. We show a field example for this supported by 3D modeling.

Figure 21 shows a typical resistivity section for a heavy-oil (HO) reservoir in the Middle East. In this case, the oil reservoir is shallow, around 250–350 m depth. The oil is routinely driven by expensive steam (conductive oil–water contact—OWC). Using CO_2 would add carbon credit to the equation and produce resistive oil–gas contact (OGC). The critical considerations are to map the HO reservoir versus the free water zone below, which is a conductive target, and to map oil leaks from the HO reservoir into the upper aquifer, which is a resistive target. Thus, all EM components are required. We modeled these scenarios using shallow borehole measurements as they provided a better image of the anomaly [66]. The results are shown at the bottom of Figure 21 with the electric field in line with the transmitter, Ex, the vertical electric field, Ez, just 10 m below the surface, and the time derivative of the magnetic field dBz/dt at the surface. The top curve represents the field values flooded and unflooded (dashed) and below the percentage variation for different times after turn-off. The water flood is clearly seen in all curves—slightly smoother for the magnetic field and most expressed for the vertical electric field. This is because of the focusing effect that shallow boreholes show.

After obtaining large anomalies with shallow boreholes, the next step in the modeling exercise is to assess the EM response when the boreholes are connected to the surface. This is because much larger source moments and signals at the surface can be more easily generated than in the borehole. Furthermore, a better signal-moving receiver in the borehole can be obtained as they are more protected from surface EM noise. Figure 22 depicts the surface-to-borehole concept with a transmitter at the surface (sketch on the bottom right of the figure). In this figure, the same components as in Figure 21 are analyzed, except this time in the frequency domain (casing effects are frequency-dependent), where real and imaginary parts of the solution can be compared, with and without steel casing, and using Finite Difference (FE) and Finite Element (FD) software codes. Here, we compared FD and FE codes. Their responses were in good agreement. The electric and magnetic fields in the borehole give additional insight into the required measurements. To a depth of 2500 m, the signal level from a surface transmitter is well above the noise floor. For magnetic field components, sensors already in production can be used. The magnetic field is in the micro-Tesla range, while typical fluxgate sensors have a 3–6 pico-Tesla noise floor. Therefore, extending the borehole sensors from a shallow borehole tool to a deep one is feasible.

shown. In this survey, the generator noise had to be filtered out during the data processing. The field tests started with careful site selection and feasibility analysis, as described in the CCUS example.

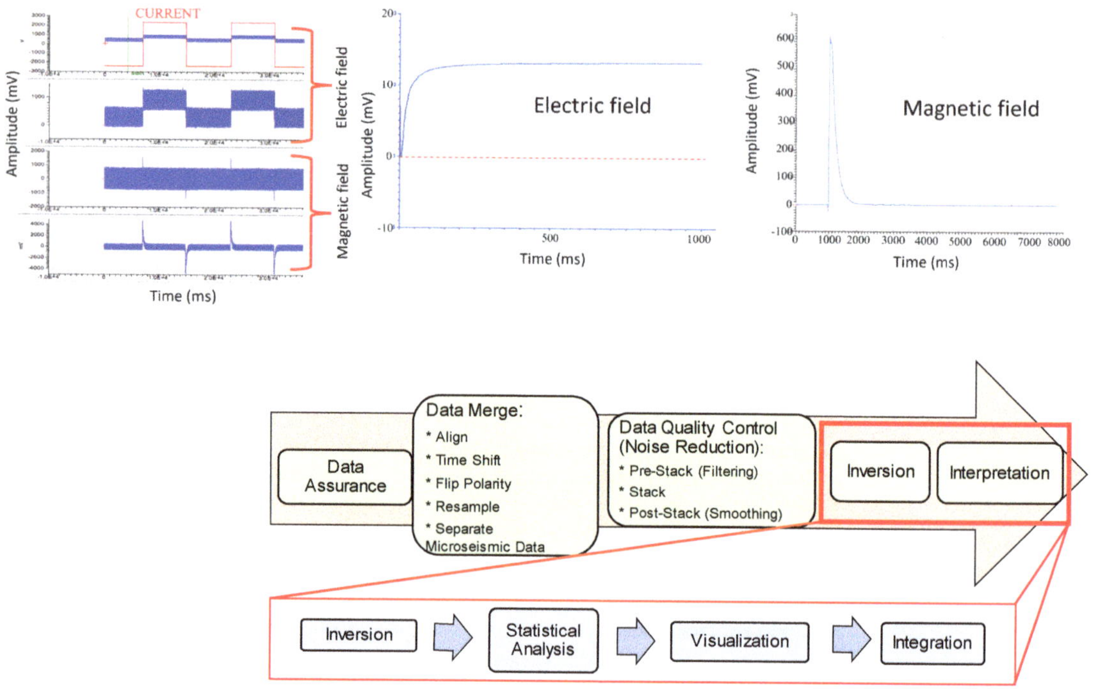

Figure 24. Data processing example from the EOR waterflood field test. The top shows (from left to right) the raw data sequences for receivers and transmitters (CURRENT). Stacked data for the electric and magnetic fields are shown. Below is the processing workflow and inversion/interpretation workflow.

The results from the field trial are shown in Figure 25. The survey layout was a 1 km long transmitter and several receivers crossing the waterflood at a 2.5 km distance. One receiver is directly above the waterflood, which consists of water injection at 2 km depth from horizontal wells. The second receiver is 400 m away from the first one. Displayed are the time derivatives of the magnetic field as a time-lapse difference in percent over approximately 50 h. Whereas the receiver directly above the water flood showed a 30% anomaly, the second only showed 2%, which we attributed to noise (our threshold in this experiment). The same responses from 3D modeling of the survey results at the bottom of the figure only showed about a 3 to 5% anomaly. Three-dimensional models, including the horizontal well casing and infrastructure, could not explain the difference via modeling. One of the two possible explanations is current channeling, where the induced transmitter current is channeled into the water flood as it covers a sizeable subsurface volume. Another possibility is the flood front related to the pore fluid mechanism described in Figure 1. When the flood front replaces fluid, it exerts forces on the old pore fluid, which causes localized alignment of the saturation water and, subsequently, an abrupt resistivity reduction. Further data on this are outstanding.

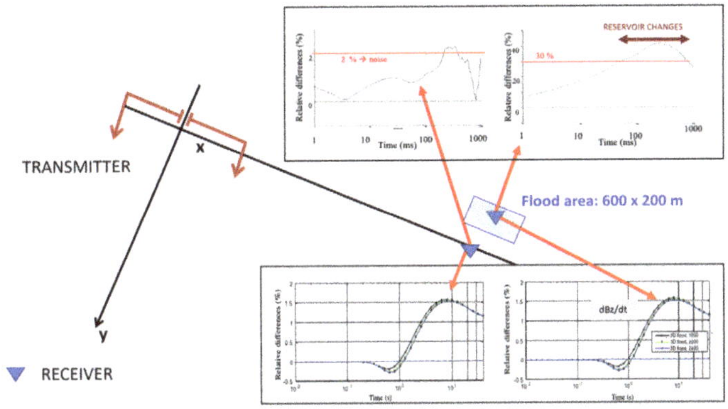

Figure 25. Survey layout and time differences derived from the field data (top curves) and comparative 3D modeling responses (bottom curves).

2.4. Critical Minerals—Lithium Exploration

As discussed above, lithium applications require the same EM technology as in the other case history examples. Mature lithium deposits can be found in Latin America, China, Australia, and the USA, and emerging deposits can be found in Europe and the Middle East [17].

2.4.1. Lithium Exploration in Argentina

In total, 65% of world reserves are in the so-called lithium triangle located in Argentina–Bolivia–Chile. This area hosts salt flats with brines that contain this mineral [69].

Due to the growth of the lithium market, the exploration frontier is changing, and currently, it extends to new concepts that drive toward the exploration of clay deposits with anomalous lithium concentrations, with much shorter production times from brine resources; fractured basements saturated with brine; and deep targets [69,70]. Figure 26 shows a summary of the process leading to lithium deposits. The minerals from mountain ranges combine with water and reach salty flats, where they build salt flats. A field setup of the equipment is shown at the top.

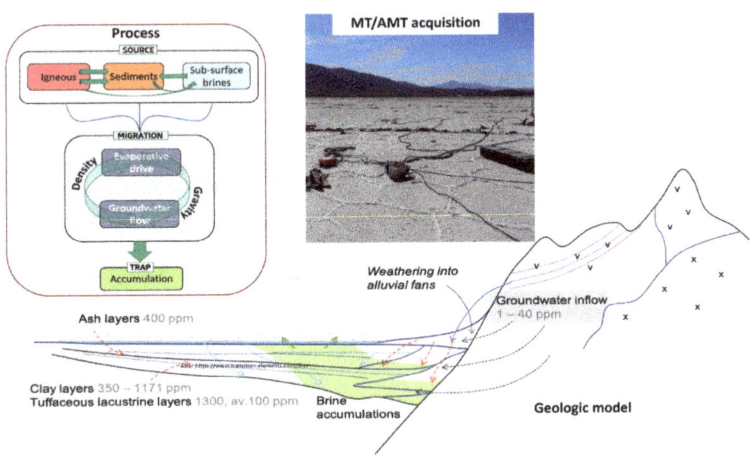

Figure 26. Schematic of lithium generation with the process flow at the top left. At the bottom is a schematic for the generation process where minerals from the mountains are carried to the bottom of the basin, where they meet evaporated salt and build salt flats.

The elements of the lithium-bearing systems require deep imaging of the endorheic (i.e., closed) basins. Geophysical variations, such as low electrical resistivity in salt concentrations, low acoustic impedances, and dynamics of the hydrogeological system, make brine exploration a complex problem. Operationally, the situation is not better because the Argentine salt flats are in sterile areas, at altitudes of 3600–4200 m and more, and, in many cases, in hostile environments with difficult accessibility and surveying. EM methods are better suited for high-conductivity targets, such as lithium brine with low resistivity (typically 0.1 to 0.5 ohm-m), than DC resistivity methods. Due to its limited dynamic range, this resistivity contrast at depth is also challenging for conventional mining and groundwater prospecting equipment.

Curcio et al. [70,71] analyzed different geophysical responses (seismic, magnetics, electric, EM, and gravity) in salt brine exploration. Various feasibility studies, field tests, and previous acquisitions were analyzed, and new acquisitions in many salt flats were made. Full-tensor MT, gravity, and electrical resistivity tomography (ERT) are currently the best combination of geophysical methods to reach the exploration objectives: characterization of the salt flat in-depth, basement delineation, definition of the main structures and main faults with the section, and detection of semi-freshwater aquifers in the edge of the salt flat providing the recharge that is key to the water balance of the endorheic basin.

MT's success in brine exploration relies on the low frequencies at which the MT method operates, compensating for the effect of the low electrical resistivity values in the skin depth relation. Sensitive to conductance, MT prospects the entire conductive column and characterizes the basement. Although the signal may be low, the equipment used in the case history here has patented amplifier behavior [18] and outstanding storage capacity that provides sufficient data statistics, resulting in an enhanced signal-to-noise ratio during advanced data processing [49]. On the other hand, a full-tensor MT provides extra interpretation attributes that cannot be achieved with other EM methods [70,71].

Figure 27 shows the MT response in salt flats. The complete study corresponds to a multi-measurement (full-tensor MT, Gravity, and ERT) survey where the geophysical model was obtained by integrating geophysical data with well data to build a 3D static model and a 3D dynamic model used in the resources and reserves estimation.

The MT frequency bandwidth is from 0.001 to 50 Hz. In both interpretations, the polar diagrams and skew indicate a 2D/3D dimensionality, and the MT section is divided into three geophysical units: the highest electrical resistivity (in red) is interpreted as the basement; the electrical resistivity in blue-violet (with the lowest resistivity of 0.1 ohm-m to 1 ohm-m) is interpreted as brine-saturated rock or wet clay (only possible—in principle); and green (mid resistivity) interpreted as rock saturated with fresh water with a lower content of salinity than the violet unit.

The first example is in the Arizaro salt flat. During 2019, approximately 65 km of full-tensor MT data with a depth of investigation of 2 km were acquired. Figure 27 (top) shows the 2D section of one of the ten lines. The MT section shows that the signal successfully passed the conductive unit that in the west has 200 m thickness, whereas the center-east model reaches 400 m depth. The model shown also delineates the shallower basement in the east (3000 m depth) and deeper in the west (2400 m depth) and main faults.

In the second example, approximately 50 km of full-tensor MT data with a depth of investigation of 2 km were acquired in the Pozuelos salt flat (bottom in Figure 27) in 2019. Figure 27 (bottom) shows the 2D section of one out of seven lines. Two conductive units, the shallower one from the surface, can reach 400 m thickness except in the NE of the area, where a fault indicates the end of the unit. According to the production wells, this unit couples a clastic or evaporite multilayer system where most of them are saturated with brine. The shallower portion of the mid-zone of the model has mid-resistivity values that indicate a zone of recharging with low-salinity water. The deeper conductive unit can reach a thickness of 500 m in the middle–NE part of the model. This unit and an important reverse fault are also detected in the crosslines. Still, we cannot affirm if its low conductivity is due to a deeper multilayer system saturated with brine or if it is due to wet clay (unlikely) because the wells are shallower than 500 m.

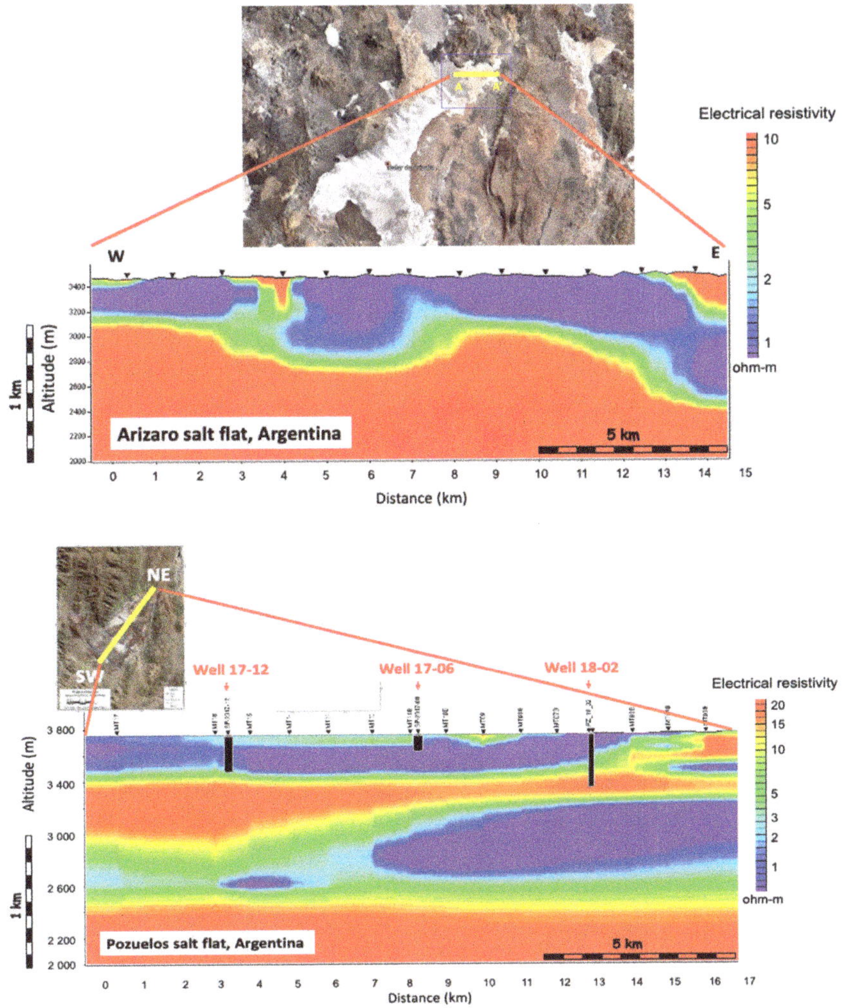

Figure 27. Two 2D resistivity cross sections, including the location of the two surveys are displayed.

2.4.2. Lithium Exploration in Saudi Arabia

Active and passive electromagnetic surveys (CSEM-LOTEM and MT) were conducted in the Half-Moon region of East Province in Saudi Arabia in January 2022 [17]. A high-power CSEM transmitter using a 1 km dipole was installed for the energy transmission, and seven receivers for both active and passive EM measurements were deployed at various distances from the transmitter, ranging from 930 m to 20 km. A 250 kVA generator was connected to the transmitter's switch box to inject current through the electrodes. The injected current varied between 40 and 200 Amperes. In addition, the duration of the injection current varied between 5 and 30 min. Four electric field electrodes (Ex1, Ex2, Ey1, Ey2) and three magnetic coils (Hx, Hy, Hz) were utilized for MT.

For the CSEM survey, two different acquisition geometries and protocols were used. The LOTEM standard CSEM layout with air loop, S20 coil, which measures the vertical time derivative of the magnetic field, and for the focused-source EM (FSEM), four electric fields with air loop. The objective of the FSEM method is to obtain deep vertical resistivity data with horizontal electric differential field measurements. FSEM enhances the conventional CSEM technique, which has significantly higher spatial resolution and provides resistivity data with greater depth [66,72]. The CSEM (active) measurements were recorded with a 1 kHz sampling frequency during the day and switched to MT (passive) measurements with a 40 Hz sampling rate at night. However, we also utilized MT data sampled at 1 kHz before and after the transmitter was turned on. Both measuring modes (MT and CSEM) displayed a clear signal for all E-field and H-field components installed.

During data acquisition, the recorded signal is the combination of the induced field for the passive source and injected transmitter current for CSEM with the effect of the signal generation process and the Earth's response [22,73–75]. Since two distinct EM measurements were collected, CSEM as an active method and MT as a passive method, the data must be separated and independently processed before they can be interpreted to create a unique subsurface model. The data were processed to perform data quality assurance and data quality control, including filtering. Since the records of the receivers contain both EM measurements with different frequency rates, these must be separated by cropping and merging before quality control processing. MT data generated the impedance, tipper, and spectral density matrices in EDI files conforming to the SEG standard. A one-dimensional (1D) analysis was performed on MT and CSEM (LOTEM) data for each sounding. The 1D inversion for MT data was performed for XY, YX, and the invariant of both data using Occam's inversion technique. The 1D inversion results for MT data with RMS values between 1.3 and 5.3 for a 3-to-5-layer model were estimated. The 1D inversion for the LOTEM-CSEM data was also performed. Both electric and magnetic field data were inverted. Ultimately, both LOTEM and MT data inversion results were compared. The final inverted models indicate that the predicted model from LOTEM and the MT results agreed, as shown in Figure 28. The fit of the CSEM data to the inversion models was generally satisfactory, falling within 1.5% root-mean-square deviation (RMS).

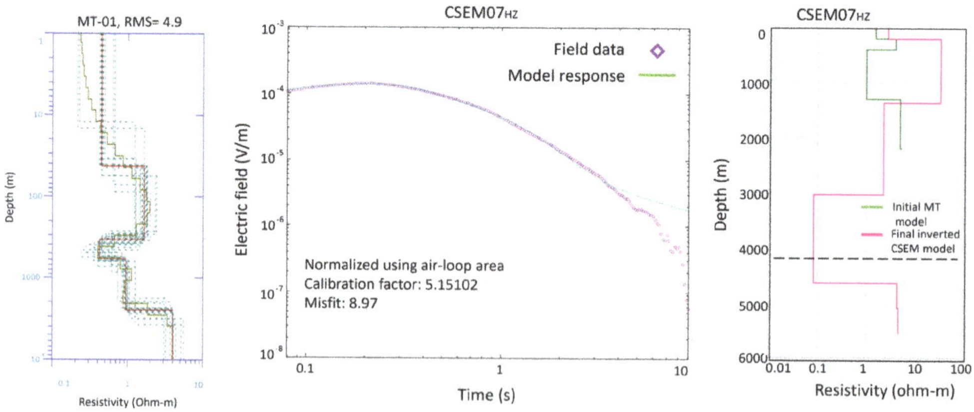

Figure 28. MT and CSEM data inversion results. Left: 1D inversion result at MT-01 location, 1 km away from the transmitter; center: the fitting between the modeled and observed E-field in CSEM-07, 20 km away from the Tx, is presented; right: the final inverted model is shown for CSEM-07 location using MT-01 location results as an initial model for the inversion. The skin depth at 4.1 km is shown as a dashed line (after [68]).

All MT data closest to the transmitter (Tx) were inverted using a five-layer model as a starting model, and the final inverted model exhibited similar behavior. In all four soundings below 300–350 m, a layer with a resistivity of less than one ohm-m was detected. The bedrock, identified as the more resistive formation, is two kilometers below the surface.

The final inverted resistivity from the MT data at the same station was used as the initial model for the CSEM inversion. At a depth of 300–350 m below the surface of the Earth, a resistivity of approximately 0.3 ohm-m was observed, and a thickness of nearly 1 km was detected.

Similarly, the processed MT data collected 20 km from the transmitter (Tx) were used as the initial model for the final 1D inversion of the collected CSEM data in ST-07. Final inverted models for ST-07 are nearly identical, indicating that the data quality, 20 km away from Tx, was surprisingly high. At a depth of 1.65 km beneath the Earth's surface, a layer with a resistivity of approximately 0.6 ohm-m and a thickness of nearly 1.9 km was detected.

The study area's high-resolution MT/CSEM survey detected a very low resistivity anomaly with values around 0.3 ohm-m, similar to seawater resistivity. As no water layer is at depth, resistivities as low as 0.3 ohm-m can be linked to lithium deposits.

Both lithium applications showed unusually low resistivity. Future drilling will confirm the measurements.

3. Discussion

The concept of using CSEM for fluid imaging is based on a carefully designed workflow supported by a sensitive acquisition system. A critical aspect of the workflow is the flexibility to select numerous calibration points. The models and data are checked against a 3D anisotropic model derived from the logs at these points. This model—verified by 3D modeling—represents the ground truth. The feasibility analysis includes on-site noise measurements to determine the optimum survey parameters and acquisition time to obtain the best signal-to-noise ratios in the data.

We selected four energy transition applications (CO_2 monitoring, geothermal, EOR, and lithium exploration) where this concept and the system architecture led to successfully reaching the objectives. At the CO_2 monitoring site in North Dakota, we thoroughly applied the workflow and, under the inclusion of all logs, derived full 3D anisotropic modeling consistent with lithology, seismic data, and all other logs. The model serves as the ground truth for verifying data acquisition, quality assurance (near-real data quality check and acquisition operations control), and inversion.

For the geothermal example in Saudi Arabia, we used integrated geology and multi-measurement (EM [MT & CSEM], gravity, and passive seismic) information to overcome the lack of deep boreholes. The initial results already explain the known geothermal system around a hot spring area, and further data and interpretation are to follow.

Applying this workflow to EOR, where numerous boreholes exist in an oil field environment, brings challenges associated with the oil field noise and operation restrictions. The feasibility allowed us to design the field test so that we could overcome these difficulties and see the injection flood front after only a few days of recording.

The same EM system was applied to lithium exploration in Argentina (MT) and Saudi Arabia (MT and CSEM). In Argentina, the system could map existing lithium deposits. In Saudi Arabia, the MT and CSEM recorded at several locations showed unusually low resistivity conductors, strongly indicating that lithium deposits are possible.

In this CCUS application, both electric and magnetic fields gave comparable information. The magnetic field was more susceptible to noise. The information content down to a depth of 3 km was verified by analyzing the eigenvalues in the inversion. Due to the extensive calibration effect of the measurements, hardware, and methodology, the data were interpreted fully automatically, with the results matching the 3D model adjusted for the seismic horizons matching all CSEM components.

The EM application to EOR is more complex than for CCUS. In most cases, both electric and magnetic components are required. Also, the signal bandwidth is larger as the depth of hydrocarbon reservoirs can vary from a few hundred to a few kilometers. Resolution can be increased by adding borehole measurements. This is important as we get closer to the ground truth of a reservoir and decision making.

4. Conclusions

The successful use of CSEM for fluid imaging for CCUS, EOR, geothermal, and lithium applications illustrates its value. The data acquisition system consisting of MT and CSEM components was based on the careful design of high-fidelity hardware coupled with a monitoring architecture that delivers repeatable measurements within an overall system repeatability of better than 0.5%. Verification of the results was conducted via integration with borehole logs. The acquisition workflow was tailored to the system so that well logs can be used to verify acquisition results in a near-real-time quality assurance mode. The system consists of MT and CSEM components, and both methodologies are complementary.

The case histories start with a CO_2 monitoring example (baseline measurements), followed by EOR applications with the same principle and workflow. While the data in the EOR example were noisier, the time-lapse anomaly clearly showed the flood front. For geothermal applications, we showed the dynamic geothermal flow cell being already mapped by the MT measurements that were used to design a detailed CSEM follow-up survey. Finally, the changes made to the hardware lead to a high acquisition dynamic that can map lithium brine.

The results of all examples are very significant:

- CCUS plume monitoring: We showed that we could certify the data and data quality within the context of the 3D anisotropic log from the beginning. This process requires extreme care during acquisition and quality assurance. As a result, potential risks can be mitigated, and cost savings can be achieved for the time-lapse measurements. By ensuring the accuracy and quality of data, researchers, practitioners, and decision-makers can make informed decisions based on sound data and analysis.
- Geothermal: The initial MT data have demonstrated a coherent match with all geological models derived for the hot spot area, attesting to the accuracy and reliability of the data obtained. This suggests that additional measurements, including CSEM measurements, will facilitate future drilling. The high data quality gives us a strong sense of optimism about the future success of the EM geothermal application.
- EOR case: We worked in existing oil fields and logs, and the noise test yielded a field test design, subsequently confirming that we could see the water flood front and confirmed the anomaly with 3D models.
- The lithium exploration examples from Argentina and Saudi Arabia have revealed a highly conductive layer positioned slightly above 1 km in depth. The data consistency indicates that this layer is probably a lithium brine (a typical finding in Argentina). We expect these results will stimulate more interest in the possibility of lithium exploration in Saudi Arabia.

Author Contributions: Conceptualization and lead writing, C.B.-O., A.C., P.S. and K.S.; methodology, K.S., Y.M. and A.Y.P.; 3D modeling, S.D.; Case histories—A.C., C.B.-O., D.C.A., R.J.K., K.S., P.S. and W.D.P. All authors have read and agreed to the published version of the manuscript.

Funding: The North Dakota case study research was funded by the U.S. Department of Energy (DOE) National Energy Technology Laboratory (NETL) as part of Cooperative Agreement DE-FE0031889 and supported by Minnkota Power Cooperativeand the Energy & Environmental Research Center (EERC) of the University of North Dakota. King Fahd University of Petroleum and Minerals (KFUPM) funded the work in Saudi Arabia. Litica Resources SA supported the work in Argentina. KMS Technologies funded field tests, equipment, and interpretation development.

Data Availability Statement: The data are not available. The DOE-funded data will be made available in due course.

Acknowledgments: The authors thank the U.S. Department of Energy National Energy Technology Laboratory for funding the North Dakota case study research. The authors also thank Minnkota Power Cooperative for supporting the North Dakota case study and our colleagues at EERC, KMS Technologies, and KFUPM for contributing to this paper. PROINGEO SA performed the work in Argentina.

Conflicts of Interest: The authors declare no conflict of interest.

Appendix A

Some complementary information related to field data acquisition and data processing, interpretation, and inversion activities in the case histories is presented in this appendix: Cloud-connected quality assurance is described in detail in [76]. MT data were processed using methods described in [49] and the CSEM processing described in [50]. The 1D, 2D, and 3D inversion using inversion statistics are described by [28,77–80]. MT and CSEM inversion was conducted using an algorithm from [80]. The 3D anisotropic CSEM modeling is based on ([25,81] for reference summary). The MT inversion module is based on [77–80].

References

1. IEA. Energy Technology Perspectives 2023, IEA, Paris. 2023. Available online: https://www.iea.org/reports/energy-technology-perspectives-2023 (accessed on 1 February 2023).
2. IEA. Energy Technology Perspectives 2020, IEA, Paris. 2020. Available online: https://www.iea.org/reports/energy-technology-perspectives-2020 (accessed on 1 February 2023).
3. Azzolina, N.A.; Nakles, D.V.; Gorecki, C.D.; Peck, W.D.; Ayash, S.C.; Melzer, L.S.; Chatterjee, S. CO_2 storage associated with CO_2 enhanced oil recovery: A statistical analysis of historical operations. *Int. J. Greenh. Gas Control* **2015**, *37*, 384–397. [CrossRef]
4. Benson, S.M.; Cole, D.R. CO_2 Sequestration in Deep Sedimentary Formations. *Elements* **2008**, *4*, 325–331. [CrossRef]
5. Available online: https://www.netl.doe.gov/carbon-management/carbon-storage/faqs/carbon-storage-faqs (accessed on 15 January 2023).
6. Barbier, E. Geothermal energy technology and current status: An overview. *Renew. Sustain. Energy Rev.* **2002**, *6*, 3–65. [CrossRef]
7. Babaei, M. Integrated Carbon Sequestration–Geothermal Heat Recovery: Performance Comparison Between Open and Close Systems. *Transp. Porous Med.* **2019**, *126*, 249–273. [CrossRef]
8. Randolph, J.B.; Saar, M.O. Combining geothermal energy capture with geologic carbon dioxide sequestration. *Geophys. Res. Lett.* **2011**, *38*, L10401. [CrossRef]
9. Randolph, J.B.; Saar, M.O. Coupling carbon dioxide sequestration with geothermal energy capture in naturally permeable, porous geologic formations: Implications for CO_2 sequestration. *Energy Procedia* **2011**, *4*, 2206–2213. [CrossRef]
10. Buscheck, T.A.; Elliot, T.R.; Celia, M.A.; Chen, M.; Sun, Y.; Hao, Y.; Lu, C.; Wolery, T.J.; Aines, R.D. Integrated Geothermal-CO_2 Reservoir Systems: Reducing Carbon Intensity through Sustainable Energy Production and Secure CO_2 Storage. *Energy Procedia* **2013**, *37*, 6587–6594. [CrossRef]
11. Weiss, C.J.; Jones, A.G. Introduction to this special section: Critical minerals exploration. *Lead. Edge* **2023**, *42*, 236. [CrossRef]
12. Barajas-Olalde, C.; Mur, A.; Adams, D.C.; Jin, L.; He, J.; Hamling, J.A.; Gorecki, C.D. Joint impedance and facies inversion of time-lapse seismic data for improving monitoring of CO_2 incidentally stored from CO_2 EOR. *Int. J. Greenh. Gas Control* **2021**, *112*, 103501. [CrossRef]
13. Boerner, J.H.; Herdegen, V.; Repke, J.-U.; Spitzer, K. The electrical conductivity of CO_2-bearing pore waters at elevated pressure and temperature: A laboratory study and its implications in CO_2 storage monitoring and leakage detection. *Geophys. J. Int.* **2015**, *203*, 1072–1084. [CrossRef]
14. Carlson, M.R. An Analysis of the Caprock Failure at Joslyn. In Proceedings of the SPE Heavy Oil Conference in Calgary, Calgary, AB, Canada, 12–14 June 2012. [CrossRef]
15. Strack, K.; Davydycheva, S.; Hanstein, T.; Paembonan, A.Y.; Smirnov, M. An array multiphysics acquisition system with focus on reservoir monitoring for the energy transition. *Earth Environ. Sci. Res. Rev.* **2022**, *5*, 237–268. [CrossRef]
16. Sheriff, R.E. *Encyclopedic Dictionary of Applied Geophysics*; Society of Exploration Geophysicists: Houston, TX, USA, 2002; p. 429.
17. Ashadi, A.L.; Martinez, Y.; Kirmizakis, P.; Hanstein, T.; Xu, X.; Khogali, A.; Paembonan, A.Y.; AlShaibani, A.; Al-Karnos, A.; Smirnov, M.; et al. First High-Power CSEM Field Test in Saudi Arabia. *Minerals* **2022**, *12*, 1236. [CrossRef]
18. Jiang, J.; Aziz, A.A.; Liu, Y.; Strack, K.M. Geophysical Acquisition System. U.S. Patent 9,057,801, 2015.
19. Peck, W.D.; Ayash, S.C.; Klapperich, R.J.; Gorecki, C.D. The North Dakota integrated carbon storage complex feasibility study. *Int. J. Greenh. Gas Control* **2019**, *84*, 47–53. [CrossRef]
20. Sarnoski, A.H. The Stratigraphy and Depositional History of the Deadwood Formation, with a Focus on Early Paleozoic Subsidence in the Williston Basin. Master's Thesis, Grand Forks, University of North Dakota, Grand Forks, ND, USA, 2015.
21. Le Fever, R.D. Sedimentology and Stratigraphy of the Deadwood-Winnipeg Interval (Cambro-Ordovician), Williston Basin. Paleozoic Systems of the Rocky Mountain Region; Golden, The Rocky Mountain Section (SEPM), 1996; Available online: https://archives.datapages.com/data/rocky_sepm/data/034/034001/11_rocky_mount340011.htm?q=%2BtextStrip%3Adeposition+textStrip%3Aenvironment+textStrip%3Astratigraphy+textStrip%3Anorth+textStrip%3Awestern+textStrip%3Adesert+textStrip%3Aegypt+-isMeetingAbstract%3Amtgabsyes (accessed on 1 September 2023).
22. Strack, K. *Exploration with Deep Transient Electromagnetics*; Elsevier: Amsterdam, The Netherlands, 1992; ISBN 978-0444895417.
23. Strack, K.-M. Future directions of Electromagnetic Methods for Hydrocarbon Applications. *Surv. Geophys.* **2014**, *35*, 157–177. [CrossRef]

80. Smirnova, M.; Shlykov, A.; Asghari, S.F.; Tezkan, B.; Saraev, A.; Yogeshwar, P.; Smirnov, M. 3D controlled-source electromagnetic inversion in the radio-frequency band. *Geophysics* **2022**, *88*, E1–E12. [CrossRef]
81. Available online: https://kmstechnologies.com/Publications.html (accessed on 1 September 2023).

Disclaimer/Publisher's Note: The statements, opinions and data contained in all publications are solely those of the individual author(s) and contributor(s) and not of MDPI and/or the editor(s). MDPI and/or the editor(s) disclaim responsibility for any injury to people or property resulting from any ideas, methods, instructions or products referred to in the content.

MDPI
St. Alban-Anlage 66
4052 Basel
Switzerland
www.mdpi.com

Minerals Editorial Office
E-mail: minerals@mdpi.com
www.mdpi.com/journal/minerals

Disclaimer/Publisher's Note: The statements, opinions and data contained in all publications are solely those of the individual author(s) and contributor(s) and not of MDPI and/or the editor(s). MDPI and/or the editor(s) disclaim responsibility for any injury to people or property resulting from any ideas, methods, instructions or products referred to in the content.

www.ingramcontent.com/pod-product-compliance
Lightning Source LLC
Chambersburg PA
CBHW040753250926
51850CB00007BB/1488
* 9 7 8 3 7 2 5 8 0 5 4 5 7 *